AF764170

Courtship and Mating in Butterflies

Reproduction, Mating Behaviour and Sexual Conflicts

Courtship and Mating in Butterflies

Reproduction, Mating Behaviour and Sexual Conflicts

Raymond J.C. Cannon

CABI

CABI is a trading name of CAB International

CABI
Nosworthy Way
Wallingford
Oxfordshire OX10 8DE
UK

Tel: +44 (0)1491 832111
Fax: +44 (0)1491 833508
E-mail: info@cabi.org
Website: www.cabi.org

CABI
745 Atlantic Avenue
8th Floor
Boston, MA 02111
USA

T: +1 (617)682-9015
E-mail: cabi-nao@cabi.org

© Raymond J.C. Cannon 2020. All rights reserved. No part of this publication may be reproduced in any form or by any means, electronically, mechanically, by photocopying, recording or otherwise, without the prior permission of the copyright owners.

A catalogue record for this book is available from the British Library, London, UK.

Library of Congress Cataloging-in-Publication Data

Names: Cannon, Raymond Julian Conway, author.
Title: Courtship and mating in butterflies : reproduction, mating behaviour and sexual conflicts / Raymond Julian Conway Cannon.
Description: Boston, MA : CAB International, 2020. | Includes bibliographical references and index. | Summary: "The aim of this book is to present a readable account of butterfly behaviour, based on field observations, great photographs and the latest research. The main focus is on courtship and mating - including perching, searching and territorial behaviour - but to understand these subjects it is necessary to explain how mates are chosen and this requires sections on wing colours and patterns. A chapter on butterfly vision is also essential in terms of how butterflies see the world and each other. There have been exciting discoveries in all of these fields in recent years, including: butterfly vision (butterfly photoreceptors), wing patterns (molecular biology), wing colouration (structural colours and nano-architecture), mating strategies and female choice (ecology and behaviour)"-- Provided by publisher.
Identifiers: LCCN 2019026041 (print) | LCCN 2019026042 (ebook) | ISBN 9781789242638 (hardback) | ISBN 9781789242645 (ebook) | ISBN 9781789242652 (epub)
Subjects: LCSH: Butterflies--Behavior. | Courtship in animals.
Classification: LCC QL562.4 .C36 2020 (print) | LCC QL562.4 (ebook) | DDC 595.78/9156--dc23
LC record available at https://lccn.loc.gov/2019026041
LC ebook record available at https://lccn.loc.gov/2019026042

ISBN-13: 9781789242638 (hardback)
9781789242645 (ePDF)
9781789242652 (ePub)

Commissioning Editor: Ward Cooper
Editorial Assistant: Lauren Davies
Production Editor: Tim Kapp

Typeset by SPi, Pondicherry, India
Printed and bound in the UK by Severn, Gloucester

Contents

Preface

I am not a lepidopterist by profession, but I have spent my working life studying insects: starting with the population dynamics of cereal aphids for my doctorate; moving on to the cold hardiness of Antarctic invertebrates; and ending with the impacts of invasive invertebrates. My passion for butterflies emerged late on in my life and developed out of an interest in photography. The more time I spent observing and photographing butterflies in the field, the more I became interested in their biology, and particularly in their behaviour. So, I embarked on a project to find out as much as I could, both from the published scientific literature, and from other diverse sources of natural history information, including the Internet. Much of what has been discovered about butterfly courtship and mating is scattered in minor journals, but fortunately many of these are now available to everyone via portals such as the Biodiversity Heritage Library. I have also benefited from having access to the British Library in London.

My inclination is to absorb, synthesize and present the information in a, hopefully, readable and accessible style. I am firmly of the opinion that it is possible to express ideas and discoveries in plain English. Research papers, for reasons of brevity and professionalism, tend to be rather daunting for the non-expert, and they are in any case not readily accessible to the curious layperson. So what I have tried to do is to take a deep dive into the literature on butterfly courtship and mating, to soak up the knowledge, and to bring up what I think will be of interest and delight to both the student and the proverbial, intelligent layman.

I want to emphasize that I have tried to present the main ideas and findings, to paint a picture of the love lives of butterflies, but not to include every last detail and fact from what is a highly voluminous literature. Aware that I need to demonstrate the solid academic foundation of what I have produced, I have included notes (at the end of each chapter) referencing the sources for everything I describe and report on. These act as both a support for the assertions made in the text and as pointers to where more details can be found.

In presenting the research findings in a shorter and hopefully more readable format than the academic literature, I hope I have not lost any of the subtleties and caveats of the original work. It has occurred to me that some researchers might not take kindly to seeing their painstaking and time-consuming research summarized in this way. Yet, most things, in my experience, boil down to a few key facts, ideas or memes. I hope that I have been able to communicate this in an interesting and revealing way. Summarizing such a vast field is an audacious and possibly foolhardy task, which probably would never be attempted by a genuine lepidopterist! I have read and reviewed literally thousands of research papers and this book is my interpretation of the literature. Science has become more and more specialized, and researchers are often accused of working in narrower and narrower fields: the silo effect. So, the task of presenting an accurate account of a broad field becomes ever more difficult for one individual. Nevertheless, I have attempted to understand and explain the subject myself. No doubt there are a few relevant publications that I have missed and should have included, but I am confident that I have captured the main ideas associated with butterfly sexual behaviour, and have sought the help of relevant specialists as I went along. All endeavours must have their limits.

In a subject much concerned with sex and gender relations, I have tried not to anthropomorphize or draw any conclusions on or parallels with the human condition, but it is hard, as a living organism oneself, not to do so somewhere in the back of your mind! The truth is, the literature itself – particularly some of the older papers – is littered with anthropomorphic, and occasionally even dubious terms, which a modern editor might take issue with. Indeed, I expect a psychologist would wonder at the emphasis, and even obsession, with which some topics of butterfly sexual behaviour are researched and discussed! Yet, the sexual lives of butterflies *are* endlessly fascinating, and whilst their behaviour cannot be compared in any meaningful way with that of humans, the diverse goings on in the butterfly realm do resonate with our own lives. Butterflies are creatures that are evolutionarily distant from ourselves, yet at the same time full of overtly similar goings on: courtship rituals, sexual conflict, mate choice, contests for access to females, the provision of gifts, and even coercion and 'rape'.

Thus, however strange it may seem to the casual reader, butterfly sexual behaviour has a huge amount to teach us in terms of biology and evolution, including: the so-called, 'battle of the sexes', the roles of males and females, the choice of mates, sexual conflicts, the costs and benefits of reproduction, and so on. Some of these subjects are now very large and highly active fields of academic research; I hope I have managed to give a flavour of the nature of this research and the elements of the debates. I have tried to avoid straying too far into subjects like sexual conflict, whilst at the same time giving a background to the type of research that is being carried out to test hypotheses regarding such central issues of evolutionary biology, such as sexual selection.

In the course of writing this book, I have concentrated more on some topics than others: essentially the things that I personally found interesting and intriguing, and which hopefully, the reader will too. Subjects relating to butterfly reproduction that I have not attempted to cover, or only referred to briefly, include development, oviposition, life histories and gynandromorphs. I have also not included very much on butterfly basking, thermoregulation, mimicry, genetics and molecular biology. Instead, I have tried to focus on the biology and behaviour of butterflies that is relevant to the field of natural history, which hopefully will enlighten and inspire readers into carrying out their own observations of butterflies in the field. There is a huge amount that the ordinary person can contribute to the understanding of butterflies, particularly in the Internet era. Citizen science projects exist in many countries, and butterfly conservation societies boast a growing membership. There is much that the amateur, or non-professional butterfly enthusiast, can bring to the table, and there is a tremendous amount that remains to be discovered concerning butterfly behaviour. Some terrific research is being carried out that has raised the bar in terms of understanding butterfly behaviour, often building on classical studies carried out many years ago. Alas, many butterflies are in danger of disappearing before we can even name them, let alone understand the rich complexity of their lives and biology.

Photographing butterflies is an excellent way to discover butterflies, and simply following them and trying to take good pictures can lead to interesting discoveries and observations. I have often noticed something in a photograph, when reviewing it on a computer afterwards, that I had not noticed in the field. Capturing interesting behaviour is as rewarding in butterflies as it is in large mammals! Perhaps more so, because their lives are often hidden from plain sight. Another wonderful thing about butterflies is that one will never run out of new things or new species to discover. Even within Europe there are hundreds of species, with new ones to watch and discover in many common holiday locations, as well as at home. We have many beautiful and marvellous species in our own backyards, although many of these are suffering, and struggling to maintain their foothold in our modern landscapes.

The Internet is awash with interesting and insightful posts and blogs about butterflies. The problem is there is no great God of the Web to filter, organize and order all this information. Some of the most detailed and professional sites – for example, I would refer readers to the Butterflies of America (https://www.butterfliesofamerica.com), Butterflies of Singapore (http://butterflycircle.blogspot.com), UK Butterflies (https://www.ukbutterflies.co.uk/index.php) and Learn about Butterflies (https://www.learnaboutbutterflies.com/index.htm) – contain a wealth of information on biology, behaviour, identification, photography and much else of use to the interested amateur. There is also a lot of fascinating information in field guides or texts on the butterflies of counties, countries and regions, too numerous to mention.

I have spent over 3 years engaged on this project, and have tried to capture the excitement and awe I felt when I first witnessed or discovered some of the amazing facts and facets of butterfly behaviour: like proboscis waving in wood whites, courtship bowing in graylings, mating refusal postures in pierids and so on. Witnessing these phenomena takes a considerable amount of time and effort – matings are rarely observed – so one must focus on species that are close at hand. I have not seen many of the species, from around the world, referred to in this book, although most of the photographs are my own. I hope that this book will inspire readers to look at butterflies with new eyes, and perhaps to spend time trying to photograph their interesting behaviour.

Acknowledgements

I am extremely grateful to Ronald L. Rutowski, Emeritus Professor in the School of Life Sciences, at Arizona State University – whose numerous publications on many aspects of butterfly behaviour are described in this book – for commenting on a draft of Chapter 3 and suggesting improvements in both detail and presentation. Professor Rutowski made the point that with more than 20,000 butterfly species, the variation in behaviour within this group is huge and perhaps we should not reach too eagerly for blanket generalizations. A suggestion I have tried to follow, although the temptation to extrapolate beyond a single species is often overwhelming!

I am very grateful to Professor Antónia Monteiro of the National University of Singapore[1] for commenting on sections on male and female choices in the squinting bushbrown butterflies, *Bicyclus anynana*, in the chapter on sexual selection and mate choice (Chapter 2), and for allowing me to use her photographs illustrating the eyespots on the wings of these butterflies (e.g. Fig. 2.13). Similarly, I am most grateful to Dr David Stella of Charles University in Prague, for commenting on Chapter 2, and for providing photographs demonstrating UV reflectance in butterflies, such as *Pieris rapae* (Fig. 2.31). I am also most grateful to Professor H. Frederik Nijhout for the loan of his slide illustrating the diversity of female forms in *Papilio dardanus* (Fig. 2.20) used in the section on female variation (Chapter 2). Also, to Shinichi Nakahara and the Florida Museum of Natural History, for allowing me to use their photograph of the sunburst cerulean-satyr (*Caeruleuptychia helios*) (Fig. 2.26) from Nakahara *et al.* (2018).

I am most grateful to Szabolcs Sáfián for commenting on an early draft of the chapter on mate-finding (Chapter 3), and for permission to use photographs of the ethereal pierid, *Pseudopontia mabira* (Fig. 5.9a, b). Likewise, to Annika Hillers and the RSPB for permission to use a photograph of a related species, *Pseudopontia gola*, from the shaded forests of West Africa. I would also like to thank Professor Roger Dennis for kind words of encouragement and for giving me papers and advice relating to hill-topping and mate location.

I am also most grateful to Professor Adriana Briscoe, University of California, Irvine, and to Professor Dr Doekle Stavenga, University of Groningen, NL, for reading and commenting on Chapter 4; and to Professor Stavenga for helping me to understand pseudopupils in butterfly compound eyes (Figs 4.9 and 4.10).

I am most grateful to Dr Manuela Pinzari, University of Rome 'Tor Vergata', Italy, for commenting on Chapters 5 and 6, and for providing advice and information about courtship behaviours in graylings and fritillaries.

I am very grateful to Dr Ana Paula S. Carvalho, of the University of Florida, Gainesville, USA, for providing fascinating photographs of sphragis in butterfly species from Carvalho *et al.* (2017) used in Chapter 8 (Figs 8.3 and 8.4).

I am very grateful to Dr Kathy Darragh, University of Cambridge, for commenting on a paragraph on *Heliconius* pheromones in Chapter 9, and for agreeing to let me use a figure from Darragh *et al.* (2017) (Fig. 9.15). I am also grateful to Dr Ananda Martins for allowing me to use his photograph of the male secondary sexual organs of the Hemon blue hairstreak (Fig. 9.8) from Martins *et al.* (2018).

I am extremely grateful to Ádám Gór, a highly talented young Hungarian photographer,[2] who granted me permission to use some of his beautiful images (Figs 3.5, 3.8, 7.20, 9.5); and to S.C. Yam for providing me with two beautiful photographs of the Asian swallowtail (*Papilio xuthus*) (Fig. 2.2a, b).

I am grateful to Professor Dick Vane-Wright, University of Kent, and Dr Blanca Huertas, Senior Curator of Butterflies at the Natural History Museum, for answering my occasional butterfly-related queries.

Finally, I would like to thank all those researchers whose works I refer to in this book. Without their painstaking work, observations and experiments we would know very little about the lives and loves of these iconic species. I hope that I have managed to pass on my enthusiasm, and a little harmless obsession, in the following pages!

Notes

[1] Professor Antónia Monteiro's laboratory‘s amazing research on butterfly eyespots and wing patterns is available online (http://lepdata.org/monteiro).

[2] The wonderful photographs of Ádám Gór can be accessed via his website (http://jasius.hu).

1 Introduction

> I thought of the long ages of the past during which the successive generations of these things of beauty had run their course … with no intelligent eye to gaze upon their loveliness, to all appearances such a wanton waste of beauty… This consideration must surely tell us that all living things were not made for man… Their happiness and enjoyments, their loves and hates, their struggles for existence, their vigorous life and early death, would seem to be immediately related to their own well-being and perpetuation alone.
>
> Alfred Russell Wallace, *The Malay Archipelago*, 1869, vol. 2, p. 119

This is a book about butterfly behaviour. Specifically, the behaviours associated with courtship rituals (described in Chapters 5 and 6) and mating and reproduction (described in Chapters 7 and 8). To understand the ways in which butterflies act as adults in search of mates, it is necessary to digress a little and wander off into associated fields, including evolutionary ecology (Chapter 2), mate seeking and territoriality (Chapter 3), vision (Chapter 4), scents and chemical communication (Chapter 9), and, finally, wing morphology and visual signalling (Chapter 10). In this chapter, we take a brief look at the origins of butterflies, how they are classified, and survey some of the pioneering studies on courtship and mating that were carried out in the 20th century. Hopefully, this will set the scene for the following chapters detailing more recent investigations.

There are many excellent tomes – and websites – on butterflies, including their biology and classification, so there is no point in duplicating them here, other than to include a brief introduction (below) to the main families and their origins and organization. Wikipedia contains species descriptions of most of the species referred to in this book, together with their scientific classification. Some old school lepidopterists may balk at relying on this resource, but I am firmly of the belief that most people now

© Raymond J.C. Cannon 2020. *Courtship and Mating in Butterflies: Reproduction, Mating Behaviour and Sexual Conflicts* (R.J.C. Cannon)

want easily accessible, online information, and it is a pity that more entomological societies and specialist journals do not make their publications available online. Another excellent online resource on butterflies is Adrian Hoskins' 'Learn about Butterflies' (https://www.learnaboutbutterflies.com).

Welcome to the Superfamily!

Origins and evolution

Butterflies have been flying about in this world for a very long time; long before sentient apes like us evolved to appreciate their functional beauty. The Lepidoptera, i.e. moths and butterflies, are thought to have first appeared in the mid Late Triassic (Norian) epoch, about 212 million years ago (mya). The development of a proboscis (Figs 1.7a, b and 1.8) is thought to have played a major role in their evolution, and the ability to suck up free liquid must have given them a huge advantage in the hot, dry conditions in which they evolved, and this organ remains to this day, a remarkable adaptation. Day-flying Lepidoptera, i.e. butterflies, probably appeared much later on, perhaps in the Early Cretaceous, c. 100–110 mya, but it was only after the cataclysmic event at the end of this epoch that butterflies, like mammals, started to diversify into many of the species we see today. So, unlike the dinosaurs, for some reason – perhaps to do with the resilience of their eggs and pupae – many of these early butterflies survived the probable asteroid impact that occurred 65.5 mya. They did not escape this major extinction event completely unscathed however – c. 60% may have gone extinct – but compared with some other groups, theirs was a relative success story, and enough survived to carry the group forward and radiate into the diversity of butterfly groups we see today.[1]

Evolution does not always proceed at a steady rate, and almost certainly occurs in leaps and bounds, as conditions change, and new niches appear. One such period of rapid change in the evolutionary history of butterflies occurred during the Oligocene epoch (34–23 mya): a period of gradual cooling and drying following a time of high temperatures during the Eocene (56–34 mya) epoch, which came before. Many extant modern plants also evolved during this time. These changes in the environment allowed butterflies to take advantage of the new niches that appeared. During the Oligocene, decreasing CO_2 levels and increased aridity favoured the expansion and radiation of grasses, which replaced many of the dicotyledonous plant forests that existed before. The (C4) photosynthetic pathway used by these monocotyledonous plants was better adapted to the lower CO_2 levels, so grasses increased and expanded in ecosystems around the world, producing extensive grasslands by c. 25 mya. This provided a huge opportunity for any species that could cope with the high level of silica in the leaves of grasses, which wears down the mandibles of feeding insects. Satyrine and hesperiid grass feeding butterflies – which had already been around since the Late Cretaceous (c. 80 mya ago) – underwent a rapid radiation, with an explosive evolution of new species adapted to feeding on grasses. This also occurred in concert with the diversification of the higher plants. Indeed, the processes driving the evolution of both flowering plants and butterflies almost certainly involved interactions between insect herbivores – butterfly caterpillars in this case – and the plants on which they fed. An evolutionary contest that has seen plants evolve a vast array of so-called, secondary plant substances – chemicals produced to deter caterpillars from eating them – and ways and means of getting around these defences by the butterflies (and other insect herbivores).[2]

So, the little skippers (Fig. 1.1) probably appeared very early on, about 100 mya, but only really began to evolve into the large number of families we know today after the end of the Cretaceous (66 mya), undergoing a four-fold increase by c. 50 mya. The oldest butterfly fossil found to date is a hesperiid – *Protocoeliades kristenseni* gen. et sp. n. – from the early Eocene, c. 55 mya, in what is now Denmark. Other families of butterflies, such as the Nymphalidae (Fig. 1.2), also diversified during this period, although the subfamily Nymphalinae had probably already split into different types well before the end of the Cretaceous period.[3]

Fig. 1.1. Hesperiidae. Linna palm dart (*Telicota linna*) male, Thailand.

Fig. 1.2. Nymphalidae. Grand duchess (*Euthalia patala*), Thailand.

Butterfly Families

The so-called 'true' butterflies occupy a taxonomic superfamily – called the Papilionoidea – that is made up of five families: Papilionidae (Fig. 1.3), Pieridae (Fig. 1.4), Nymphalidae (Fig. 1.2), Lycaenidae (Fig. 1.5) and

Fig. 1.3. Papilionidae. Great windmill (*Byasa dasarada*), Chiang Mai, Thailand.

Fig. 1.4. Pieridae. Chocolate albatross (*Appias lyncida*), Thailand.

Fig. 1.5. Lycaenidae. Lesser grass blue (*Zizina otis*), Bali, Indonesia. (From: Creative Commons – Flickr.)

Riodinidae (Fig. 1.6). Recent phylogenetic studies have suggested including the closely related Hesperiidae (c. 3500 species) and Hedylidae (butterfly moths) in this superfamily as well, so I will treat them all as families, as shown below, based on the classification by van Nieukerken and others, published in 2011.[4]

Papilionoidea Latreille, 1802 (7 families).
Family **Papilionidae** Latreille, 1802 (32 genera, 570 species).
Family **Hedylidae** Guenée, 1858 (1 genus, 36 species).
Family **Hesperiidae** Latreille, 1809 (570 genera, 4113 species).
Family **Pieridae** Swainson, 1820 (91 genera, 1164 species).
Family **Riodinidae** Grote, 1895 (1827) (146 genera, 1532 species).
Family **Lycaenidae** Leach, 1815 (416 genera, 5201 species).
Family **Nymphalidae** Rafinesque, 1815 (559 genera, 6152 species).

Names and numbers

It is conventional to refer to a family, such as the Nymphalidae (brush-footed butterflies) with a capital letter, but when referring to members of this family – i.e. nymphalids – to use lower case. There are 12 subfamilies in this family, including the Heliconiinae for example, which are referred to as heliconians. Within this subfamily, there are four tribes, including the Heliconiini, the passion-vine butterflies, which contains 10 genera and about 100 species. Species, such as the common postman, *Heliconius melpomene* (Linnaeus, 1758) (Fig. 9.15), are usually described using genus, species and authority – the authority being the person credited with the first formal use of the name for this species. In this book I have dispensed with using authorities, in the interests of keeping it as readable and non-technical as possible, but these can always be obtained by referring to the publication that is quoted in a note for each paragraph. I have also tried to include the common name – where these exist, because some species do not have a widely accepted common English name – where possible.

Fig. 1.6. Riodinidae. Common punchinello (*Zemeros flegyas*) male, Thailand.

I have also, frequently included names for the family (ending in -idae), subfamily (-inae) and tribe (-ini) in parentheses after the species name, where I have considered it necessary to indicate the taxonomic classification and relatedness of a butterfly. Very occasionally, I have also included the subtribe (-ina). For example, the plain tiger, also known as the African queen, *Danaus chrysippus* (Fig. 1.16), is a member of the brush-footed family, Nymphalidae, in the subfamily, Danainae. N.B. milkweed butterflies such as this are now classified as the subfamily Danainae (milkweed butterflies), but the previous family name, Danaidae, is still occasionally used. Scientific names are continually being revised and changed; many of the butterflies called by a given name in the original publication now have a new name! I have tried to use the most up-to-date scientific name wherever possible, so readers may notice that the species (or genus) referred to in the original paper appears to be different. That's taxonomy! Hopefully, the rate of change will slow down, but don't bank on it, as new classifications and reorderings are happening all the time, particularly as a result of molecular genetic investigations. Returning to the plain tiger (Fig. 1.16), butterflies in the genus *Danaus* – which include many well-loved and well-studied species (tigers, milkweeds, monarchs, wanderers and queens) referred to in the following pages – are in the subtribe Danaina. So, we can describe this species as follows: *Danaus chrysippus* (Nymphalidae, Danainae, Danaina). There are also subspecies, like this butterfly from Bali, which give it a triple-barrelled name: *Danaus chrysippus bataviana*. There are a few other such examples of subspecies in this book, including *Pieris rapae rapae* and *P. r. crucivora* (note the abbreviation). N.B. the term *sensu lato* (s.l.) is used to describe all the different types, subspecies or forms of polytypic butterflies like these.

How many butterflies are there in the world? No one knows for sure. New species continue to be discovered and described – perhaps 100 or more every year – but others are disappearing, going extinct, before anyone has even discovered them. Higher-end estimates suggest there may be as many as 20,000 different butterfly species, but most current estimates place the figure closer to c. 18,000.[5]

Pioneering Investigations of Butterfly Courtship Behaviour

Publications with observations on the courtship and mating behaviour of butterflies go back many centuries. For example, Maria Sibylla Merian (1647–1717) was an early naturalist and superb illustrator, with an interest in lepidopteran metamorphosis and behaviour (Fig. 1.9). Nevertheless, I have not attempted to do a historical review, but instead have concentrated on some seminal studies carried out in the 20th century, which provide significant foundations for further, more modern studies, described in the chapters on courtship behaviour (Chapter 5) and body language (Chapter 6).[6]

Fig. 1.7. (a) Stately Nawab (*Polyura dolon*) with yellow proboscis, Thailand. (b) Brimstone (*Gonepteryx rhamni*) with proboscis in flower.

Fig. 1.8. Red Admiral (*Vanessa atalanta*) showing vestigial forelegs, Spain.

Tinbergen and the grayling

The pioneering Dutch ethologist Niko Tinbergen and his students famously studied the courtship behaviour of the grayling, *Hipparchia* (=*Satyrus*) *semele* (Fig. 1.10), at Hulshorst – a sandy site now located in the De Hoge Veluwe National Park, in the province of Gelderland, in the Netherlands – during the summers of 1938 to 1942. They were studying what was then called *releasing mechanisms*, i.e. stereotypical behavioural responses to stimuli. In a classic series of experiments, they investigated the effects of colour, shape, size, distance and movement, using butterfly models mounted at the end of a rod and line and dangled in front of, mostly, male graylings. Shape was shown to be of little consequence, but size mattered: larger than normal models were followed more often by male graylings. Movement was especially important, with a dancing – i.e. continually turning – motion, eliciting the greatest response. The function of androconial areas on the forewings of the males (see Chapter 9, this volume) was also studied in relation to mating success.[7]

Fig. 1.9. Maria Sibylla Merian Metamorphosis insectorum Surinamensium, (From: Maria Sibylla Merian, public domain, plate LX. 1705 Wiki.)

Fig. 1.10. Grayling (*Hipparchia semele*) male, Galicia, Spain (9 September 2018).

The grayling, *H. semele* is a so-called, lateral basker – meaning it holds its wings and body at right angles to the sun to warm up – and rarely opens its wings, other than during courtship attempts and flight. Tinbergen famously called the grayling, 'bark with wings' and it is remarkable how well camouflaged they are, especially when resting on trees (Fig. 1.11). Tinbergen also observed the so-called, 'startle' response in graylings, i.e. when the forewings are suddenly lifted and flicked upwards, to reveal the apical eyespot, or ocellus (Fig. 1.12a, b). This is a defensive move, probably intended to startle, frighten or at least distract would-be

Fig. 1.11. Grayling (*Hipparchia semele*) male on a tree looking like 'bark with wings', Galicia, Spain.

Fig. 1.12. (a) Grayling (*Hipparchia semele*) showing the apical eyespot, or ocellus, Galicia, Spain (19 September 2017). (b) Grayling (*Hipparchia semele*) showing the apical eyespot, or ocellus. Rather worn individual. Galicia, Spain (19 September 2017).

predators. Sometimes both apical and anal eyespots are revealed by this motion (Fig. 1.13). The forewings are then lowered, and the spot is hidden when the butterfly feels that it is no longer threatened. The sudden appearance of the eye-like eyespot – with its white pupil – breaks up the camouflage and probably has a startling effect for this reason. Tinbergen observed that smaller animals, like flies and wasps, are driven away or warded off by wing fluttering, and females also flap their wings in order to ward off males. Tinbergen noticed that the eyespots of females are much blacker, and their centres more striking, than those of the males. This was remarkably prescient because these butterflies are quite variable in form, and such differences are not always obvious to the naked eye. Indeed, it has only very recently been proven that there are such subtle differences in the sizes of the eyespots – including the white pupil, black spot and halo – between the sexes. Females have disproportionately larger wing eyespots than males, perhaps because, being larger – and hence more desirable to a predator – they need to produce a bigger startle response in order to deter their attacker![8]

Fig. 1.13. Grayling (*Hipparchia semele*) female showing both apical and anal eyespots, Galicia, Spain (14 September 2017). (From: Creative Commons – Flickr.)

Tinbergen and his students also famously studied the courtship behaviour of graylings, and meticulously recorded the sequences involved, using words and drawings. The detailed courtship behaviour of *H. semele* and other grayling species is described in Chapter 6, this volume. Suffice to say here that the courtship sequence is divided into five main stages: quivering, fanning, antennal spinning, bowing and clasping. Of these, bowing (Fig. 6.12) is the most iconic and widely quoted behaviour. Essentially, the male grayling stands directly in front of the female and flicks his wings forwards, thus bringing the female's antennae into contact with his androconial (scent) organs. Using Tinbergen's 1948 terminology, this movement by the male acts as 'the releaser' necessary to 'secure the female's cooperation in coition'. This pioneering study has now been supplanted by more recent studies of courtship behaviour in several grayling species and is presented in detail in Chapter 6, this volume.[9]

Crane and 'irrelevant actions' in tropical heliconiids

Jocelyn Crane from the New York Zoological Society – now known as the World Conservation Society – carried out research on butterfly behaviour at the Tropical Research Station in Trinidad, between 1954 and 1957. Using outdoor insectaries, Crane studied what she referred to as 'releasing mechanisms' and 'social behaviour' (essentially courtship and mating) in six heliconiid species – *Dryas iulia* (Fig. 6.2a, b), *Heliconius melpomene euryades*, *Heliconius sara rhea* (Fig. 7.15), *Heliconius ficini insulana* and *Heliconius isabella isabella* – in large, open-air, wire-mesh, insect cages. The terminology used, i.e. releasing behaviour, followed that current at the time and originally formulated by Konrad Lorenz. Social releasers were described as being external sensory stimuli – visual, chemical and tactile – which could release innate response chains in the organism that received them. Crane found that the similarities in the courtship patterns of these species were greater than the differences, and she described their behaviour as consisting of several distinct stages. In the first (*aerial* phase), flying males typically approach perching females from the rear and induce them to take off, without touching (but called *nudging*). There then follows a *flight* stage, where the males chase and then fan the females from above with rapidly vibrating wings. In *Heliconius erato* (Fig. 2.6), the male may brush the wings of the female on the downbeats of his wing flapping, and 'spread the products of his scent scales'. The second (*sedentary*) phase of courtship begins with *primary fanning*, where the male fans the female on the ground, and she flutters her wings and elevates her abdomen. The motion, colour and air currents produced by the fanning wings of the male are important visual and sensory cues, or releasers, to which the female responds. Next follows *secondary fanning*, where the male changes position and the female extrudes her sub-terminal scent glands. Odours released by both sexes were recognized as chemical cues. Finally (in *alighting and engaging*), the male alights beside the female and, as she closes her wings, he curls his abdomen around and engages her abdomen with his 'harpes' (=claspers, see Chapter 7, this volume). Crane continued her investigations of courtship behaviour with a later study of the crimson-patched longwing, *Heliconius erato hydara*, where she noted that the latter phases of courtship may be curtailed, or even omitted, if the female is highly motivated. She also gives a detailed description of the abdominal scent glands of *H. erato* – which are bright yellow in the female – and tiny stink clubs (see Chapter 9, this volume).[10]

The notion of so-called *irrelevant actions* – or displacement activities – was also in vogue at the time and was thought to occur as a result of a conflict, e.g. when an organism is motivated to carry out two conflicting behaviours (such as fight or flee). Crane identified disorientated *wing fanning* and *antennal palpation* as irrelevant actions – i.e. not part of the normal behavioural repertoire of the organism. Crane was aware of the importance of antennal chemoreception but chose to categorize these behaviours as 'irrelevant actions' because they occurred in the latter stages of courtship, when males were thwarted by unresponsive females, and thus engaged in such displacement activities. I think we would now consider these as being stereotypical courtship behaviours, which are, in many cases, frequently repeated in sequences that often do not result in copulation (see Chapters 5 and 6, this volume). Crane also identified what she thought were two other, so-called displacement activities: proboscis uncurling and 'excessive, rapid, undirected flight'. She considered that these behaviours occurred when a 'strongly activated sex drive' was denied expression. This sounds rather Freudian to me! Nevertheless, uncoiling of the proboscis, and using it to touch the head and body of the female, probably occurs in many species – for example, in skippers (Fig. 1.14) – and is particularly well-known in wood white butterflies (Fig. 6.15a, b). Prolonged courtships in *Leptidea* species occur as a result of a failure by males to recognize females as being of their own kind (i.e. in cryptic species) (see Chapter 6, this volume).[11]

Fig. 1.14. Lulworth skipper (*Thymelicus acteon*) courting pair male with extended proboscis, Galicia, Spain.

Crane also investigated the spectral composition of the colours present on a wide range (n = 41) of butterflies from Trinidad. She showed that whilst females, as well as males, of black and red race of *Heliconius erato*, were stimulated by the colour red on the wings of these butterflies, females were in general less influenced by colour changes than were males (see Chapter 2, this volume). It was already known in the 1950s that butterflies were sensitive to ultraviolet (UV) light, but Crane found that only a minority (5/41) of her test species – those like *Morpho* butterflies (Fig. 10.1 and 10.15), with iridescent areas – had a significant level of UV reflectance from their wings and bodies, which suggested that this property might be involved – or in her words, had an 'adaptive value' – in intraspecific behaviour. She also investigated how the spectral reflectance of iridescent *Morpho* wings, changed with the angle of incidence – i.e. the viewing angle – relative to the sun, and remarked on how this would generate different colours in the eyes of other butterflies, passing from blue-green to blue to UV, as the butterfly flapped its wings. These subjects – butterfly colour vision (Chapter 3, this volume) and the generation of colours (Chapter 10, this volume) – have progressed enormously since these early, pioneering investigations.[12]

Stride and the danaid eggfly

George O. Stride carried out behavioural experiments in Ghana in the 1950s on the danaid eggfly, or diadem, *Hypolimnas misippus* (Fig. 1.15a, b), very much in the tradition of Tinbergen. *Hypolimnas misippus* males – whilst perching on the ground – fly up to investigate passing objects (birds as well as insects!) in a similar manner to that of the common grayling (above). Darker subjects, i.e. similar in colour to that of the female, are almost invariably investigated and 'frequently harried for some distance' by the males, according to Stride. In other words, a behaviour more suggestive of mate searching, or mate seeking, than aggressive territorial defence,

Fig. 1.15. (a) Danaid eggfly (*Hypolimnas misippus*). (From: Creative Commons – Flickr, Tarique Sani.) (b) Danaid eggfly (*Hypolimnas misippus*) male, Madayipara, Kannur District, Kerala, India. (From: Uajith [CC BY-SA 4.0 (https://creativecommons.org/licenses/by-sa/4.0)].) (c) Danaid eggfly (*Hypolimnas misippus*) Female, form *inaria*. (From: Ajith U [CC BY-SA 2.0 (https://creativecommons.org/licenses/by-sa/2.0)].)

as Denis Owen suggested in his 1971 book, *Tropical Butterflies*. The females of this species are mimetic, closely resembling different forms of the African queen, also called the plain tiger in Asia, *Danaus chrysippus* (Fig. 1.16). There are three different forms of *H. misippus* in Africa, which are mimics of the three forms of its model, *D. chrysippus*. Although not all models and mimics may be present in a given location, the existence of this multitude of morphs implies – as Denis Owen suggested – that there might be a considerable level of complexity in terms of the courtship behaviour amongst these mimetic assemblages. N.B. the monomorphic male of *H. misippus*, is very similar to that of *Hypolimnas bolina* (Fig. 2.1).[13]

Stride used cardboard models on the end of a long bamboo cane to investigate the ability of male *H. misippus* to recognize and court females, similar to the method of Tinbergen. The painted models were moved in front of free-flying males 'with a gentle dancing motion' in order to test their response to different visual stimuli. He had already demonstrated that the males respond to the females with a 'characteristic sequence of courtship reactions', notably a courtship flight, where the female ascends high into the air with wings quivering (see Chapter 5, this volume). When the female is ready to mate, she lands and holds her wings open; closed wings are a sign of rejection or unreceptivity in this species. Stride's experiments were remarkably detailed, and he came to a number of conclusions regarding the preferences of males and the importance of visual

Fig. 1.16. Plain tiger (*Danaus chrysippus bataviana*) male nectaring, Bali, Indonesia.

stimuli during the early stages of courtship. He concluded that 'stimuli arising' from the hindwings of females had a greater effect on the behaviour of males than those from the forewings, and furthermore, that the presence of white on the female hindwing has an inhibitory effect on male courtship behaviour. Thus, negative sexual selection was thought to be the reason for the rarity of the white hindwinged form of *H. misippus*. Stride has, however, been criticized for his small sample sizes and consequent lack of statistical analysis. Nevertheless, his observations of courtship and mating in *H. misippus* remained amongst the most detailed for some time. Researchers have continued to use coloured model butterflies in choice experiments; for example, to investigate male and female preferences and choices of wing colours.[14]

Magnus and the silver-washed fritillary

In a series of experiments, over nearly a decade, German ethologist Dietrich Magnus investigated how male silver-washed fritillary butterflies, *Argynnis paphia* (Fig. 1.17a, b), responded to optical stimuli. Magnus recognized that optical stimuli attracted the male during the zigzag flight, carried out whilst searching for females. The courtship behaviour of the silver-washed fritillary is particularly striking: the female flies in a straight line and the male loops over and under her, showering her with pheromone scent-laden androconial scales (see Chapter 5, this volume). Magnus reported that similar zigzag seeking behaviour occurred in other species, including the dark green fritillary, *Argynnis aglaja*, and the brimstone, *Gonepteryx rhamni* (Fig. 3.2).[15]

There are two, differently coloured, female morphs of *A. paphia*: the normal (or nominate) fulvus-coloured form, *paphia*, and the much darker, rarer, greenish-grey coloured form, *valesina* (Fig. 1.18). It is only the females who are dimorphic: male *valesina* are indistinguishable from normal males. Male *paphia* form silver-washed fritillaries are a brighter orange than the *paphia* females, but both sexes are fundamentally the same colour (or hue). Although, the *valesina* morph is relatively rare in

Fig. 1.17. (a) Silver-washed fritillary (*Argynnis paphia*) male, Dereköy, Kırklareli, Turkey. (From: Zeynel Cebeci [CC BY-SA 3.0 (https://creativecommons.org/licenses/by-sa/3.0)].) (b) Silver-washed fritillary (*Argynnis paphia*) males. (From: Pixabay.)

Fig. 1.18. Silver-washed fritillary (*Argynnis paphia*) female, form *valesina*. Bentley Wood, Hampshire, UK. (Photo by Charles J. Sharp [CC BY-SA 4.0 (https://creativecommons.org/licenses/by-sa/4.0)].)

the UK, c. 5%, it becomes commoner towards the east, in Europe and Asia. Magnus found that the *valesina* morph was preferred by males only 10% of the time, but this lack of visual interest in the darker form appears to be overcome by the attractive effect of a female pheromone. The advantages of the *valesina* female forms – which presumably maintain the presence of this rarer morph in natural populations – are not, as far as I am aware, definitively known, but might include: (i) a thermoregulatory advantage in some warmer climates, (ii) decreased predation due to better camouflage and (iii) an avoidance of harassment by males. For example, *valesina* form, *A. paphia* females, studied in Corsica, appeared to have a higher heat tolerance; e.g. remaining for longer at feeding sites during the midday heat, whilst *paphia* females departed to the shade.[16]

Magnus investigated the attraction of male *A. paphia* to stationary pieces of coloured paper, in an effort to determine flower colour preference. To attract the male, the target had to be coloured, and of a similar orange/yellow colour to that of the base colour of the butterfly wing. One colour – Ostwald colour paper No. 4 PA (a now discontinued colour system) – a beautiful, bright orange colour, proved to be most effective.

Male silver-washed fritillaries were attracted by the coloured paper models, but they lost interest when they got close (within 10 cm) to them and did not attempt to copulate with them. They did, however, remain interested – and were sufficiently aroused to attempt to copulate with the models – if a female was hidden under a stationary dummy, so that her scent was present. The response to this bright, orange/yellow colour was thought to be innate.[17]

These experiments with silver-washed fritillaries were repeated in a PhD study by Alexandra Freeman, at the University of Oxford, to investigate whether males are innately attracted to orange, *paphia*, females or whether they learn a preference between the two morphs by experience. Freeman's results confirmed the earlier studies by Magnus, that male *A. paphia* were attracted to very bright orange, and at close quarters appear to be able to differentiate between the sexes (presumably via the female pheromone). Additional experiments carried out by Freeman in Crab Wood, near Winchester, UK, in 1997 – using, pinned, dead females of each form – confirmed that males significantly preferred *paphia* females to *valesina* females (which were present at about 7% in the natural population of *A. paphia* in Hampshire at the time).[18]

Magnus discovered that the males were most powerfully attracted to what he called 'super-optimal' – what we would now call supra-normal – stimuli. For example, when the colour surface was as large as possible, and appeared and disappeared – like the flapping wings of a butterfly in flight – as quickly as possible. In other words, the males were responding to a very large, fluttering patch of colour, in preference to butterfly-sized models. The key finding was that male butterflies responded more strongly to brightly coloured, strongly reflecting 'models' with simple black markings, than to the normal, realistic wing colours of females (and males) of the *paphia* morph of the same species. In other words, the natural black, fritillary wing pattern, and wing shape, are not key stimuli for the males, and there were no shape preferences.[19]

To investigate the effects of flapping or wing fluttering (optical flickering), Magnus attached paper models to the end of 6-ft-long arms, on a rotating merry-go-round, so that the butterfly models appeared to be flying. Males responded to the size, colour and flicker frequency of the models. In another experiment, he used a rotating cylinder, orange on one half side, black on the other. Male *A. paphia* preferred flashing stimuli, at wing strokes of 20 Hz, increasing to 140 Hz; beyond the flicker-fusion rate of the eye and much faster than the normal flapping frequency of this species. Another supra-normal signal it would seem.[20]

Brower and the queen butterfly

The courtship behaviour of the queen butterfly, *Danaus gilippus* (Fig. 5.17), was systematically investigated by Lincoln Pierson Brower (1931–2018) and colleagues in a landmark study, carried out in Florida and Trinidad, in the summers of 1959 to 1961, and 1962 to 1964, respectively. Clearly a remarkably comprehensive investigation, carried out over 6 years, and published in one large publication. Lincoln *et al.* used laboratory-reared females and wild caught male *D. gilippus* butterflies. The investigation was comprehensive, with 187 courtships, involving 40 different females, analysed in one year (1960) for example. The courtship sequence was broken down into four main components, consisting of nine different phases. These can be summarized as follows (from table 1 of Brower *et al.*, 1965).

1. First aerial component: aerial pursuit and aerial hairpencilling.
2. First ground component: ground; hairpencilling; hovering and striking; copulation attempt; copulation.
3. Second aerial component: post-nuptial flight.
4. Second ground component: insemination; termination of copulation.

Details of some of these courtship phases will be discussed with reference to other danaids in Chapters 5 and 9 of this volume. Brower *et al.* describe the courtship sequences as following a 'stimulus–response' chain: one behavioural act by the male – such as brushing the female with extruded hairpencils – followed by another behavioural act from the female, e.g. folding her wings, if she is receptive. Absence of a positive response from the female can induce the male to initiate the courtship sequence once again. Queen butterflies do not mate until the male has brushed his abdominal hairpencils over the female's head and antennae, transferring the 'love dust' particles containing the male pheromone (see Chapter 9, this volume). This pattern of repetition of one or more elements of the courtship is a feature of sexual interactions in many species, as will be shown in Chapter 5, this volume. Female queen butterflies can reportedly mate up to 15 times – a record for butterflies.[21]

Notes

[1] Wahlberg (2006), Wahlberg *et al.* (2009), Zhan *et al.* (2011), Heikkilä *et al.* (2012) and van Eldijk *et al.* (2018). N.B. the Cretaceous-Paleogene (K-T) mass extinction occurred 65.51 +/– 0.3 million years before the present (Jablonski and Raup, 1995).
[2] Ehrlich and Raven (1964). Butterflies in the subfamily Satyrinae feed on monocotyledonous plants (Ackery, 1988) and the bulk of species (tribe Satyrini, c. 2200 species) almost exclusively use grasses (Peña *et al.*, 2006; Peña and Wahlberg, 2008). Hesperiids also diversified as a result of evolving an ability to feed on monocots (Sahoo *et al.*, 2017).
[3] Wahlberg *et al.* (2005, 2009), Wahlberg (2006, 2009), Heikkilä *et al.* (2012), De Jong (2016) and Sahoo *et al.*, 2017).
[4] Peña and Wahlberg (2008), van Nieukerken *et al.* (2011) and Heikkilä *et al.* (2012).
[5] There are close to 20,000 described species according to Kristensen *et al.* (2007), c. 18,768 known butterfly species according to van Nieukerken *et al.* (2011), 17,698 according to Hoskins (see Annex 1), 17,280 species according to Shields (1989), 13,750 species of true butterflies, i.e. excluding skippers (Robbins and Opler, 1997), 15,900–18,225 butterflies including skippers (Robbins, 1982) and 17,500 (Ackery *et al.*, 1999).
[6] Maria Sibylla Merian began a study on the insects and plants of Dutch Surinam – now called the Republic of Suriname – in 1699, at the age of 52, and published a beautifully illustrated book in 1705, called *Metamorphosis insectorum surinamensium* (Etheridge, 2011).

[7] Kruuk (2003).
[8] Tinbergen (1941, 1972), Tinbergen *et al.* (1942), Tinbergen *et al.* (1972) and Dapporto *et al.* (2018).
[9] Tinbergen (1948, 1972) and Tinbergen *et al.* (1972).
[10] Crane (1955, 1957). Lorenz (1935) is described by Tinbergen (1948) as putting forward the idea of social releasers. These are stimuli sent out by one organism (the 'actor') that are received by another (the 'reactor'). The original William Beebe Tropical Research Station, located at Simla in the lower Arima valley, Trinidad, was donated to Asa Wright Nature Centre (AWNC) in 1974.
[11] Crane (1957).
[12] Crane (1954).
[13] Owen (1971).
[14] Edmunds (1969), Stride (1956, 1957 and 1958) and Owen (1971). The different colour forms of female *H. misippus*, and its supposed model, *Danaus chrysippus*, are shown in Smith (1976; Imafuku and Kitamura, 2015, 2018).
[15] Magnus (1950, 1963).
[16] Magnus (1958) used males that originated from populations in south-west Germany, where the *valesina* morph was, and perhaps still is, rather rare. Vane-Wright and Boppré (1993) commented that the involvement of a female pheromone in the courtship sequence of this species is probably a rare exception amongst butterflies, generally (Brunton *et al.*, 1990; Freeman, 1998; Smith, 2014).
[17] The Ostwald 'scientific colour organ' was a collection of 680 pigment powders, created by Wilhelm Ostwald in 1925 (Bridarolli *et al.*, 2016).
[18] The experiments of Magnus (1950, 1954 and 1958) – published in German – are described to some extent by Freeman (1998) in her PhD, and comprehensively by Bastock (1967). I must admit that I used Google Translate to read the German!
[19] The 'fluttering patch of colour' was described by Kral (2016). Male *A. paphia* followed conspecific females (with wing areas of c. 20 cm^2) less frequently than much larger (2–4×) models with wing areas of up to 88 cm^2 (Magnus, 1958).
[20] The flicker-fusion rate for butterflies is around 100 Hz (White *et al.*, 2014, quoting Rutowski, 2003). The flapping frequency of *A. paphia* wings is about 8–10 Hz according to Kral (2016).
[21] The work on the queen butterfly by Brower *et al.* (1965) was also undertaken at the same laboratory in Trinidad where Jocelyn Crane had carried out her investigations a decade or so earlier (Crane, 1957). Brower and Jones (1965) described the complete sequence of behaviour involved in the extrusion of the hairpencil glands in the queen butterfly, *Danaus gilippus xanthippus*.

2 Sexual Selection and Mate Choice

> As the laws of Nature must be the same for all beings, the conclusions furnished by this group of insects must be applicable to the whole organic world; therefore, the study of butterflies – creatures selected as the types of airiness and frivolity – instead of being despised, will some day be valued as one of the most important branches of Biological science.
>
> H.W. Bates (1863)

Choosing a Mate – the Female Prerogative?

In butterflies, as in most animals, females produce many fewer gametes than males. However, their eggs are larger and more costly to make than sperm, which are very small, and cheap to produce. It is for this reason, that females generally get to make the selection of whom to mate with, because they have invested so much more, in terms of material contribution to the next generation. When females choose mates based on the form and appearance of secondary sexual characteristics, the benefits can be direct – e.g. in the form of a large spermatophore (see Chapter 8, this volume), or indirect, in terms of the genetic fitness of the male, which will be expressed and realized in their offspring. In some species however, such as the monarch butterfly, it is not altogether clear which sex is making the most determinative choices, and two-way choices – i.e. preferences for traits or ornaments by both sexes – can occur simultaneously. Choices and decisions regarding mates will also depend on a range of internal and external factors, such as the age and reproductive history of the chooser, and the range and availability of potential mates. Choices based on a preference for a particular trait or ornament may be beneficial in terms of selecting mates with good genes, i.e. better than average alleles that are

© Raymond J.C. Cannon 2020. *Courtship and Mating in Butterflies: Reproduction, Mating Behaviour and Sexual Conflicts* (R.J.C. Cannon)

linked to the expression of the ornamentation trait, but these are not necessarily the most genetically compatible individuals.[1]

Choice of a mate may be made before or after mating, based on *objective* criteria, such as the size of a nuptial gift, the quantity of sperm (cryptic female choice) or even the size and location of male territories. Choices may also depend on *subjective* assessments of quality – for example, based on courtship behaviour – including visual and chemical signals. There is also a theory – the *genetic compatibility hypothesis* – that females, rather than choosing males with the best or most superior genes in a population, may choose individuals with whom they are the most genetically compatible. In other words, because the survival and fitness of the offspring will result from the combination of both male and female genotypes, it might be better to choose a mate who best complements your own make up, rather than choosing the best or highest-ranking individual in the field of candidates. An example of a bad choice would be a closely related individual, who would produce deleterious effects in offspring as a result of the coming together of recessive genes. A good choice would be a genetically variable individual, who will – in combination with the choosers' own genes – produce genetic diversity in their offspring.[2]

Sexual selection

Charles Darwin has dominated thinking on sexual selection for nearly 150 years, ever since he published his book, *The Descent of Man, and Selection in Relation to Sex*, in 1871. He was however, mainly focused on vertebrates and on aspects of sexual selection that occur in animals prior to copulation, sometimes called pre-ejaculatory, or Darwinian, sexual selection. Darwin speculated that male ornamentation and displays, like the peacock's tail, were selected by the aesthetic senses of the female, her 'sense of beauty' and not necessarily as utilitarian functions indicating 'good genes' or male fitness. In other words, they were *non-adaptive* characteristics. We now tend to think of such elaborations however, as indications of some sort, either overt or covert, displaying some measure of material or genetic fitness. Darwin believed that differences – in form, function and behaviour – between males and females of a species could be explained by the operation of two mechanisms: female choice and competition between males. That is, competition between males for territory and/or access to reproductively available females, and female choice of aesthetically pleasing features or ornaments, which would, over evolutionary time, produce the male weapons and adornments seen on animals, from peacocks to stag beetles, as well as butterfly wings![3]

For much of the time after Darwin however, his views of sexual selection, and especially the importance of female choice, were eclipsed by other theories emphasizing the role of the male in stimulating the female into mating. Other Victorian scientists, such as Alfred Russel Wallace, thought that *all* traits were subject to natural selection, and male ornamentation had come about as a result of natural selection for some purpose, such as camouflage. Many scientists simply could not accept that animals could make rational, or even aesthetic choices, which were thought to be something only human beings could do. For cultural, and perhaps even chauvinistic reasons, the emphasis was placed firmly on the male, and on *his* role in stimulating the female. Instead of a female *choosing* a male, biologists suggested that their selections were a result of the male overcoming the natural reticence of the female. The more stimulated a female was by a given male, the more likely she was to mate with him. In other words, it was the male's action in inducing the female response, rather than the female's in making choices or selections based on the male phenotypes that were at play, or on display. Explanations of courtship displays usually sought to place them in the realm of natural, rather than sexual selection. There has however, been a belated acceptance of Darwin's proposition that female mate choice is widespread in nature, and there is now widespread agreement that Darwin was correct in assigning such an important role to sexual selection – as a driving force in evolution.[4]

Sexual conflicts

Darwin focused on events that occurred before copulation; e.g. courtship displays and contests between males, but in recent decades however, scientists have discovered that sexual selection also occurs during and after mating. Such *post-copulatory* sexual selection, such as sperm competition, occurs inside the female reproductive tract during and after copulation. Attention has also focused on sexual conflicts, which occur when the evolutionary interests of the two sexes differ. The male and female genomes are different in unrelated individuals and sexual conflicts can occur as a result of their mutual investment in offspring. Conflicts arise as a result of differential investment, where one sex – usually the male – reduces their investment – of protein, time and energy, for example – in producing offspring so that they can try and maximize their fitness via another reproductive act with a different individual.[5]

This emphasis on sexual conflict highlights a paradigm shift in the way biologists think about reproduction. For much of the 20th century, sexual reproduction was mostly seen as an exercise in cooperation, viewing mating as a harmonious venture. The consensus now, however, is that conflicts of interest will always arise between individuals that are not genetically identical, and that conflicts between the sexes are an important evolutionary force: an evolutionary arms race of adaptation and counter-adaptation, as the sexes struggle to maximize their own benefits from reproduction, and minimize the costs resulting from the on-going conflict. Different types of interaction – which have been categorized as persuasive, coercive and forcing – may occur between the sexes as they compete to gain control of both fertilization and reproduction. It has been said that it is a wonder that the benefits of sexual reproduction are great enough to have allowed it to have evolved at all! Yet such 'sexually antagonistic coevolution' has, for example, produced the vast range of features and armaments seen in the genitalia of insects.[6]

Sexual selection occurs via competition for mates, the outcome of which produces differences in reproductive success. Competition can occur between members of the same sex, for access to the opposite sex, or via members of one sex choosing a mate from amongst the opposite sex. In practice there may be a continuum between competition and choice. The intensity of sexual selection usually differs between males and females, often being stronger in males as a result of their promiscuity, i.e. sexual selection operates at each mating event. Female butterflies can store sperm and can usually obtain enough sperm from one mating to fertilize all of the eggs they can produce. Thus, females may only mate once, and often exhibit considerable resistance to further matings, although variations occur in long-lived species. Males on the other hand, can mate many times and therefore come under sexual selection on each occasion.[7]

Good genes and handicaps

A key question for biologists is whether the ornamental traits exhibited by males – i.e. secondary sexual characteristics – have evolved as a result of arbitrary female choice, or are adaptive, in the sense of having been produced by sexual selection as genuine indicators of male quality. Arbitrary female choice could, in theory, result in the overelaboration of a trait until it became such a burden that it would be weeded out by natural selection. The other alternative is that certain traits are 'honest advertisements'; in other words they are genuine indicators of quality or good genes. These characters are sometimes called 'handicaps' because they impose a cost on the bearer, who will have to spend surplus energy or resources producing such traits, which are therefore, an indication of the quality of the male. The fact that such sexual ornaments are costly to produce is an essential element in their persuasive power. Implicit in this theory is the idea that the costs incurred in producing a sexual trait – like highly iridescent wing scales (Fig. 10.1), for example – are less for high-quality males than they are for those of low quality. Another way of thinking about it is that the costs of producing sexual ornamentations are higher for low-quality individuals in poor condition than for high-quality males in good condition. Finally, the costs themselves may be direct – perhaps being predated upon as a result of being too shiny or flashy – or indirect, as a result of spending too much energy and becoming depleted or compromised in some way. There is, however, very little empirical evidence in the literature to verify the assumption that so-called sexual traits really are costly to produce (see Chapter 10, this volume).[8]

Identifying a Potential Mate

In general, male butterflies are notoriously poor at recognizing a potential mate at a distance: perching males often pursue any butterfly, or other flying insect, which intrudes into their territory. These males are it seems, not so much exhibiting territorial behaviour as just being rather poor-sighted in terms of determining whether the intruder is a male or female! For example, great eggfly, *Hypolimnas bolina* (Fig. 2.1), males establish and defend territories in order to help them locate receptive females: perching males launch themselves at any object that flies past, but they are only able to detect females at about 4–5 m, or less. Nevertheless, this tactic of launching themselves at anything that comes within their visual field is probably the best way they have for locating suitable mates. Males of the Xuthus, or Asian swallowtail, *Papilio xuthus* (Fig. 2.2a, b), are initially attracted to the striking black and yellow pattern of the wings of this species, which act as a sort of 'visual releaser', but they probably only determine the sex of a conspecific individual by making actual physical contact with it.[9]

Once males are close to a female however, they can usually distinguish between small differences in wing colours, marks and patterns, and often have specific preferences in terms of female wing coloration. For example, males of two lycaenid species – the green hairstreak, *Chrysozephyrus smaragdinus* (Theclini) and *Neozephyrus japonicus* (Theclini) (Fig. 2.3) – were shown to use wing colours to tell the sexes apart, and prefer the cryptically coloured female wings. Cabbage white butterflies, *Pieris rapae*,

Fig. 2.1. Great eggfly (*Hypolimnas bolina jacintha*) male, Thailand.

Fig. 2.2. (a) Asian swallowtail (*Papilio xuthus*) dorsal view. (b) Ventral view. (Photos by S.C. Yam.)

Fig. 2.3. *Neozephyrus japonicus japonicus*, in Mie pref., Japan. (Photo by Alpsdake [CC BY-SA 3.0 (https://creativecommons.org/licenses/by-sa/3.0)].)

use the small but conspicuous black wing spots on their dorsal forewings to tell each other apart – i.e. male, one spot; female, two spots (Fig. 2.4) – and to discriminate their own species from other pierids. Leopard lacewing males, *Cethosia cyane* (Fig. 2.5a–c) also use wing coloration – as well as olfactory cues – to identify females, rather than wing size; this even though females are on average, considerably larger. The reflectance spectra of the dorsal (upper) surfaces of male and female *C. cyane* wings differ markedly in certain portions of the visible and UV spectrum, allowing the sexes to identify each other. The wing *patterns* themselves are, however, relatively similar in both males and females, and the beautiful ornate markings on the ventral wing surfaces (Fig. 2.5c) are probably the result of natural selection for crypsis. Female *C. cyane* also emit a scent containing cedrol, an aromatic compound also found in the oils of conifer trees.[10]

Visual signals

Visual signals produced by butterflies as a way of advertising their qualities to potential mates – or vigour to potential rivals – need to be highly conspicuous to the receiver, but at the same time, not too obvious to would-be predators. This seemingly impossible requirement has been achieved in many butterflies by a combination of 'design' structure and behaviour. Structural colours (see Chapter 10, this volume) are one way in which butterflies can produce very bright and vivid colours, which can in effect be aimed or directed at the receiver by the signaller. By using a combination of behaviour and targeted reflectance – almost like we might use a mirror to communicate across long distances – butterflies can reflect bright, sometimes UV colours, often only visible within a narrow angular

Fig. 2.4. (a) *Pieris rapae* female mounted specimen. (Photo by Sarefo [CC BY-SA 3.0 (http://creativecommons.org/licenses/by-sa/3.0/)].). (b) *Pieris rapae* male mounted specimen. (Photo by Ypna [CC BY-SA 3.0 (https://creativecommons.org/licenses/by-sa/3.0)].)

Fig. 2.5. (a) Leopard lacewing (*Cethosia cyane*) male upside, Thailand. (b) Leopard lacewing (Cethosia cyane) female upside, Thailand. (c) Leopard lacewing (*Cethosia cyane*) male underside, Thailand.

range, during courtship displays. Reflections can be combined with movements, such as wing flickering, which amplify the signal. Evolution has selected behaviours and 'designs' that enable highly targeted displays, which can be switched on and off, with the flick of a wing! The advantages of signalling using UV light for example, is that it is scattered more efficiently in air than light of longer wavelengths, and therefore may be invisible to at least some predators that might lack UV photoreceptors. For example, the yellow bars present on the hindwings of *Heliconius erato*

(Fig. 2.6), and the yellow patches on the forewings of *Heliconius numata* and *Heliconius ethilla* (Fig. 2.7), contain a yellow pigment that reflects both UV and long-wavelength (i.e. visible) light. This remarkable pigment, which also fluoresces, is used by these butterflies to recognize members of their own species, and to tell themselves apart from very similar looking mimics that do not have this type of pigmentation. It's their own unique colour signal, called UV yellow, and we can only guess what it looks like to them, with their enhanced spectral discrimination.[11]

Polarized light generated by iridescent wing scales is widely used for mate recognition in some butterflies. For example, males of the cydno longwing, *Heliconius cydno chioneus* (Fig. 2.8) were significantly less

Fig. 2.6. *Heliconius erato petiveranus* (taken at Krohn Conservatory butterfly show). (Photo by Franz Xaver – own work, CC BY-SA 3.0, https://commons.wikimedia.org/w/index.php?curid=10271252)

Fig. 2.7. Ethilla longwing (*Heliconius ethilla narcaea*), Argentina.

Fig. 2.8. Cydno longwing (*Heliconius cydno chioneus*). (Photo by Greg Hume [CC BY-SA 3.0 (https://creativecommons.org/licenses/by-sa/3.0)].)

interested in females when the light from their polarized, blue, iridescent wings was filtered out by depolarizing filters. Bright colours on the wings of male butterflies may also have an inhibitory effect on other (conspecific) males in some species, such as the shiny blue (male) lycaenid, *Favonius taxila*, and may have evolved through male–male competition. Iridescent colours do not always work in both ways however, i.e. for attracting mates *and* competing with other males. Individual butterflies will have evolved their bright colours for specific purposes, which need to be elucidated for each species. Patches of colour on a butterfly's wing are complex features, generating powerful signals via several mechanisms – structural and pigmentary – which may be perceived in different ways in different situations. For example, large greenish-blue patches on the hindwings of female *Cheimas opalinus* (Satyrinae), a Venezuelan species, fade away and disappear during mating. The highly reflective blue scales, used for attracting mates by both sexes, conveniently get rubbed off the females' wings – brushed off by hairs on the males' wings – during mating and nuptial flights, rendering the female inconspicuous and cryptically coloured. The shiny, structurally coloured scales are no longer needed once mating is over, and are a disadvantage, when once they were an advantage![12]

The evolution of costly sexual ornamentation in females does not usually occur because of the competing demands of egg production, but it could occur via sexual selection if the benefits of such displays outweighed their costs, e.g. in terms of loss of fecundity. The sulphur butterflies, *Colias eurytheme* and *Colias philodice*, are sexually dimorphic in terms of their iridescent wing coloration, as are many coliads (sulphurs or yellows). Over half of studied taxa (62%) of Coliadinae exhibit some form of UV reflectance pattern. The UV patterning is found more frequently in males than in females, but the females of many species also display UV coloration, in either a full (e.g. *Eurema candida, E. herla*) or partial form (e.g. *Eurema hecabe, E. sana, E. laeta*); whilst the dorsal wing surfaces of some species, such as *Eurema lisa*, reflect little or no UV light. There is, therefore, a great deal of variation in UV reflectance, between species, but also, between individuals, which suggests that this property is unlikely to be used for species recognition, at least in *Colias* species, but this is an active area of research and it is probably premature to make generalizations about the functional significance of UV patches. In *Gonepteryx* butterflies, the dorsal UV patches on both fore- and hindwings are also variable – more so than their visible coloration – but it is possible that they play a role in species recognition. However, it is also likely that these UV patterns are involved in female mate choice, and perhaps, male–male competition.[13]

It is generally thought that male *Colias* butterflies are attracted to the yellow ground colour and ultraviolet *absorption* of the female hindwings, whereas females choose mates using the brilliant UV *reflectance* of the males' dorsal wing surfaces. However, male sulphur butterflies court females with and without a basal UV patch with equal intensity, suggesting that UV patterns are not a mating cue for males in these species. Female large grass yellows, *Eurema hecabe* (Fig. 2.9), for example, have smaller, duller, less bright, iridescent patches at the base of their dorsal forewings, compared with males, which are not used as mating cues, although there is a suggestion that males preferentially court females with larger UV patches. The different UV reflectance properties of the wings of male and female grass yellows, *E. hecabe*, correlate with the density of nanoscale, lamellar-bearing ridges, i.e. multilayer UV reflectors (see Chapter 10, this volume), on their wing scales. Male *E. hecabe* have been shown to be attracted to pupae in which pharate adults with yellow wings are developing. The males hover above, and sometimes alight on a pupa, but do not appear to discriminate between male and female ones, suggesting that the attraction is visual rather than chemical. N.B. the metallic, UV reflectivity of the pharate males' wings is not visible through the pupal cuticle.[14]

Subtle differences in wing colour patterns have been used by taxonomists to separate closely related species – such as the North American lycaenids, *Lycaeides* (=*Plebejus*) *idas* (Fig. 2.10) and *Lycaeides melissa* –

Fig. 2.9. Common grass yellow (*Eurema hecabe*), Thailand. Sex undetermined.

Fig. 2.10. Idas blue (*Plebejus idas baldur*) female underside, Bulgaria. (Photo by Charles J. Sharp – own work, from Sharp Photography, sharpphotography, CC BY-SA 4.0, https://commons.wikimedia.org/w/index.php?curid=61829458)

which look very similar yet occupy separate ecological niches. The males use some of these detailed patterns – the size of certain hindwing spots and orange crescent-shaped areas (aurorae) – to distinguish between females of the two closely related species, in favour of their own kind (i.e. species). Whilst this is just one example, it does lend support to the intuitive idea that butterfly patterns are inherently meaningful and informative, both as a means of communication between individuals in the same species, and – as in this case – as a visible template for making choices and distinguishing between different species.[15]

Choices, by either sex, may be direct or indirect, active or passive (see below). In its purest form, a choice may be made directly by comparing and judging one or more traits, or characters, from a pool of potential mates. Choices can, however, occur indirectly, for example by being attracted to the brightest, or loudest member of the opposite sex; or by mating with the one – usually a male – who has removed or excluded all of the other contenders, either by means of a contest, or by gaining control of a territory or lekking site (see Chapter 3, this volume). The extent of the 'choice' clearly depends on circumstances, contingent factors such as the size of the pool of prospective candidates and the extent to which contests have been resolved between competitors.[16]

Female Choice

The essential elements of Darwin's definition of sexual selection are that individuals vary in terms of reproductive success, and the advantage of one male over another is worked out in terms of competition and female choice. The concept was further elaborated into what was termed the Darwin–Bateman Paradigm (DBP), after experiments on fruit flies by the English geneticist, Angus John Bateman (1919–1996) suggested that the reproductive success of females was limited by the number of eggs they produce, whereas in males it is a function of the number of times they can mate.

This supposedly fundamental variation in gamete production was used as a basis to suggest that the reproductive success of males is more variable than that of females; males are more eager and indiscriminate in their choice of mates; females are coy and more discerning; and that males gain more from repeated mating (polyandry) than females. Some of these ideas are now seen as rather quaint 'gender stereotyping', and whilst they may hold for some cases, the role of chance, and the ability of individuals to make mating decisions based on their circumstances, has led to a more 'gender-neutral' theory. The characteristics and behaviour of males and females in terms of reproduction are probably much more flexible and *contingent* – i.e. upon internal and external factors – than was previously realized.[17]

It is now generally accepted that females exert considerable control over their own reproduction and there is no reason to think that this does not apply to butterflies. Females of most species are much more promiscuous and active in terms of soliciting copulations than was once thought, and polygamous females have a greater reproductive success as a result (see Chapter 7). This more 'gender-neutral model' of sexual selection focuses on the behaviour of, and constraints operating upon, *both* sexes. Nevertheless, even though studies on sexual selection and female choice have undoubtedly been biased to some extent in the past by prevailing, stereotypical views of gender roles, the contributions of pioneers such as Darwin and Bateman nonetheless remain valid in a general sense. Fundamental differences between the sexes, particularly relating to their relative investment in reproduction – technically described as 'genetic expenditure' – shape the roles and behaviours of the sexes. Thus, females are generally the choosier sex – evaluating a repertoire of male features, including coloration, pheromone bouquets and courtship ardour – but there are many examples where males are also choosy (see below, 'Male Choice'). It is this dynamic of sexual selection that has, to a large part, produced the wonderful array of butterfly colours, patterns and behaviours we see in the natural world.[18]

Eyespots, ornamentation and ultraviolet light

Eyespots are the bull's eye-shaped roundels found on so many butterfly wings, particularly species in the subfamily Satyrinae, which are masters of this sort of patterning (Fig. 2.11). Their function appears to be mainly one of predator deflection – i.e. deflecting predator attacks away from the body – or startle. These spots are remarkably invariant over time within a given species – despite considerable genetic potential for variation – which suggests that they are maintained, or stabilized, at a certain size and shape by forces of selection. These forces may, however, be different on the two sides of the butterfly wing, and in different seasons. Ventral eyespots, mainly used for predator deflection, are probably under natural selection, whereas the dorsal eyespots – often used by females as a signal of male quality (see below) – can be under sexual selection. However, we

Fig. 2.11. Common fivering (*Ypthima baldus*), Chiang Mai, Thailand.

must remember that these generalizations are based on a limited number of experiments carried out on a few pierid and nymphalid species and may not apply to all butterflies. Nature loves to create exceptions to any rule![19]

In the African butterfly, *Bicyclus anynana* (Nymphalidae, Satyrinae) (Fig. 2.12, dry season (DS) form), females choose males based on the size and UV reflectivity of the central white pupil in the eyespots on their dorsal (upper) wings (Fig. 2.13, wet (WS) and dry (DS) forms). Butterflies have UV-sensitive photoreceptors (see Chapter 4, this volume) and UV reflections are important in discriminating both conspecifics and mates. Other factors, such as the size of the males' wings do not appear to play a role in the females' choice of a mate in this species. Specialized wing scales in these small, UV-reflective patches may be energetically costly to make and could therefore, be a measure – an honest badge or signal – of the genetic potential of the males. This is not to say that preferences for other morphological features – i.e. wing patterns, which were not tested in the experiments – may not also be at play in the females' choice of a mate, but their predilection for males with white pupils that are not too large, or not absent from the eyespot, is thought to exert a moderate amount of control in maintaining spots of a certain size and reflectivity in this species. However, mutual mate choice occurs in *B. anynana*, albeit in different seasons, with females (Fig. 2.14, WS female) choosing males in the WS – based on their wing patterns, amount of wing damage, age and pheromones – and males choosing females in the DS, based on their wing patterns and ornamentation (see below, 'Male Choice'). WS males approach 'choosy' WS females and engage in vigorous courtship (see Chapter 5, this volume), whereas the males become the choosier sex in the DS, and the females do the courting.[20]

In the orange sulphur, *Colias eurytheme* (Fig. 2.15), females are presented with both visual (UV-reflecting scales) and olfactory (pheromone) cues by males, but the brightness of the UV reflectance is the best predictor of male courtship success. In other words, females prefer brighter males, perhaps because they are more likely to be younger and therefore more likely to transfer larger spermatophores (see Chapter 7). Similarly, in the checkered white butterfly, *Pieris* (=*Pontia*) *protodice* (Fig. 2.16),

Fig. 2.12. Squinting bushbrown (*Bicyclus anynana*) female, dry season form from a lab strain (Louvain-La-Neuve, Belgium). (Photo by Gilles San Martin – own work, CC BY-SA 3.0, https://commons.wikimedia.org/w/index.php?curid=17649104)

Fig. 2.13. Wet and dry seasonal forms of *B. anynana*. (a) WS ventral female, (b) DS ventral female, (c) WS dorsal female, (d) WS dorsal male, (e) DS dorsal female, (f) DS dorsal male. (From Prudic *et al.* (2011), fig. 1. With permission of Antónia Monteiro.)

Fig. 2.14. Wet-season morph of *Bicyclus anynana.* (With permission of William H. Piel and Antónia Monteiro.)

Fig. 2.15. Orange sulphur, or alfalfa butterfly (*Colias eurytheme*), Cincinnati, Ohio, USA. (Photo by Greg Hume [CC BY-SA 3.0 (https://creativecommons.org/licenses/by-sa/3.0)].)

both species use UV reflectance to discriminate between the sexes, and in *Pieris rapae* females prefer brighter – i.e. higher colour and luminance contrast – males who stand out more against green foliage. Males are in effect advertising themselves to females. Unfortunately, these males are also more conspicuous and likely to be attacked by birds, suggesting that natural selection exerts a brake on the extent of the ornamentation – bright, UV-reflecting wings – evolving as a result of sexual selection.[21]

In the pale grass blue, *Pseudozizeeria maha*, virgin females have a marked preference for blue or bluish coloured male wings – dorsal surfaces – rather than their own black or dark greyish-blue wings. Female western white butterflies, *Pieris occidentalis*, however, prefer males with darker areas – increased melanization – on the middle and margins of their dorsal forewings. It is possible that this factor is indicative of male quality, but wing melanin patterns are also subject to the forces of natural selection – for thermoregulatory ability – which will favour darker morphs in colder regions. Male *P. occidentalis*, on the other hand, appear to prefer the conventional, or typical, melanin patterns of their species and discriminate against heavily melanized females. Quite why females choose males with dark forewing markings is not altogether clear, but the melanin pattern could be an indicator of male quality if it correlated with some physiological factor(s). This example illustrates the difficulty in relating sexually selected features or ornaments, which have been identified as a mate preference, with genuine biological indicators of some desired quality. The extent to which such features can be expressed and developed as a result of sexual selection may also be constrained by other factors, which are under natural selection. For example, if butterflies become too black, they may be at a disadvantage in some other way, like being more apparent to predators. In the closely related checkered white, *P. protodice* (Fig. 2.16), UV reflectance was more important for sexual discrimination, for both sexes, than melanic wing markings.[22]

Active and passive choices

To some extent at least, female butterflies may benefit from a sort of passive mate choice, whereby the males which manage to contact them have already been selected – or filtered out – because some of them lacked the ability to find and locate females. Finding a mate is not necessarily a trivial thing in a large and complex habitat, where females may be hidden and inactive, only found in some specific locations, e.g. on or near host plants. Chance may also play a role, but some of this filtering could be described as active, in the sense that the place where the female decides to position herself will determine the ease with which males can locate her. This may be a trade-off on the part of the female, i.e. between exposure

Fig. 2.16. Checkered white (*Pieris protodice*) male underside. (Photo by Megan McCarty – own work, CC BY 3.0, https://commons.wikimedia.org/w/index.php?curid=7242082.)

to potential harm, risk of being spotted by predators, against the possibility of failing to be discovered and mated. In terms of selecting the 'best' males, once again some winnowing may have already occurred – via passive selection – as a result of male-to-male competition: selecting and defending the best territories, encounter sites or leks, for example. Active selection, by directly evaluating and choosing a mate, is probably what we primarily think of as female choice. This will depend on female preferences for age, size, colour, etc. as well as on the behavioural repertoire of the male, including persistence, and perhaps even the quality of his courtship performance. In other words, whatever properties best convey, overtly or covertly, the genetic qualities of the male. The choices themselves may be direct – e.g. females actively soliciting courtship – or indirect, where females adopt rejection postures.[23]

Iridescent ornaments and a preference for flashy males!

Females of the great eggfly, *Hypolimnas bolina*, prefer males that display bright, iridescent ornamentation in the form of bluish-UV dorsal wing patches (Fig. 2.1). This choice by the females, of a bright UV signal – which is an honest signal of male quality – explains the evolution of sexual dimorphism in this species. The males invest in physiologically 'costly' scales, which generate these bright UV colours via complex nanoscale surface structures, only visible over narrow viewing angles (see below). In other words, the preference for bright, exaggerated UV colours in *H. bolina* has driven the evolution of the male wings, and exaggeration of the signal we see in this species. This preference – i.e. for 'limited-view' UV markings – has presumably evolved as a way of enabling female choices, in terms of male quality, via structural coloration. In another, closely related species, *Hypolimnas alimena*, females have no such preference for bright UV coloration. Instead, non-iridescent, blue bands – on the dorsal surfaces of both fore- and hindwings of male *H. alimena* – appear to be enough for female choice in this species.[24]

Research has shown that male great eggflies, *Hypolimnas bolina*, place themselves in positions, usually beneath females, that display to best effect the visibility and UV brightness of their ornaments (Fig. 2.17, fig. 6 from White *et al.*, 2014). During courtship, the males increase their wing beat frequency, yet flap their wings through a shallower amplitude, which has the effect of producing a sharply flashing signal at a frequency of about 11 times/s. The intensity and duration of such flashes may well be used by the females to evaluate the quality of the males. Producing such signals is not something that the males can fake, although they can enhance the effect by their choice of perching site, body position and time of day, e.g. under direct sunlight. They can also maximize the effect by using all four wing patches together, in a dazzle of UV reflectance. It is tempting to think that some males may be more adept than others in using their physical attributes in a mating display.[25]

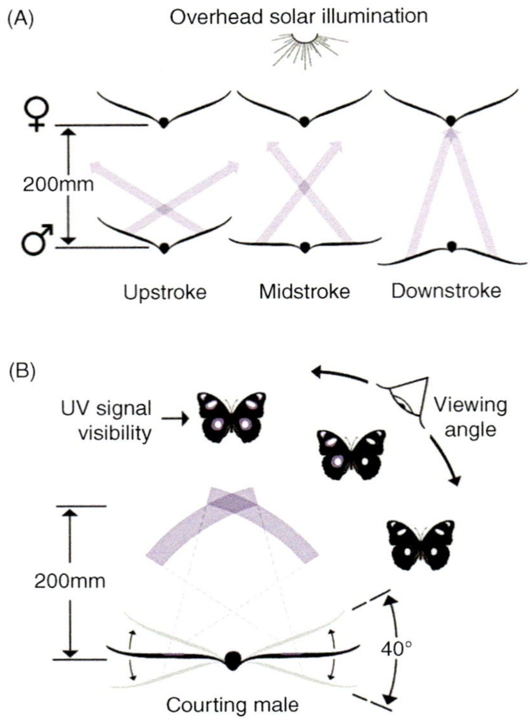

Fig. 2.17. Summary dynamics of male UV visibility during courtship. Panel (A) traces the angular visibility of maximally bright UV reflectance (violet arrows) from the wings of a courting male over a complete wingstroke. A female situated 200 mm overhead would see this bright UV flash only at the point of the male's downstroke, but from all four wings simultaneously. Male UV would be invisible to the female for most of the remainder of each wingstroke, contributing a distinct series of flashes if this geometry can be maintained by the male. Panel (B) expresses the same courting male scenario but represents the visibility of maximally bright UV as a function of viewer orientation (and summarized for a viewing distance of 200 mm). The violet bands trace the orientations for UV visibility during a single male wing sweep, and the butterfly icons indicate which (of any) UV patches would be visible (indicated for viewers off to the right-hand side only). UV reflectance from both wings would be visible from overhead orientations (where the violet bands intersect), whereas viewers from more oblique angles would see UV from only one wing (one violet band visible) or from neither (no bands visible). Both panels assume identical 'plane of depth' for the signalling male and viewer, and directly overhead illumination. (From White *et al.*, 2014, fig. 6)

The idea that the bright iridescent colours displayed by many male butterflies on their upper wing surfaces are a measure of their fitness accords with the theory that females choose mates based on such reliable, or honest, indicators of quality. Studies have revealed that in male pipevine swallowtail butterflies, *Battus philenor* (Fig. 2.18), the more chromatic individuals – those with the most intensely blue iridescent colour on the dorsal surfaces of their hindwings – produce the largest spermatophores, on average. In other words, these bright iridescent blue patches predict qualities that are highly attractive to females; namely, the mass of the ejaculate transferred by the male during copulation. The male is therefore investing both in the brightness of his scales and the size of his nuptial gift. The relationship is not perfect – there is rather a lot of variation in the relationship between spermatophore size and dorsal chroma – but presumably, it is a good 'rule of thumb' for the female to follow. Female, *B. philenor* butterflies appear to be preferentially attracted to males with this highly chromatic dorsal hindwing coloration. Take away these bright colours, as researchers did in one experiment where the wings of male *B. philenor* were blackened out, and the females were no longer interested. It is worth mentioning that these bright blue iridescent colours, together with the lovely black and orange hindwing spots, also function as a warning – aposematic coloration – to predators that this species is highly distasteful.[26]

The finding that the quality of a nuptial gift is related to some obvious feature, such as the intensity of iridescent wing patches in the male, is very interesting and profound, but it does not appear to be a universal rule. In other species, such as the orange sulphur, *Colias eurytheme*, the UV coloration of male ornamentation provides no guide as to quality of the spermatophore provided. Perhaps there are a repertoire of available cues, visual or chemical, which *could* be evolutionarily selected to communicate such information regarding male fitness? Discovering which feature, or combination of signals, is used in a given species as a possible measure of male quality, is challenging.[27]

Females of the large grass yellow or common grass yellow butterfly, *Eurema hecabe* (Fig. 2.9), evaluate structurally coloured visual signals – which emanate from the yellow or orange areas on males' dorsal wing surfaces – and they prefer males bearing bright UV iridescence. Female *E. hecabe* also have iridescent patches but these are smaller, less bright than those of the males and restricted to the base of their dorsal forewings.

Fig. 2.18. Pipevine swallowtail butterfly (*Battus philenor*) male.

Nevertheless, males of this species prefer to court females with larger UV-reflecting patches and appear to use the sight of small, UV-reflecting patches at the base of the dorsal forewing as a signal that the individual is a female. It is a moot point however, as to whether there are two-way choices going on in this species. Females choose males with brilliant UV iridescence – presumably as an indicator of their condition – and males preferentially court females with larger UV patches. But males also court females with or without basal UV patches, so the results are mixed, and the responses of the male to these female ornaments may not be strongly related to choosing a mate. Such ornamentation may be produced almost incidentally, as a result of the expression of 'highly penetrating genes' with no direct benefit to females. This example clearly illustrates the complexities of determining whether an ornament is being used as a sexual signal for the opposite sex. It also shows the potential for two-way mate choices, especially regarding ornamentation that is expressed in both sexes.[28]

Females with a tendency to vary

Alfred Russel Wallace (1865), an inimitable observer of nature and a great field naturalist, is thought to have been the first biologist to propose that female butterflies are inherently more variable than males – a phenomenon that is now thought to apply to Lepidoptera in general, although some researchers have questioned its generality. Basically, male butterflies tend to be less variable in terms of patterns – particularly on dorsal wing surfaces – than females. For example, females of the Australian butterfly, *Heteronympha merope merope* (Satyrinae), were found to have consistently more variable wing patterns than males. Similarly, the patterns on the dorsal surface of the wings of female eastern black swallowtails, *Papilio polyxenes*, were more variable than those on the males, perhaps as a result of the suppression of phenotypic variation in the males.[29]

Female-limited polymorphisms, including female-limited Batesian mimicry – where several different female forms, mimetic and non-mimetic, exist – are examples of both female variability in general, and reduced male variability in particular. However, such polymorphic mimicry is not limited to female butterflies, as there are species with polymorphic males as well, e.g. the common mime, *Papilio clytia* (Fig. 2.19). Nevertheless, female-limited polymorphism is much more prevalent, with some impressive examples of variability, such as the mocker swallowtail, *Papilio dardanus* (Fig. 2.20), where there is just one male form, but at least 14 different female morphs, throughout sub-Saharan Africa. Most of these morphs are Batesian mimics of other, distasteful, butterflies, and remarkably, there can be up to six different female forms coexisting within the same population. The mimetic patterns within individual *P. dardanus* forms are also more variable than those within non-mimetic ones.[30]

There are various hypotheses as to why males remain non-mimetic in these female-limited, polymorphic species. Females may prefer males

Fig. 2.19. Common mime (*Papilio clytia* form *dissimilis*) underside, Kadavoor, Kerala, India. Mimics the blue tiger (*Tirumala limniace*). (Photo © 2009 Jee & Rani Nature Photography (Licence: CC BY-SA 4.0) [CC BY-SA 4.0 (https://creativecommons.org/licenses/by-sa/4.0)].)

Fig. 2.20. African swallowtail, mocker swallowtail (*Papilio dardanus*) females (homozygous forms). (With kind permission of Professor H. Frederik Nijhout.)

with non-mimetic wing patterns, usually the ancestral condition, but this has rarely been tested. Males prefer mimetic vs. non-mimetic females in *Papilio polytes* (Fig. 5.21), but females showed no preferences for males with white spots on their dorsal forewings, which are not present in mimetic female forms. Further experiments are required to rule out the possibility that females are using some other feature or marking that is only present on the wings of males to make their choice in this species. Nevertheless, it leaves open the possibility that intrasexual selection, via male–male competition is playing a more important role than female choice in constraining the evolution of male wing patterns and limiting mimicry variations to females – a situation that appears to exist in a congeneric species, the eastern black swallowtail, *Papilio polyxenes*.[31]

Mr Normal – stabilizing selection

Females of the eastern tiger swallowtail, *Papilio glaucus*, prefer males with a natural, or normal, tiger-stripe pattern (Fig. 9.5). Males who deviate from the norm are less likely to be successful at mating: an example of stabilizing sexual selection, which may explain why females are typically more variable than males. In alfalfa butterflies, *Colias eurytheme*, it seems that exceptionally large individuals of either sex are unattractive to their opposite gender, and butterflies that are successful at courtship usually exhibit less variability in size than those which are rejected by the opposite sex. Males accepted by *C. eurytheme* females are less variable in size than males rejected by females, and, likewise, the size of females accepted by males is less variable than that of rejected females. In other words, there appear to be widespread preferences for mates of a normal or standard size, which maintains the population around a mean, perhaps an optimal size for each species in a given habitat.[32]

Male Choice

Whilst males are far more promiscuous than females in most butterfly species – and are not usually limited by their ability to produce sperm – there are costs associated with producing ejaculates, particularly so when these are very large. There are also costs and risks associated with searching for females, as well as with activities such as nectaring and puddling, carried out in order to replenish their reproductive powers. As such, males might be expected to be discriminating and to choose mates who will provide the highest reproductive return. Females of high reproductive value are likely to be young and large, with high potential fecundity. So, it is unsurprising that some male butterflies, including orange tips, *Anthocharis cardamines* (Fig. 3.5), court young virgin females more intensively than older mated ones. Male checkered white butterflies, *Pieris protodice* (Fig. 2.16), also court younger and larger females for longer than older and smaller ones, and male *Colias philodice* and *Colias eurytheme* prefer large rather than small females as mates. There are good reasons to explain male preferences for young females, and vice versa. As well as being easier to find, for example, in the case of virgin females eclosing from their pupae – there are often chemical signals emitted by the females prior to eclosion – young females are preferable in terms of their life expectancy and vigour, e.g. in terms of evading predators and locating host plants.[33]

It has been argued that researchers in the past were perhaps too eager to project stereotypical views of the roles of the sexes onto the organisms they studied. Thus, the conventional view of males as being more dominant and active – e.g. in terms of searching for mates – and females being passive and reactive, needs challenging. A shift to a more female-orientated perspective has produced examples of both male choice and female–female aggression in butterflies, and an awareness that mate choice is not restricted to one sex. Mate choice by males may not be as common as that by females, but it may be more common than is generally supposed. Males can be both competitive – with each other – *and* choosy, in terms of selecting a female with which to mate. However, the ability to make choices depends, for both males and females, on how often they are spontaneously presented with multiple members of the opposite sex from which to choose a mate.[34]

Mate preferences by male *Heliconius* butterflies are thought to be based on specific colour patterns and spectra, seen in the coloured patches, bars and rays found on both surfaces of the fore- and hindwings (Fig. 2.21, *Heliconius doris*). In particular, male *Heliconius* butterflies are strongly attracted to the colours found on the forewing bands of their own species and will spend much longer courting virgin females of their own race than they do with females of another species. The colour patterns most favoured by males are also the most effective ones in terms of deterring attacks by predators; i.e. warning colours. So, the males seem to be choosing the patterns that best confer a greater chance of survival for their offspring; i.e. sexual and natural selection working in parallel. Their strong preferences for their own, local type of colour pattern is what keeps these species reproductively isolated – preventing hybridization – despite often

Fig. 2.21. Doris longwing (*Heliconius doris*) Amsterdam Zoo.

occurring together in the same habitat. The colours of their wings, and the colours of the butterflies they are attracted to, and prefer to mate with, are very tightly linked genetically. There is also some evidence to suggest that female *Heliconius* butterflies also make mate choices, although this aspect has been less well studied than male choices.[35]

Males and females of the Japanese cabbage butterfly, *Pieris rapae crucivora* (Fig. 7.11), differ in terms of their reflectance spectra: female wings reflect more UV light because they lack the UV-absorbing pigments found in the male. Males use this difference in UV reflection as a cue to recognize a potential mate and preferentially approach females that have more colourful wings – to their eyes – that is, ones with a higher pteridine content (see Chapter 10). N.B. pteridine pigment granules absorb UV light, but scatter light in the visible wavelength range, making them appear bright and colourful. Studies have shown that male *P. rapae crucivora* prefer to locate mates in the shade, where the *relative* UV reflection of female wings is much higher, even though the *absolute* UV reflection is much (10×) stronger in direct sunlight. This is a convenient adaptation, since virgin females can be found on the shaded undersides of leaves, close to where they have eclosed in the early mornings. The females become receptive to mating about 30 min after emerging but remain close to their cast-off pupal cases until they are ready to fly off, some 2–3 h later, when their wings have hardened. So, it seems that females are more apparent, i.e. visible, to males in these shaded, UV-rich locations. In contrast to the Japanese subspecies, the wings of females of the British subspecies *P. rapae rapae* reflect little UV light and there is only a very slight sexual dimorphism in terms of visible wing colour. So, discriminating between the sexes is more difficult for male *P. rapae rapae*, which probably make more mistaken approaches to other males. Fortunately, resting males – of both subspecies – have a 'flutter response' when approached by flying males, which in effect says, 'I am not a female'. This behaviour saves the active male from having to land, investigate the sex of the conspecific individual and take off again.[36]

Flavonoids are naturally occurring plant compounds that absorb ultraviolet light. Many butterflies – including the common blue, *Polyommatus icarus* (Fig. 2.22a–c), and the chalkhill blue, *Polyommatus coridon* (Fig. 5.2) – sequester flavonoids whilst feeding as caterpillars on their host plants, and these pigments are transferred to the wings during metamorphosis. Females sequester larger amounts of flavonoids than males and, in *P. icarus*, these flavonoid-rich, strongly UV-absorbing females are attractive to males, perhaps because of their more colour-saturated wings. The undersides of the wings of flavonoid-rich butterflies are slightly more creamy or yellowish in colour – in visible light – but in ultraviolet light, flavonoid-free butterflies reflect much more UV light than flavonoid-rich individuals. In the latter, the white wing patterning almost disappears completely, showing that flavonoids have a profound influence on the wing coloration of *P. icarus*, especially in the ultraviolet range. There is a large intraspecific variation in the wing coloration of common blues, and

Fig. 2.22. (a) Common blue (*Polyommatus icarus*). Male dorsal wing surfaces in *P. icarus* exhibit an iridescent bright blue to violet coloration due to structural coloration of wing scales. (b) Females have dark brown upper sides with a row of more or less pronounced submarginal, highly ultraviolet-absorbing, orange spots. (c) Females may have varying degrees of additional iridescent blue coloration, 'dusted' over a smaller or larger part of the wings.

it is possible that males are attracted to flavonoid-rich females because they have had the benefit of a better diet in their larval stages. Their quality, and hence desirability as a mate, is perhaps reflected – although in this case the UV light is not reflected (!) – by their coloration.[37]

Males of the little yellow, or little sulphur butterfly, *Eurema lisa*, can recognize females of their own species, and can tell that they are not males, solely based on their UV reflectance. Or rather, their lack of UV reflectance, as the dorsal wing surfaces of *E. lisa* reflect few or no UV wavelengths. It is thought that because males can distinguish females so rapidly and precisely, both mate selection and courtship (see Chapter 5) can proceed very rapidly in this species. However, the coloration of the ventral wings and the presence or absence of a wing bar affect the attractiveness of an individual female to *E. lisa* males, so presumably there is still scope for qualitative decision-making in the otherwise extremely short (c. 3.5 s) courtship sequence in this species. The females are probably also capable of making sex discriminations as rapidly as the males, although their evaluations of the male may involve pheromones as well.[38]

In the section on 'Female Choice' (above) it was shown that female squinting bushbrown butterflies, *Bicyclus anynana*, which occur in the WS are choosy and prefer to mate with males with their *dorsal* eyespots intact. Similarly, DS males – who become the choosier sex in this season – also prefer to mate more readily and stay *in copula* longer, with females that have their dorsal eyespots intact. In other words, the *dorsal* eyespots (Fig. 2.13) have a crucial role to play in mate choice by both sexes, albeit at different times. However, *B. anynana* butterflies also have a series – of even more conspicuous – eyespots on the *ventral* sides of their wings, which become particularly apparent in the WS. Those on the ventral side of the forewing include a particularly large and prominent (Cu1) eyespot (Fig. 2.14), which is often fully or partially concealed by the hindwing when the butterflies are at rest. It has recently been shown that these ventral wing eyespot markings also play an important role in sexual signalling as, once again, in experiments, males preferred to mate with females that had eyespots with their centres intact, rather than obscured by black paint. This is exciting because it demonstrates that males notice – and are attracted to – these white, UV-reflective, ventral forewing elements, which are being used as sexual ornaments by DS females. The DS females probably display both their ventral and dorsal eyespots during their courtship, and it is possible that these ornaments may be used by the males as an indication of female quality. The extent to which these findings can be extrapolated to other species is a tantalizing, but largely unknowable proposition.[39]

The banded peacock

Anartia (Nymphalidae) is a small genus of so-called peacock butterflies – but the European peacock (*Aglais io*) is not in this genus – found in the Americas. In the neotropical, banded peacock, *Anartia fatima* (Fig. 2.23a, b), there is a colour dimorphism, with yellow-banded and white-banded morphs of both males and females. Yellow morphs are slightly more abundant (55–64%) in lowland areas of Costa Rica, where they have been studied, whilst white-banded forms tend to predominate in cooler and higher locations, where they presumably have a physiological advantage. The bands of both yellow males and females – whose colours can be in a continuum from yellow to white – fade to white after about 2 weeks. Mate selection experiments carried out in Costa Rica demonstrated that both white- and yellow-form males had a marked preference (by 2:1) for white-banded females, which curiously are older females. Interestingly, the white bands were highly UV-reflective, whilst the yellow-pigmented bands were UV absorbent, so the males were preferentially attracted to the UV-reflecting forms. N.B. the *A. fatima* males are probably using female

Fig. 2.23. (a) Banded peacock (*Anartia fatima*) white-banded morph, Selva Verde Lodge, Costa Rica. (Photo by Jerry Oldenettel – originally posted to Flickr as DSC_4900a.jpg, CC BY-SA 2.0, https://commons.wikimedia.org/w/index.php?curid=3883984) (b) Banded peacock (*Anartia fatima*) yellow-banded morph, San Francisco Reserve, near Torti, Darien Province, Panama. (Photo by gailhampshire from Cradley, Malvern, UK, CC BY 2.0, https://commons.wikimedia.org/w/index.php?curid=73442010)

pteridine-based coloration – i.e. of the yellow bands – as the basis for choosing mates, but in the opposite direction to that in *P. rapae crucivora* (Fig. 7.11), described above. The *A. fatima* females, on the other hand, did not appear to mind which sort of male approached them, so the first male to approach a virgin female is generally successful, in terms of mating. Females mate soon after eclosion – a few hours after sunrise – whilst males generally need 3 days before they commence mating.[40]

Role Reversals

Experiments with insects other than butterflies – crickets and bush crickets (Tettigoniidae) – have shown that a reversal of sex roles can occur in species where males invest more in reproduction than females. In such cases, it is the males who are the choosiest in terms of selecting mates, and the females who engage in competitive behaviour. The choosiness of males usually increases where there is a high degree of variation in terms of female quality, e.g. differences in fecundity. Male choosiness is expected to be strongest in situations where female fitness increases cumulatively (i.e. with every mating); for example, in species where the males provide food or nuptial gifts (see Chapter 8) for the females. So, not surprisingly, male choosiness is widespread in butterfly species that invest heavily in producing large spermatophores, e.g. *Pieris* and *Colias* species.[41]

Males would not be expected to be choosy, and reject mating opportunities, unless potential mates were abundant and easy to find, which is the opposite of most natural situations. There are, however, butterfly populations where males are very rare as a result of the male-killing bacterium, *Wolbachia*. For example, populations of the African butterflies *Acraea encedon* and *Acraea encedana*, studied in southern Uganda, were extremely female-biased (>90% infected with *Wolbachia*). This bacterium only kills its host if it finds itself in a male during the egg stage, so in a population where its prevalence is high, there is a very low proportion of males. These highly female-biased populations often aggregate at prominent landmarks – lekking sites – with the aim of attracting males. The females appear to compete, by chasing each other, for the few males that find their way to the lek site, i.e. a reversal of the usual sex roles. There is some evidence to suggest that males preferentially choose uninfected females within the swarm: a behaviour that – if correct – would prevent the bacterium from completely wiping out the population.[42]

Another consequence of a shortage of males in female-biased populations is increased female solicitation. Females that have received smaller nuptial gifts may be more eager to receive additional benefits. For example, female checkered white butterflies, *P. protodice* (Fig. 2.16), with small spermatophores – from previous matings – spent more time approaching and chasing males than those with larger spermatophores. Clearly, they were willing to invest more energy in order to obtain greater benefits from males. Similarly, female *H. bolina* butterflies in populations where there was a pronounced female bias as a result of male-killing *Wolbachia* bacteria were promiscuous and increasingly willing to mate as the proportion of males decreased. Surprisingly perhaps, these extremely female-biased populations are remarkably enduring. A Samoan population of *H. bolina* survived for over 100 years (around 400 generations) with a sex ratio of 100 females to every male. One reason for the persistence of these nearly all-female populations is the remarkable reproductive activity of the males, who reportedly, can mate more than 50 times in their lifetimes. Despite this heroic effort by a few males, many females remain unmated and there is a significant overall loss of female fertility within these populations. In other words, the mating capacity of the males is finite, and fatigue eventually sets in, at least in extremely female-biased populations. Sexual selection via male–male competition still occurs in moderately female-biased populations, via sperm competition within female reproductive tracts. But when males are severely limited (rare), such competition breaks down.[43]

The *alba* polymorphism in *Colias* species

Many species within the genus *Colias* (sulphurs and yellows) have a sex-limited – i.e. only found in females – wing colour polymorphism called *alba*. The creamy-white *helice* form of the clouded yellow, *Colias croceus* f. *helice* (Fig. 2.24b), which occurs at relatively low frequencies (e.g. c. 5%) in European populations, is a well-known example of the *alba* aberration. These white or greenish-white *alba* forms express

Fig. 2.24. (a) Clouded yellow (*Colias croceus*) female with tiny wasp, Galicia, Spain. (b) Clouded yellow (*C. croceus*) form helice, Galicia, Spain.

a dominant allele (A), which produces colourless wing pigments – pteridine molecules (see Chapter 10) – rather than the orange or yellow ground colours seen in 'normal' (aa) females, and in males. The effect of the *alba* polymorphism is to free up more pupal nitrogen for allocation into other parts of the body – such as fat reserves – rather than into pigments for wing colours. *Alba* females of *Colias eurytheme* produce more eggs per day than orange females. The physiological advantages of the *alba* forms are, however, offset by some behaviour disadvantages. *Alba* females occur more frequently in the colder parts of the species' ranges where they have a fitness advantage, often reaching levels of 80–90% of the population in mountain regions, where they use the nitrogenous pigment compounds to help them to develop faster and produce more eggs. The price these pale females pay, however, is that they are less attractive to males![44]

Male *Colias* butterflies are more attracted to the normal, orange or yellow forms in a population, and approach *alba* females much less frequently. As a consequence, they usually suffer less harassment from males, compared with their colourful sisters, who sometimes emigrate as a result of excessive male attention. This lower level of interaction with males can be an advantage – e.g. allowing them more time for oviposition – but not at low population densities, when they may fail to mate altogether. The *alba* females also must spend more time resting and basking, because they lack the more efficient, heat-absorbing pigments of the yellow morphs. The reason why males are less attracted to the whiter *alba* forms is probably because there are usually a lot of other white butterflies – i.e. pierine species – flying around in their habitats. Given the relatively poor ability of butterflies to discriminate species over long distances, they would waste a lot of valuable time chasing the wrong sort of white butterfly! So evolutionary pressures will favour those males limiting their courtship investigations to yellow butterflies.[45]

The Japanese sulphur butterfly, *Colias erate* (Fig. 6.5), presents a rather different situation, in that the yellow morph is generally *less* common than the white *alba* morph in Japan, e.g. at a frequency of 15% or less at a site in central Honshu. The yellow males heavily outnumbered the polymorphic females at this location, but this was found to be an artefact caused by their much greater mobility, and the true sex ratio was estimated to be 1:1. Male *C. erate* were said to 'persistently' court *alba* morph females and to sometimes harass mating pairs comprising an *alba* morph female in preference to yellow morph females. Such harassment has the effect of increasing the duration of copulation. Male *C. erate* also change their mating preference depending on the 'copulation experience' of the female. Males preferred yellow morphs if the females were virgin, but showed no morph-related bias towards experienced, i.e. mated, females. Furthermore, courting males appeared to persist longer with *alba* females than with yellow ones, perhaps as a result of underlying preferences.[46]

Sexual Dimorphism

Differences in size

Differences between the sizes of adult male and female butterflies are quite common – as they are in most organisms – and relate to the fact that there are different selection pressures operating on each sex. There will in practice be a mix of different sources and directions of selection: some operating in favour of getting bigger, others driving a diminution of size. For example, larger females will be advantaged in terms of increased fecundity, whilst larger males will usually benefit in male–male competition – such as for access to mates – and in sperm competition. Smaller adults, on the other hand, may benefit from shorter development times and a decrease in associated risks, e.g. from predation or starvation. The observed size differences between males and females in different species can be assumed to reflect the balance of such forces – frequently that between sexual and natural selection – at play in their evolution and development. Sexual size dimorphism is, however, plastic, i.e. variable, and subject to environmental influences.[47]

There is also evidence to suggest that sexual size dimorphism is influenced by the resource contribution of each sex to the next generation, and therefore the male-to-female size ratio tends to increase as the relative ejaculate size contributed by the male increases. Larger males are also reproductively more successful in some, but not all species, because of producing larger ejaculates, which can result in more offspring and delayed remating by the female. This correlation between the size of the spermatophore and reproductive success does not, however, apply to all species. In some *Heliconius* butterflies, body length is negatively correlated with mating success (see Chapter 7, 'Pupal Mating'). The ability of a female to rely on nuptial gifts from males in the case of developmental food shortages probably puts less selection pressure on females than on males to grow larger. In other words, variability in the availability of food may be an important factor in determining sexual size differences.[48]

Female butterflies are often the largest sex; for example, Baltimore checkerspot (*Euphydryas phaeton*) females are about twice the size of males, and most *Morpho* females are bigger than males, but the reasons why this is the case and precisely what selection pressures determine body size and wing shape are not fully understood. A variety of hypotheses have been put forward to try and explain why there are sometimes such large differences in size and shape between the two sexes. One clear advantage in being large is that of greater fecundity: large females can deposit more eggs, both within a day, and in a lifetime. Differences between the sexes might also be a way of dividing up resources, i.e. in favour of the female, who is usually the one investing the most in the next generation. The fact that males are often smaller may be a direct consequence of protandry: early emergence as a result of rapid development in order to take advantage of the high receptivity of females just after eclosion. There is, however, some evidence of faster development by males to compensate for early emergence, implying that the 'early birds' need not always be small. Clearly, the advantages of early emergence are greater in species where generations are discrete and the females only mate once. Consequently, the male-to-female size ratio tends to shift in favour of relatively larger males as the degree of polyandry increases. For example, males are slightly larger (male:female forewing length ratio of 1.03) than females in the strongly polyandrous green-veined white, *Pieris napi*, where females mate at least twice, on average.[49]

Differences in shape

There can also be marked difference between the sexes in terms of wing and body shape. Female butterflies tend to have a larger wing area compared with males, probably as a result of having greater body weight, on average, but there are exceptions. For example, the forewings of male clearwing swallowtails, *Cressida cressida* (Fig. 2.25) are considerably larger on average, than those of the female – males have a wingspan of about 80 mm, in females it is about 70 mm – although there is a great deal of variation within each sex. In birdwing butterflies – like, the New Guinea birdwing, *Ornithoptera priamus poseidon* – the situation is reversed, with the female having larger wings (forewing 80 mm) than the male (70 mm). Differences in wing shapes between the sexes probably relate to differences in lifestyles, with smaller relative wing dimensions and increased wing loading being adaptations to perching in the males of some species, which fly up rapidly to investigate passing females. Males that spend longer in the air, patrolling for example, tend to have wings with a higher aspect ratio; i.e. long and narrow wings. In *Morpho* butterflies, differences in the shape of the forewing between males and females are most pronounced in species that also have a strong colour dimorphism, like *Morpho zephyritis*, whose females are white, and males are blue.[50]

Fig. 2.25. Clearwing swallowtail or big greasy (*Cressida cressida*). (Photo by JKDw – own work, CC BY-SA 4.0, https://commons.wikimedia.org/w/index.php?curid=41812381)

Differences in colours and patterns

As well as differences in size and shape, male and females of the same species can appear very different in terms of colours and patterns. There is in nature a wide spectrum in terms of the extent or degree of sexual dimorphism in butterflies, from the negligible to the extreme. Indeed, the sexes can be so different in appearance that it may not be obvious that they are even of the same species. It took DNA barcoding to reveal that two butterflies described and placed in different genera were in fact males and females of the same species! The iridescent blue sunburst cerulean-satyr, *Caeruleuptychia helios* (the male), had been described in 1911, whilst the drab, brown *Magneuptychia keltoumae* (now known as the female *C. helios*) was only described in 2012 (Fig. 2.26). On the other hand, differences in wing patterning between the sexes can be slight and subtle; for example, limited to small areas of the wing where sexual dimorphic features – like the presence of a white spot on the dorsal hindwing of the scarlet peacock, *Anartia amathea roeselia* (Fig. 5.13), for example – occur. Female *A. amathea roeselia* have a prominent white spot near the costal margin of both hindwings, which is largely absent from male wings. Other areas of the wing may be very similar, with very little variation, which suggests perhaps that butterflies must somehow focus on the sexually dimorphic areas, or markings, to identify members of the opposite sex. However, chemical signals may also be involved in mate recognition. The ability and facility with which butterflies recognize, or not, tiny intra- and interspecific differences in wing patterns is an area ripe for more research.[51]

Fig. 2.26. Sunburst cerulean-satyr (*Caeruleuptychia helios*) sexual dimorphism (Nakahara *et al.*, 2018). Male (top), female (bottom), dorsal (left), ventral (right). (From: Florida Museum of Natural History, photo by Shinichi Nakahara.)

Sexual dimorphism is strongly expressed in birdwing butterflies, both in terms of wing size and coloration. For example, in *Ornithoptera* species, males are black combined with bright green, blue, orange or yellow markings, whilst the females are larger and less colourful with black or dark brown coloration, together with white, pale-brown or yellow markings (Fig. 2.27a, b, Goliath birdwing). The female is considerably larger than the male and is various shades of brown on the dorsal wing surfaces, rather than the iridescent green of the male. The common archduke (*Lexias pardalis dirteana*) is a good example of a sexually dimorphic species. The females are larger, with cryptically coloured upper wing surfaces of dark brown with rows of yellow spots (Fig. 2.28b). The dorsal wing surfaces of the males on the other hand, are black with shimmering greenish-blue margins, especially in the hindwings (Fig. 2.28a). This suggests some structural coloration may be on display on the male wings.[52]

Lycaenid butterflies, such as the common blue butterfly, *Polyommatus icarus*, exhibit marked sexual dimorphism on their dorsal wing surfaces, with considerably more structural blue colours in males (Fig. 2.22a) and pigment brown colours in females (Fig. 2.22b and c). The patterns on the ventral wing surfaces, however, are almost identical, although the background hue is slightly darker in the female (Fig. 2.29). There is also sexual dimorphism in terms of UV reflectivity in many species, especially

Fig. 2.27. (a) Goliath birdwing (*Ornithoptera goliath samson*) male (above) and female (below). (Photo by Anaxibia [CC BY-SA 3.0 (https://creativecommons.org/licenses/by-sa/3.0)].) (b) Goliath birdwing (*Ornithoptera goliath samson*) male, GLAM muséum, Lille. (From: Musée d'histoire naturelle de Lille [CC BY-SA 4.0 (https://creativecommons.org/licenses/by-sa/4.0)].)

amongst pierid butterflies, but also in lycaenids, such as the common copper, *Lycaena salustius*, found in New Zealand. Male *L. salustius* have extensive UV-reflecting regions on the dorsal sides of both fore- and hindwings, but on females these are reduced to tiny, submarginal spots. To our eyes, under normal lighting conditions, male and female *L. salustius* look rather alike (Fig. 2.30), but under UV light, the males display a striking

Fig. 2.28. (a) Common Archduke (*Lexias pardalis dirteana*) male resting on the ground, Chiang Mai, Thailand. (b) Common Archduke (*Lexias pardalis dirteana*) female resting on the ground, Chiang Mai, Thailand.

Fig. 2.29. Common blue (*Polyommatus icarus*) female, Galicia, Spain.

pattern of highly UV-reflecting veins. In other words, the two sexes look strikingly different to each other, in their own eyes, even if we primates cannot see it![53]

Although there are no significant differences in the overall size of the forewings between male and female *Heliconius erato phyllis* – and the co-mimicry species, *Heliconius besckei*, *Heliconius melpomene burchelli* and *Heliconius melpomene nanna*, in the so-called *Heliconius* postman Müllerian mimicry ring – there are small differences in the size and shape of most of the yellow and red colour elements on the dorsal and ventral wing surfaces. These fine-scale differences in the size and shape of colour

Fig. 2.30. *Lycaena salustius*, male, female, set specimens, Sumner, Christchurch, New Zealand. (Photo by Door Accassidy – own work, CC BY 3.0, https://commons.wikimedia.org/w/index.php?curid=9419040)

bands are evidence for the operation of sexual selective forces; otherwise one would expect natural selection to have resulted in no differences between species.[54]

Female butterflies can be darker – i.e. with a greater degree of wing melanization – than their male counterparts, because they need to generate higher body temperatures in order to mature and deposit eggs, particularly at higher elevations and at cold temperatures. Melanization – an increase in the density of black wing scales – is an important factor in determining the reproductive success of females in the butterfly, *Colias philodice eriphyle*. Surprisingly, males seem to prefer lighter coloured, more yellow females in this species, i.e. less melanized individuals. An example of natural and sexual selection appearing to be pulling a phenotypic feature – melanization of the wings – in opposite directions. In the powdered oakblue, *Arhopala bazalus* (Arhopalini, Lycaenidae), it is the male that appears mostly black – in visible light – whilst the females have striking iridescent purple-blue patches on their dorsal surface fore- and hindwings. Under UV light however, the whole dorsal surfaces of both fore- and hindwings reflect UV light, and would appear shiny and reflective when viewed with UV photoreceptors as butterflies do. Only the iridescent patches of the female reflect UV light, so there is a clear sexual dichromatism in this species, which is reversed under UV light. So, one must always bear in mind that butterflies see each other differently from the way we do.[55]

When observed under UV light, male and female *Pieris rapae* appear markedly different: the wings of the female appear white – because they reflect UV light – whilst those of the male are much blacker, because they

Fig. 2.31. Photograph demonstrating the difference in UV reflectance between male (a) and female (b) specimen of *Pieris napi*. Fig. 5 from Stella *et al.* (2018b).

absorb the UV light (see Chapter 10). Under daylight, or human visible light, the reflectance of the *Pieris napi* female wings is 25% more intense – on average, across the Palaearctic region – than that of the males. There is, similarly, a marked difference in the UV-reflectivity of male and female *P. napi* wings (Fig. 2.31). Curiously, females prefer males with a higher degree of reflectance in the visible spectrum, i.e. whiter specimens, like themselves. The whiteness of the male's wings in visible light (and their darkness under UV light) is due to higher concentrations of pterin granules (see Chapter 10, this volume).[56]

Differences in ornamentation

Male *Bicyclus anynana* butterflies are slightly more ornamented than females, having one more hindwing *dorsal* eyespot, on average. In nymphalids more generally, both sexes can have more dorsal eyespots than the other, but females are slightly more likely to have more. In a survey of 278 nymphalid species with at least one eyespot on their wings, Christopher K. Tokita and colleagues found that 22% had no difference in total eyespot number between males and females, 44% had females with more eyespots than males and 34% had males with more eyespots than females. Ventral wing surfaces, and ventral *hindwings* especially, display the least amount of sexual dimorphism regarding eyespots, probably because eyespots in these places function to deflect predatory attacks and as such are under natural selection, which applies equally to both males and females. Researchers have discovered that the traits on the forewing used for mate signalling – communication between the sexes – are evolving rapidly as a result of sexual selection, whereas those on the hindwing are more evolutionarily stable, presumably because they serve a vital function that confers survival value and hence the trait is unlikely to vary. Similarly, characters on the *dorsal* wing surface – including eyespots – are used for mate signalling and are evolving at a higher rate than those on *ventral* surfaces, such as eyespots used in predator avoidance.[57]

Fig. 2.32. Spangle butterfly (*Papilio protenor demetrius*) male. (Photo by Hsu Hong Lin from 南投縣集集鎮, 中華民國 – 92.1黑鳳蝶8(洪振龍攝), CC BY 2.0, https://commons.wikimedia.org/w/index.php?curid=64361095)

Differences in smell and taste

There is, of course, a marked sexual dimorphism in the chemical composition of pheromone secretions, with males usually emitting significantly stronger scents than females (see Chapter 9, this volume), although many odoriferous compounds are shared by both sexes. For example, males of many papilionid species possess an odour that can be detected by humans, and in spangle butterflies, *Papilio protenor demetrius* (Fig. 2.32), these volatile scents are significantly stronger in the male. The same is the case in *Papilio polytes* (Fig. 5.21). Additionally, there can be marked differences between the sexes in terms of the chemical composition of cuticular lipids, e.g. long-chain aliphatic hydrocarbons (CHCs), and other epicuticular compounds, some of which – as in the sulphur butterfly, *Colias erate poliographus* (Fig. 6.5) – only occur in males. Large numbers of these cuticular compounds, composed of up to 40 carbon atoms, have been identified in *B. anynana*, with some being associated with specific body parts. Researchers think that these unique compounds (cuticular lipids) might provide a library of information – communicated between individuals during courtships by gustatory (i.e. taste) contacts – concerning the identity (sex), status (age) and quality of the other partner.[58]

Notes

[1] Dewsbury (1982), Møller and Jennions (2001), Tinghitella *et al*. (2013), Davis *et al*. (2007) and Rodríguez (2015).

[2] Brown *et al*. (1997), Tregenza and Wedell (2000), and Mays and Hill (2004). Mays and Hill (2004) point out that genetic dissimilarity of potential mates can produce fitness benefits for their offspring, but the effects may be subtle and only evident in large samples.

[3] Racevska and Roberts (2018) and Kreutzer and Aebischer (2015). Clutton-Brock (2009) writes that, since 1871, a substantial body of research has confirmed Charles Darwin's explanation of the evolution of secondary sexual characters in males. According to Jones and Ratterman (2009), Darwin was correct about almost every topic related to sexual selection.

[4] Milam (2010) and Rodríguez (2015).
[5] Arnqvist and Rowe (2002).
[6] Chapman *et al.* (2003), Andersson (1994), Tregenza *et al.* (2006), Andersson and Simmons (2006), Wedell *et al.* (2006), Milam (2010), Parker (2014) and Ruse (2015). Brown *et al.* (1997) identified three different types of interaction: persuasive, coercive and forcing.
[7] Turner (1978), Shuker (2010), Simmons (2014), and Racevska and Roberts (2018). Levin (1973) found that a single insemination, i.e. one mating event, provides enough sperm to fertilize all the eggs that a female eastern tiger swallowtail butterfly, *Papilio glaucus,* can produce.
[8] Zahavi (1975), Grafen (1990a, b), Hill (1994), Kotiaho (2001), Parker (2014), and Steiger and Stökl (2014). See Zahavi handicap, in Glossary. There are three different models for mate choice. It is thought that mate selections can be made based on: (i) good genes, i.e. superior alleles or combinations of alleles; (ii) high heterozygosity, meaning a high degree of genetic variability; or (iii) high genetic compatibility, which will result in heterozygous offspring.
[9] Kemp and Rutowski (2001) and Hidaka and Yamashita (1975).
[10] Imafuku and Kitamura (2015), Stoehr *et al.* (2016) and Li *et al.* (2017).
[11] Colour, patterns, signal directionality and polarization have coevolved with display behaviour to produce specialized transmission (White *et al.*, 2014). Many animals, including butterflies, have UV photoreceptors and recognize it as a separate colour (Cronin and Bok, 2016). 3-OHK yellow pigments allow *Heliconius* species to recognize conspecifics for interspecific communication and sexual selection (Finkbeiner *et al.*, 2017). *Heliconius* butterflies are better at separating 3-OHK (*Heliconius* yellow pigment) from non-3-OHK spectra compared with other butterflies, and birds (Bybee *et al.*, 2012).
[12] Sweeney *et al.* (2003), Grether *et al.* (2004), Imafuku and Hirose (2016), and Pyrcz *et al.* (2018).
[13] Silberglied and Taylor (1978), Brunton and Majerus (1995), Rutowski *et al.* (2007b) and Stella *et al.* (2018a) found a UV reflectance pattern in 62 out of the 106 taxa representing 54 species of *Colias* butterflies. Pecháček *et al.* (2019) found that UV patches are better at discriminating different *Gonepteryx* species than venation patterns are.
[14] Kato and Nakane (1989), White *et al.* (2012), and Kemp and Rutowski (2007).
[15] The detailed patterns found on *Lycaeides* spp. wings act as effective mate recognition signals (Fordyce *et al.*, 2002).
[16] Based on Alexander *et al.* (1997).
[17] Dewsbury (2005), Tang-Martinez and Ryder (2005) and Ah-King (2011). Sex roles can change in response to environmental variables (Gwynne and Simmons, 1990).
[18] Parker and Birkhead (2013) consider that the Darwin–Bateman Paradigm remains valid for most species that reproduce by internal fertilization and exhibit no or low male parental care, which would seem to apply to butterflies.
[19] Large marginal eyespots increased butterfly survival during predation encounters (Prudic *et al.*, 2015).
[20] Breuker and Brakefield (2002), Robertson and Monteiro (2005), Costanzo and Monteiro (2007), Fischer *et al.* (2008), Nieberding *et al.* (2008), Prudic *et al.* (2011), Westerman *et al.* (2012) and Dion *et al.* (2016).
[21] Rutowski (1981b), Papke *et al.* (2006), Morehouse and Rutowski (2010) and Tuomaala *et al.* (2012).
[22] Rutowski (1981b), Wiernasz (1989, 1995), and Imafuku and Kitamura (2018).
[23] Lloyd (1979) and Rutowski (1982a).
[24] Kemp and Macedonia (2006), Kemp (2007a) and Kemp *et al.* (2014). The costs of such ornamentation in *H. bolina* might also include the risk of increased predation.
[25] White *et al.* (2014). Individuals position themselves beneath females in a way that maximizes their conspicuousness: namely UV brightness, visible area and flash effect (White *et al.*, 2015).

[26] Rajyaguru *et al.* (2013) and Rutowski and Rajyaguru (2013). See fig. 1 in Rutowski and Rajyaguru (2013), or fig. 1 in Rajyaguru *et al.* (2013). *Battus philenor* females prefer older and greener males, and these individuals were more effective competitors (Sasaki *et al.*, 2017) and perhaps more successful at mating (Sasaki *et al.*, 2015a). Larval *B. philenor* sequester acids, which make them unpalatable as adults (Pegram *et al.*, 2013).
[27] In the pierid *C. eurytheme* no single aspect of the dorsal wing coloration of males – including UV, chroma or hue – was related to the size of the spermatophore (Kemp *et al.*, 2008).
[28] Kemp (2007b), Rutowski and Kemp (2017) and Stella *et al.* (2018a).
[29] Pearse and Murray (1982), Hazel (1990), and Herrell and Hazel (1995). Wallace (1865) was the first Darwinian biologist to study the different forms of papilionid butterflies (Mallet, 2004).
[30] The 14 female colour morphs of *P. dardanus* are controlled by allelic variation within a single mimicry gene (see fig. 1 in Nijhout, 2003).
[31] Low and Monteiro (2018), Westerman *et al.* (2018), and Lederhouse and Scriber (1996).
[32] Krebs and West (1988) and Rutowski (1985).
[33] Marshall (1982a), Rutowski (1982a, b, 1984c), and Wiklund and Forsberg (1985, 1991). Some male butterflies (*Hypolimnas bolina*) mate more than 50 times in their lives (Dyson and Hurst, 2004)!
[34] Based on Paul (2002), Dewsbury (2005), Bonduriansky (2009) and Ah-King (2011).
[35] Based on McMillan *et al.* (1997), Jiggins *et al.* (2001), Kronforst *et al.* (2006, 2012), Finkbeiner *et al.* (2014), and Southcott and Kronforst (2018). Both elements of the forewing colour pattern of *Heliconius heurippa* – a hybrid species – were necessary to produce courtship (Mavárez *et al.*, 2006). The pronounced preference of male *Heliconius* butterflies for red patterns is associated with a 'locus', or fixed position on a chromosome, also responsible for producing the red forewing band (Merrill *et al.*, 2011).
[36] Obara and Hidaka (1964), Obara (1970), Hirota *et al.* (2001), Giraldo and Stavenga (2007), Obara *et al.* (2008), and Obara and Majerus (2000, 2009). Male *P. rapae* use female pteridine-based coloration as the basis for mate choice (Tigreros *et al.*, 2014).
[37] Burghardt *et al.* (2000) and Knüttel and Fiedler (2001).
[38] Ghiradella *et al.* (1972), Rutowski (1977a) and Rutowski *et al.* (2007a).
[39] Huq *et al.* (2019). N.B. the different (DS and WS) forms can be produced in the laboratory by manipulating development temperature regimes.
[40] Emmel (1972, 1973), Taylor (1973a) and Silberglied *et al.* (1979). Reading these papers, especially that of Emmel (1973) suggests to me that the story – mate choice and polymorphism in *Anartia fatima* – is a complex one and has not been fully resolved.
[41] Bonduriansky (2001).
[42] Owen and Chanter (1969) and Jiggins *et al.* (2000). According to Hornett *et al.* (2006), the heavily outnumbered males preferentially select females uninfected with the bacterium, although Jiggins *et al.* (2002) could find no evidence for this type of mate choice.
[43] Rutowski (1980a), Dyson and Hurst (2004) and Charlat *et al.* (2007).
[44] Watt (1973), Graham *et al.* (1980), Gilchrist and Rutowski (1986), Ley and Watt (1989), Whiteley (1992) and Woronik *et al.* (2018). About 28 of the 90 *Colias* species have *alba* forms (Woronik, 2017).
[45] Gilchrist and Rutowski (1986) and Nielsen and Watt (1998, 2000).
[46] Watanabe and Nakanishi (1996), Nakanishi *et al.* (1996), and Hasegawa and Watanabe (2008). Watanabe *et al.* (1997) reported that 'lone males' appeared to be able to discriminate between young and old males *in copula*.
[47] Stillwell *et al.* (2010).
[48] Environmental conditions are important factors in the evolution of the mating systems and sexual size dimorphism (Karlsson *et al.*, 1997). Male nuptial gifts may reduce the value of large size in females, but increase its value in males (Leimar *et al.*, 1994). Rossato *et al.* (2018) found

no significant difference between males and females in forewing size for *H. erato phyllis* and co-mimics. In *Heliconius numata*, male forewings are significantly larger (Jones *et al.*, 2013).

[49] Singer (1982), Gilchrist (1990), Wiklund and Karlsson (1988), Wiklund and Forsberg (1991), Wiklund and Kaitala (1995), Cordero (2000) and Chazot *et al.* (2016).

[50] Orr (1999a), Orr and Kitching (2010), Rutowski (1997b), Chazot *et al.* (2016) and Le Roy *et al.* (2019).

[51] Nakahara *et al.* (2018). Most spots on the dorsal surface of *A. amathea roeselia* wings presented little or no variation between the sexes, but one white spot (no. 18) was only present in females (Francini *et al.*, 2015).

[52] Obara and Hidaka (1968) and Silberglied and Taylor (1973). The wings of birdwing butterflies exhibit brightly coloured patches combining structural and pigmentary effects in the wing scales (Wilts *et al.*, 2015).

[53] Figs 1 and 2 in Meyer-Rochow (1991) show *Lycaena salustius* males and females under ordinary sunlight and UV light, respectively. Sexual dimorphism in UV-reflectivity is also present in many other species, including: *Colias philodice* (Silberglied and Taylor, 1973), *Eurema candida* (Rutowski *et al.*, 2007a), *Eurema hecabe* (Rutowski *et al.*, 2007a), *Pieris protodice* (Meyer-Rochow and Järvilehto, 1997), *Pieris rapae* (Obara *et al.*, 2008) (Fig. 2.31), *Hypolimnas bolina* (Kemp and Macedonia, 2006), *Pieris occidentalis* (Wiernasz, 1989) and many more (David Stella, 8 April 2019, pers. comm.).

[54] Three out of four of the yellow and red colour elements on the wings of *Heliconius erato phyllis* and *Heliconius besckei* were larger in males (Klein and de Araújo, 2013).

[55] Ellers and Boggs (2002, 2003a, 2004) and Imafuku *et al.* (2002b).

[56] Makino *et al.* (1952), see Fig. 5 in Stella *et al.* (2018b).

[57] Oliver *et al.* (2009) and Tokita *et al.* (2013).

[58] Ômura and Honda (2005), Ômura *et al.* (2012), Ômura and Yotsuzuka (2015) and Barbero (2016). Heuskin *et al.* (2014) identified over 100 cuticular compounds in *B. anynana*.

3 Strategies for Locating a Mate

Perching and Patrolling

Finding a mate is one of the biggest challenges facing any animal that relies on sexual reproduction. For butterflies, the process of finding, recognizing and attracting a mate involves a combination of visual (sight) and chemical (scent) signals. Visual cues are particularly important for locating mates at long distances, and males need to stand out against the background, both whilst perching or patrolling. The task of locating a mate is often carried out by males, but females can facilitate the process by placing themselves in the vicinity, for example flying past perching males who then give chase, or making their way towards sites where males gather and/or maintain territories. So, the process of finding a mate is perhaps best seen as a cooperative venture with, as always in Nature, a huge variation in behavioural patterns between species.[1]

Perching in certain places makes sense in situations where the habitat provides resources – e.g. host plants, nectar sources or basking sites – which attract females. Perching on a strategic vantage point allows a male to survey an area across which females predictably transit (Fig. 3.1). Where the features and resources of interest and attraction to females are widely spread out throughout the habitat, or appear unpredictably, it then makes more sense for males to go looking for females by actively patrolling the environment. The density of receptive females in the habitat is also a factor – but not the only one (see below) – in determining the most efficient mate-locating strategy. Deciding whether to perch or patrol – to wait or wander in lay terms – may depend on overall population numbers, with low densities generally favouring perching, and high densities encouraging patrolling. For example, male American snout butterflies, *Libytheana carinenta* (Fig. 5.11), are classic patrolling species, searching

© Raymond J.C. Cannon 2020. *Courtship and Mating in Butterflies: Reproduction, Mating Behaviour and Sexual Conflicts* (R.J.C. Cannon)

in and around their larval host plants – the desert hackberry, *Celtis ehrenbergiana* – for females. However, their mate-searching tactics are to some extent density-dependent: they tend to patrol more when numbers are high, or at least above a certain threshold, and perch more when numbers are low. There are, however, many exceptions to this rule. For example, the brimstone, *Gonepteryx rhamni* (Fig. 3.2), is a butterfly that continues to patrol at low densities, whilst other, species – like the fiery skipper, *Hylephila phyleus* (Fig. 5.10) – perch at high densities.[2]

The theory behind the density-dependent switch from perching to patrolling seems to rest on the idea that it is only at low population densities that females become 'mate-limited', i.e. with not enough males to find and mate with them. Males are often the most abundant sex and are usually ready to mate more frequently than females. So, females will only actively search for males and visit male territories and lek sites – such as hill-tops (see below) – when numbers are low. At high population densities, there are plenty of males flying around searching, so females can rest, and wait for the males to find them. For the males, it also makes sense to give up defending their territories and switch to patrolling, as the numbers of competing males, and hence intruders, increases.[3]

Fig. 3.1. Malay Viscount (*Tanaecia pelea pelea*) male perching on a leaf, Chiang Mai, Thailand.

Fig. 3.2. Brimstone (*Gonepteryx rhamni*), Galicia, Spain.

Optimal search strategies

The concepts of perching and patrolling as mate-location tactics are well established in the literature, but there is another way of looking at mate-locating behaviour, which focuses on the location of encounter sites – i.e. places where the sexes are most likely to come across each other – and then separately considers the behaviour of males and females when they are in these places. First, butterflies must decide – or have evolved ways of determining – *where* to look for mates. Second, they must decide *what* strategy – e.g. perching or patrolling – is most efficient, at a given time and place. Third, they must decide whether it is worth defending their site, which is usually an evolved, species-specific behaviour. We can also add to this basic scheme the concepts of *when* a butterfly should visit an encounter site, and *how long* it should stay there, before moving on or switching strategies.[4]

Sites where male and female butterflies can typically locate and encounter one another, include: (i) emergence and oviposition sites, i.e. on larval host plants; (ii) flowering plants visited by adults to feed on nectar, and (iii) so-called *landmark sites*, which do not contain any obvious resources, other than a physical space in which to perch or patrol. For example, the highest density of receptive, virgin Empress Leila, *Asterocampa leilia* (Fig. 4.8), females were near the hackberry trees, *Celtis occidentalis*, on which they lay their eggs. Female butterflies – including the Anicia checkerspot, *Euphydryas anicia* – are frequently harassed by males whilst searching for larval host plants on which to deposit their eggs. There are of course subtleties and variations in behaviour within the broad categorizations of perching and patrolling, and butterflies will vary in the extent to which they remain attached to a place – partly because the environmental parameters are constantly changing – and the degree to which they will defend it. Similarly, the extent to which males patrol, and the scale of the peregrinations, will also differ over time, often in response to changes in the availability and receptivity of females.[5]

A new naming system for describing mate-locating behaviour in butterflies was recently suggested by James A. Scott, who is sceptical of the existence of true territoriality in butterflies ('a butterfly is about the least-equipped to fight of any animal on earth'). He invented some new words, which are mnemonics for the behaviour they describe. In summary, Scott applied the following terms to male behaviour: *rait* (resting to await females for mating); *flait* (fly around a small mating site waiting for females to arrive); and *fleek* (flying farther to search for receptive females for mating). These terms do not appear to have been widely adopted, however.[6]

Perching – Staying Put and Waiting for Females

Male butterflies frequently perch on prominent locations – such as on stones on the ground or on leaves in a tree – surveying the landscape for prospective mates. The flight of a female elicits an immediate response, as does any passing object in many cases. Sometimes, sitting on the ground is a better way of spotting mates and rivals as they fly past, silhouetted against the sky, and butterfly eyes have different zones for detecting different types of activity (see Chapter 4). In other situations, perching higher up, off the ground, is a better way to detect – and then chase off, or court! – members of the same species. Such intraspecific interactions – i.e. between members of the same species – are usually much lengthier and more intense than those between different species. When a rival male flies into a percher's patch, there is usually a rapid spiralling flight, with the two individuals encircling each other, followed by the intruder (usually) being chased out of the territory. For example, male–male interactions led to extended spiral flights and chases of up to 50 m – lasting up to 119 s – in the purple-edged copper, *Lycaena hippothoe*. The small copper, *Lycaena phlaeas* (Fig. 6.3), is also largely a perching species.[7]

In the European map butterfly, *Araschnia levana*, studied in the Czech Republic, perching was largely restricted to the afternoons, a behaviour in common with other nettle-feeding nymphalids. Males aggregated on patches of nettles – their food plants – perching with wings closed and chasing passing butterflies, bees and even a dragonfly. The rare appearance of a female provoked repeated bouts of chasing, often involving several males. These chases involved ascending spirals, up to 10 m in height; the males either returned to their perch site, or disappeared, perhaps in a successful copulation.[8]

Perching butterflies have also been described as employing a perch-and-sally strategy. On detecting a passing object – which might be either a mate or another insect altogether – the male flies or sallies out to investigate and pursue the object of interest. The choice of perching site is critical, as it usually determines how well the passing objects appear in contrast to the background. The term 'perceived luminance contrast' is used to describe this phenomenon: passing butterflies – or other insects – which are highly contrasted against the background are more easily detected. Body posture is also important to ensure that the zone of acute sensitivity in the compound eye (see Chapter 4) is best aligned to the likely flight paths of any items of interest. There must, however, be some instinctive behaviours that place males at locations best placed to detect passing females, which, for a given species, fly at a certain average height.[9]

Territories

A perching butterfly may or may not be territorial, but territoriality is nevertheless widely observed in butterflies which rely on mating systems that revolve around prominent landmark features or sunspots. Territories typically consist of a site that is open and sunny; roads and tracks made by man are often ideal sites in which to establish a territory. The territory may only be occupied for certain periods during the day, depending on the onset of shade or the characteristics of the species. Red admirals for example, typically set up territories in the late afternoon. The proportion of males in a local butterfly population occupying territories may vary, depending on how many suitable sites are available. Only a few males – the most superior competitors – in a population of scarce coppers, *Lycaena virgaureae*, were seen to be territorial. The best territories are those that a female is most likely to fly through, but these locations are often fluid and changeable. There are often fewer such good sites than there are males to occupy them. Male Turkish fiery coppers, *Lycaena ochimus*, aggressively defend 'territories' around perching sites on flowers, and chase other butterflies as well as other non-lepidopterous insects, passing by.[10]

Thermoregulation and site selection

Butterflies are ectotherms, and therefore rely on heat from the sun, as well as the air surrounding them, to maintain their body temperature. They are, therefore, highly skilled at positioning their bodies relative to the sun to maintain an optimum temperature. For example, graylings, *Hipparchia semele* (Fig. 1.10–1.13), orientate their bodies with respect to the angle of the sun: sun-basking at right angles to the incoming radiation when they need to warm up, but facing directly into the sun – to minimize the area exposed to the sun's rays – when they need to cool down. They can elevate their body temperature by up to 3°C, relative to ambient, in this way. Such thermoregulation can also be fine-tuned by altering the angle of the wings relative to the vertical. Small heaths, *Coenonympha pamphilus* (Figs 3.3 and 3.26), are also very good at angling themselves towards the sun. The height and location of perching places within the territory or mate-location site can also vary as temperatures change during the day or season. Male meadow browns, *Maniola jurtina* (Fig. 3.4), often perch in open locations, with short turf, in order to warm up. Favoured perches are often selected by butterflies based on their temperature relations, i.e. the best places to maintain body temperatures so that they can spring up quickly and investigate any passing objects.[11]

Fig. 3.3. Small heath (*Coenonympha pamphilus*) warming up in the grass, Galicia, Spain.

Fig. 3.4. Meadow brown (*Maniola jurtina hispulla*) male underside, Galicia, Spain.

Patrolling – Moving Around and Searching for Mates

Male butterflies can either sit and wait for a female to come to them, or they can actively search for receptive partners within a wider area. Clouded yellows, *Colias croceus* (Fig. 2.24a) are typically non-territorial, patrolling species, with males visually searching large areas for females. These yellows or sulphur species usually have large green eyes (Fig. 4.6), proportionately larger in terms of surface area than many other species, particularly so in the males. There are however, not thought to be any significant differences in eye size that can simply be related to the type of mate-locating strategy; i.e. perching species do not have larger eyes than patrolling species, on average. Patrolling flights are usually relatively slow and close to the ground, traversing habitats where females might occur, with frequent turns and investigatory behaviour based on visual cues. For example, male Japanese luehdorfia, *Luehdorfia japonica* (Fig. 8.6), butterflies patrol continuously, c. 0.5–1.5 m above the ground, searching for females. Similarly, the bog fritillary, *Boloria eunomia* (Fig. 7.1a, b), flies less than 1 m above the ground, with frequent turns, every metre or so. This type of behaviour is very different from so-called, orientated flights, which butterflies make when crossing unsuitable ground or searching for nectaring sites. These tracks are generally straighter, higher and faster than patrolling flights. Patrolling butterflies, such as the orange tip, *Anthocharis cardamines* (Fig. 3.5), for example, typically follow defined features such as woodland paths or river valleys, and have a tendency to occur along the edges of their habitat, and similarly, along the margins of any clusters of their host plants.[12]

Many species – such as the purple-edged copper, *Lycaena hippothoe*, or the danaid eggfly, *Hypolimnas misippus* (Fig. 1.15), for example – utilize both perching and patrolling behaviour, depending on the weather, or the time of day. Male *H. missipus* are territorial and commonly occur on open hill-tops all over Africa (Szabolcs Sáfián, 2019, pers. comm.). *Lycaena hippothoe* is also very territorial and flies up to investigate any passing insect;

males of the same species are chased, and the two interact with each other in spiralling flights lasting up to 2 min. The scarce copper, *Lycaena virgaureae* (Fig. 3.6), also exhibits similar behaviour, flying up to confront any invading butterflies, particularly males of the same species. Male wall browns, *Lasiommata megera* (Fig. 3.7), can also switch mate-finding

Fig. 3.5. Orange tip (*Anthocharis cardamines*) male. Biatorbágy, Hungary. (Photo by Ádám Gór.)

Fig. 3.6. Scarce copper (*Lycaena virgaureae*), Somiedo NP, Asturias, Spain.

Fig. 3.7. Wall brown (*Lasiommata megera*) female, Cotley Hill, Wiltshire. (Photo by Charles J. Sharp – own work, CC BY-SA 4.0, https://commons.wikimedia.org/w/index.php?curid=69490922)

strategies, spending less time perching when temperatures increase and wind speed decreases. According to Roger Dennis, the behaviour of male *L. megera* varies between two extremes, along a continuum: from that of territorial defence on one end, to that of patrolling – and passively accepting the presence of competitors – at the other end. Males of the Diana treebrown, *Lethe diana* – a satyrid butterfly that occurs in Japan and eastern Russia – similarly patrol over wide areas in the mornings but defend territories in the afternoon, where copulations are achieved by the owners.[13]

The apollo butterfly, *Parnassius apollo* (Fig. 3.8) (Papilionidae), is a patrolling species where males roam over many different locations that provide an opportunity for encountering receptive females. The males emerge 1 or 2 weeks before the females (protandry) and carry out both random patrols – without any obvious patterns – of likely emergence sites, as well as a more regular zigzag scanning of their habitats, often gliding downhill about 0.5 m above the vegetation. Search strategies for locating mates depend on the density and distribution of receptive females in the butterflies' habitat. When females are thinly distributed over a wide area, it makes more sense to go and look for them, rather than sit and wait for one to pass by. Thus, perching and patrolling represents a continuum, with very few butterflies engaging in all of one type of behaviour and none of the other. The satyrine butterfly, *Erebia epipsodea*, a North American alpine species found from Alaska to New Mexico, is a patrolling species that ranges almost continuously over large montane areas containing one vast randomly mating population.[14]

The large skipper, *Ochlodes sylvanus* (Figs 3.11, 6.4 and 9.13), is another species where mate-location behaviour is flexible: males typically patrol during the morning and perch in the afternoon. Patrolling flights are usually slow and extended, with the males staying close to vegetation. Males bask on prominent perching sites, although these are rarely higher than 1 m off the ground, sallying out on aerial 'sorties' to investigate any passing objects. Interactions with males of the same species are intense and violent. Researchers studying these butterflies on a site in Cheshire, England, reported that the impact of these males – as they bashed into each other –was audible![15]

Fig. 3.8. Apollo butterfly (*Parnassius apollo*), Voskopoja, Korça, Albania. (Photo by Ádám Gór.)

Hill-topping – Sexual Rendezvous Points

Hill-topping is the name given to a mate-location strategy where male butterflies aggregate on or near the highest places in their habitats – including hill-tops, mountain ridges and even mountain peaks – to wait for receptive females, which fly uphill towards them. Most females visiting these sites will presumably be virgins, but some may have mated previously. In other words, they gather at *sexual rendezvous points*. In some, but not all, cases the males become territorial, challenging and chasing away other males, defending what is best described as a *lek*, a display site where females visit to choose males with which to mate. Butterflies are by no means unique in carrying out hill-topping behaviour; many insects – including moths, flies, bees, wasps and beetles – congregate on hill-tops to 'pursue, assault, or engage in ritualistic combat with rivals' according to a leading researcher in this field, Professor John Alcock of Arizona State University, USA. A nice example of a territorial, hill-topping butterfly is the great purple hairstreak, *Atlides halesus*; the males defend entire creosote bushes and 'palo verdes' – tall cacti in the genus *Cercidium* – as perching territories from which they repel rival males, in Arizona, USA. Hill-topping species are often less common – i.e. occurring at lower population densities – than those mating elsewhere. Aggregating on hill-tops and ridges inevitably leads to an increase in density, with vastly improved chances of encountering prospective mates, as long as they are also engaging in the same, upwardly mobile behaviour. N.B. some species reverse this behaviour, by aggregating in gullies; the spatial concentration effect is the same.[16]

A very detailed study of hill-topping was carried out by Mr Oakley Shields in California in the late 1960s, which confirmed the theory that hill-tops are meeting arenas where certain species instinctively gather to mate. Alternative theories, such as the idea that butterflies simply congregate on hill-tops as a result of updrafts, or due to the presence of larval food plants, were discounted. Shields found that summit populations consisted primarily of territorial males displaying aggressively to other butterflies. A total of 21 hill-topping species – exhibiting territorial or patrolling behaviour – were categorized, whilst another 25 species were observed to fly 'up-and-over' the summit without congregating there. Another more recent study – carried out on three peaks in the Kongo-Ikoma mountain range, Central Japan – also discriminated between 'real' and 'controversial' hill-topping species. Hill-topping species were further categorized into subgroups depending on the amount of time they spent on the summits and lower slopes, and the extent to which they engaged in fighting, for example. Swallowtail butterflies, *Papilio machaon* (Fig. 3.9), were amongst the most abundant of the true hill-topping species found, exhibiting territorial perching and patrolling as well as aggressively interacting with other, conspecific males and females (see below). Hill-topping is relatively common in Afrotropical butterflies, including species in the genera *Graphium*, *Deudorix*, *Aphnaeus*, *Charaxes*, *Acraea* and *Coeliades*.[17]

Fig. 3.9. Old world swallowtail (*Papilio machaon hispanicus*) third generation brood, Galicia, Spain, 14 September 2017.

Niche segregation

There are great variations in terms of the behaviour and territoriality of hill-topping butterfly species; they often differ in response to both the environmental conditions and the population density of rival males in the area. Different species may utilize the same hill, segregating themselves in terms of time and space. For example, 18 hill-topping butterfly species were recorded on one hill near Barcelona, Spain; some perched on the ground, others used bushes; some species established territories in the morning – e.g. *Papilio machaon*, *Iphiclides podalirius*, *Lasiommata megera* and *Charaxes jasius* – others in the afternoon, e.g. *Vanessa atalanta* and *Vanessa cardui* (Fig. 3.10). The locations of these territories depended on time of day together with individual species preferences regarding factors like windiness, sunshine, insolation, temperature and so on. It is thought that stronger-flying species are better able to establish territories on the summits of hills, as they are better equipped to cope with the windiness and exposure of high elevations. Many species in the genus *Charaxes* – called rajah and pasha butterflies in Asia, and emperors in Africa – display temporal segregation on hill-tops. For example, in Sarawak, east Malaysia, the tawny rajah, *Charaxes bernardus*, was the first butterfly to appear on hill-tops at c. 10.30 h, leaving around 13.30 h; a larger species, *Charaxes repetitus*, appeared shortly afterwards, around 14.00 h and remained until 15.30 h. Presumably these species have evolved such temporal segregation to ensure species segregation, although some overlap exists.[18]

A study of hill-topping lycaenid butterflies on a 2200 m mountain in Colombia found that 20 species – all beautiful iridescent green and blue butterflies in the tribe Eumaeini – out of a total of 72 species present in the area, used a small 200 m^2 mountain top for mate location. These tropical butterflies also demonstrated a remarkable ability to segregate themselves in space and time, particularly in terms of their vertical distributions. Their stratification was such that there were three or four separate levels, or strata, within 8 vertical metres of forest, at which butterflies of different species would perch and fly. This demonstrates an impressive

Fig. 3.10. The painted lady (*Vanessa cardui*) on thistles, Galicia, Spain.

three-dimensional stratification of a diverse community within a relatively small mating arena.[19]

James A. Scott described how three *Hesperia* skippers (Hesperiidae) effectively divided up their habitat – in Chaffee Co., Colorado, USA – to avoid interspecific interference, despite sharing the same larval host plant. The Pahaska skipper, *Hesperia pahaska*, perched on hill-tops and ridge-tops; the green skipper, *Hesperia viridis*, perched on stones and bare spots at the bottom of gullies; and the Uncas skipper, *Hesperia uncas*, perched on 'slight rises'. The males of *H. pahaska* were described as not being territorial, but needed to space themselves out, about 3 m apart, to avoid interference. Clearly species have evolved to partition the physical space in the habitat, to minimize the costly effects of interference – e.g. having to fly up to investigate or chase off males of a different species.[20]

Leks – arenas for contests

Butterflies often fight over so-called lek sites – areas such as rocks or bare ground, devoid of any resources such as food plants or oviposition sites – which are used to attract females and engage in courtship behaviour. The snow apollo, *Parnassius imperator* – a high-altitude species found in western China and north-east India – defends such territories and the owners always defeat intruders, regardless of their relative sizes. Males fight over areas of bare rocks, which are used as lek sites to attract and court receptive females. The perching order – i.e. when they arrived on the site – determined the outcome of the contests, not body size. How contests are settled in different species is a topic of active research with studies looking at relative body sizes, manoeuvrability, fat reserves, stamina and so on of the competing butterflies. Resident males of the Mexican cycadean, *Eumaeus toxea*, a butterfly whose host plants are cycads, were larger and fatter than intruders and, usually – but not always – won contests. However, one rather intangible feature that might determine the outcome of territorial contests, is sheer persistence. The residents may just be more motivated to hang on to their territories, because ownership is the key determinant in attracting females.[21]

Pushing and shoving!

The neotropical rhiodinid, *Charis cadytis*, engages in what can only be called 'pushing and shoving' contests, as a form of male–male competition. Dominant males defend small core areas of a lekking arena, located in sunny patches along the borders of forest trails, which females visit looking for mates. In Serra do Japí, São Paulo state, Brazil – where they were studied by researcher Gabriela W. Chaves and colleagues – they engage in three different types of interactions with each other: (i) frontal flights; (ii) pursuits or chases; and (iii) 'sumo-type' pushing contests. The agonistic interactions usually commence when rival males confront each other in face-to-face ascending flights (frontal fights). They ascend vertically, only about 3 cm apart, with their bodies remaining almost perpendicular to the ground; sometimes almost hovering, as they seem to be sizing each other out. At some point, between 1–5 m up, they break off from facing each other, and a pursuit starts. They do not touch, or spin around each other, in these aerial pursuits, which often punctuate bouts of sumo-type behaviour. The apparent loser, or defeated individual, who is chased away, frequently returns to the core area – on the same or subsequent days – to engage in new contests with the resident.[22]

The contests took place within one or two small, fixed areas, within a larger lekking arena. These core areas (3–5 m long) contained one or two resident males, perched on the top of leaves – with their wings spread open – in wide, sunny sites along trails or roadsides. Outside of these disputed patches, males were observed to be very tolerant of each other. The shoving contests usually started after a short pursuit flight in the core area, when the antagonists landed close to each other on the same leaf.

They start by facing each other, then move around to an oblique position, where they push against one another with partially opened wings, in contests lasting 22 s, on average (longest was 149 s). These pushing contests took place between pursuit flights, with the butterflies rapidly shifting from one behaviour to another, in sometimes prolonged interactions, which in one case, lasted nearly 5 min (292 s). The loser eventually starts to tilt sideways and then flies off; remarkably this was usually the largest one of the two. Other, non-combatant, more submissive males were observed to occupy scattered perches around the periphery of the lek sites, avoiding entering the contested areas, but perhaps attempting to intercept passing females, suggests Chaves. Females were seen much more rarely, and then only fleetingly, unless observed *in copula*.[23]

The defence of landmark encounter sites by dominant males is very unusual for butterflies, although it might be more prevalent than we realize. The sumo-like contests are rather low-impact shoving matches – in other words, they do not result in injuries – and appear to be a form of ritualized behaviour that may have replaced the more common, spiralling and circling interactions seen in other species. Interestingly, it was the smaller males who came out as winners in these contests, and the dominant males residing in the centres of the arena were smaller than non-residents on the periphery. Somehow, they manage to unbalance the larger ones and force them to fly away. Perhaps this a skill where agility is more important than size. N.B. smaller males were also more successful in contests between males in the Sara longwing (*Heliconius sara*) (see above).[24]

Landmark mating systems

By far the most common sort of territory however, at least in temperate regions, is where butterflies defend a territory on or near a feature in the landscape – a so-called landmark point. Here the only resource is the presence of the male himself and females only visit these lek sites for the purpose of copulation. Typically, these landmark points are hill-tops, sunspots, sunny gaps in a wooded area, roads or prominent perches along forest margins, but they could be any prominent feature in the environment of the species concerned. The brown Arctic grayling, or chryxus Arctic, *Oeneis chryxus*, is a subalpine satyrid found in the far north-west regions of North America, as far north as southern Alaska and the Yukon Territory. The males aggregate on lek sites associated with certain visual landmarks, including dirt roads, bare ground, ridges and rocky knolls, where they consistently defend territories and engage in spiralling flights with other males. Such 'landmark mating' systems may have developed in response to the scarcity of females and widely dispersed resources across their sparsely vegetated habitats. So-called 'drifters' – males without territories – are less successful than those who do hold territories.[25]

Examples of Perching and Patrolling by Family

Hesperiidae

Skippers are some of the most characteristic perching butterflies, and often occupy prominent vantage points on conspicuous landmarks where they can survey the scene for the arrival of receptive females (Fig. 3.11). Many skippers show very similar behaviour: darting out to investigate passing

Fig. 3.11. Large skipper (*Ochlodes sylvanus*) male perching, Galicia, Spain.

objects, with the aim of locating a female or challenging another male. The mournful duskywing, *Erynnis tristis* (Erynnini), typically perches on a wide range of shrubs, trees and cacti on peaks and ridges in the Sonoran Desert, in the south-western USA. They defend their perches against rival males and occasionally engage in spinning-wheel flights, where the intruder is forced down towards the ground. The large sprite, *Apallaga safiani* (Pyrginae), is an African hill-topping sprite, which occurs on the forested summit of Mount Beeton, Nimba Mountains, Western Range, Liberia. It flies very rapidly, up and down along linear display areas, such as footpaths, chasing away other males. Many skipper species have been found to be crepuscular, or semi-nocturnal in Africa – including those in the generas, *Caenides*, *Leona*, *Pteroteinon*, *Zophopetes*; and *Coeliades chalybe* – flying at dusk and well into the night (Szabolcs Sáfián, 2019, pers. comm.).[26]

Lycaenidae

The grey hairstreak, *Strymon melinus* (Strymonini) (Fig. 3.12) – a North American butterfly – also switches from vigorously defending its perching site, to a non-aggressive patrolling mode, when the numbers of intruders get too great: adaptability in the face of changing circumstances. Another hill-topping lycaenid, the mottled scrub-hairstreak, *Strymon mulucha*, engages in relatively lengthy male–male contests, lasting up to 21 s, to defend their hill-top territories.[27]

Nymphalidae

Some species, like the California patch butterfly, *Chlosyne californica* (Melitaeini), have more than one arrow in their quiver when it comes to searching for mates. They can sit on hill-top perches and wait for females to find them, or, if the competition gets too fierce, they can fly off on

Fig. 3.12. Grey hairstreak (*Strymon melinus*), Virginia, USA. (Photo by Judy Gallagher [CC BY 2.0 (https://creativecommons.org/licenses/by/2.0)].)

searches, patrolling the downslopes for incoming females. In other words, they switch from territorial behaviour to non-aggressive patrolling; i.e. from 'contest' competition to 'scramble' competition (see Glossary). Dotted checkerspot, *Poladryas minuta* (Melitaeini), males also switch strategies, from perching on ridges and hill-tops in the mornings – waiting for females to come to them – to patrolling hillsides and flatter ground near flowers in the afternoon, searching for females. The wall brown, *Lasiommata megera* (Elymniini) (Fig. 3.7), and the large wall brown, *Lasiommata maera* (Fig. 3.13), are both hill-topping species, but this behaviour is highly dependent on favourable weather conditions, namely high temperatures and low wind speeds. Male *L. megera* switch from stationary behaviour – waiting on hill-tops for females to come to them – to a more active, vagrant type of activity, as conditions become more favourable for patrolling. Virgin females of this species have a natural tendency to fly up hill, but this desire for meeting males in high places wanes once they have mated. Similar behaviour is seen in Edith's checkerspot, *Euphydryas editha* (Melitaeini), where only virgin females hill-top.[28]

The forest mother-of-pearl, *Protogoniomorpha* (=*Salamis*) *parhassus* (Junoniini), a large (male wing span: 65–80 mm) tropical nymphalid that occurs in African forests, uses sunlit patches as landmark sites in which to locate mates, in a manner similar to other sunspot perchers like *Hypolimnas bolina* (Fig. 2.1) and *Pararge aegeria* (see below). However, males change their mate-location strategies, from perching to patrolling, in more open, disturbed forests, where larger areas of the forest floor are bathed in sunlight.

James Lawrence and Michael Samways studied hill-topping behaviour in *Acraea encedon encedon* (Acraeini) in Pietermaritzburg, South Africa, and compared individuals at the top of an 896 m local summit, with those at the bottom. There were no differences in the density of males in the two locations, but there were significant differences in their behaviour. The hill-topping males perched and investigated passing insects, including other members of the species, whereas those that stayed at the base of the hill patrolled and investigated other flying insects, but *not* conspecifics.

Fig. 3.13. Large wall brown (*Lasiommata maera f. adrasta*), Somiedo, Asturias, Spain.

In other words, the hill-topping males spent much more of their time in – interspecific and intraspecific – interactions with other individuals and in attempting to mate with females. The hill-topping males were also significantly older, but no larger, than the non-hill-topping males.[29]

Riodinidae

Males of many metalmark butterflies – they often have small metallic-looking spots on their wings – typically perch in small well-defined areas, along streams and forest edges, but also on hill-tops and ridge-tops. They exhibit marked niche segregation in terms of their choice of when and where to perch. A 5-year study by Curtis J. Callaghan of neotropical riodinids – from southern Brazil to Panama – found that different species within the same genus avoid interbreeding and remain isolated despite occurring within the same habitat. They manage to achieve this by dividing up their forest habitat into separate rendezvous sites, based on time of day, location – using tree fall and forest edges, for example – and segregation of spaces using light and shade. Riodinid males usually only perch for a few hours each day, and the females actively search out these rendezvous sites. The choice of perch site varies considerably between different metalmark genera, with some perching conspicuously on upper leaf surfaces and others hiding under the leaves. Species-specific pheromones probably play a major role in isolating congeneric species, which are often remarkably similar in terms of morphology, including their genitalia.[30]

Papilionidae

Swallowtail butterflies frequently defend mating territories at the top of prominent hills. They typically patrol a relatively small, well-defined patch, searching around a prominent feature, like a rock, and darting out to chase away any intruders. I monitored such a patch in 2017, occupied by European swallowtails, *Papilio machaon*, in Galicia, Spain (Fig. 3.9). The site was occupied by male swallowtails over three different generations, in a summer. The phenology of the swallowtails changes slightly as the summer progresses: the butterflies show progressively more yellow scales, a fine dusting of yellow, on the background black colour, presumably in order to blend in with the drying, yellowing vegetation.

The giant African swallowtail, *Papilio antimachus* (Fig. 3.14) – the largest butterfly in Africa, with a wing span between 18 and 23 cm – is a rare and elusive species that has been studied by Hungarian lepidopterist Szabolcs Sáfián in the Nimba mountains of Liberia, west Africa. The males maintain well-defined territories, c. 100 m in radius, on hill-tops (c. 1000 m), chasing away other large butterflies and even swifts! A hill-top, in effect concentrates the density of a thinly populated species such as this and gliding at heights of 25–30 m above the ground gives the males

Fig. 3.14. (a) The giant African swallowtail (*Papilio antimachus*), Musée zoologique de Strasbourg. (Photo by Ji-Elle [CC BY-SA 3.0 (https://creativecommons.org/licenses/by-sa/3.0)].) (b) Illustration of male (below) and female (above) of *Papilio antimachus*, upper and lower-lateral view, with its former name: *Druryia antimachus*. (By Robert Henry Fernando Rippon (died January 1917) (public domain).)

a good chance of spotting females. Courtship and mating occur high off the ground in the canopies of large trees. This iconic species is of conservation concern and is threatened by habitat fragmentation and forest clearance.[31]

Male pipevine swallowtails, *Battus philenor* (Fig. 2.18), patrol mountain peaks in Arizona and defend their patrolling sites against other intruding males. They engage their male competitors in ascending flights that appeared to the researchers to be 'ritualized aerial combat'. Remarkably, the two interacting males can ascend to 300 ft (91 m) in the air, as they spiral around one another.[32]

Pieridae

Observations on two *Pereute* butterflies – *Pereute callinira* and *Pereute telthusa* – in montane tropical forests of the Chanchamayo district of Peru, South America, revealed that the males were territorial and frequently patrolled the valley floors of ravines and gorges, as non-resource-based encounter sites. The males of both species typically flew at a height of about 5–8 m above the ground, or water, in small territories located along the course of a river, at intervals of about 30 m (*P. telthusa*) and 50 m (*P. callinira*). The males patrolled up and down over small territories (10–20 m in diameter) for about an hour and a half in mid-morning, chasing out rival males and interacting with other species. The males flew slowly, 'sailing or gliding' with their forewings outstretched and held at angles between 135° and 180°, which exposes a conspicuous bright red median wing band (Fig. 3.15). The visitation times varied with ambient temperature and the patrolling was carried out by older males. The younger males – freshly emerged with no wing damage – were seen puddling in the river, in cool, shaded habitats; presumably building up the nutrients required for spermatophore production.[33]

Fig. 3.15. *Pereute callinira* from Peru. Mounted specimen. (Photo by Hectonichus – own work, CC BY-SA 3.0, https://commons.wikimedia.org/w/index.php?curid=17755960)

Territoriality

Territoriality in butterflies usually involves the defence of strategic sites – places where females can reliably be encountered – rather than obvious food resources. It should not be confused with just perching; a high degree of site fidelity must be present, together with a high frequency of putative contests. Defended sites may contain plants used for oviposition; they may just be prominent landmarks used for aggregation, or they may be places that females regularly visit. For example, territories function as mating stations in the xami hairstreak, *Callophrys* (= *Xamia*) *xami* (Fig. 5.3). There are also instances of males defending territories containing immature females – i.e. pupae – waiting for them to emerge, as seen in the Taylor's checkerspot butterflies, *Euphydryas editha taylori*, in Washington and Oregon, USA.[34]

Some researchers have objected to the use of the word territorial in terms of butterfly behaviour because they lack any obvious offensive weapons and do not appear to be aggressive. It is also difficult to determine whether interactions are genuinely aggressive contests, or simply instances of two butterflies attempting to ascertain the sex of the other one (discussed below). Nevertheless, there are examples of physical contact between male butterflies, with some species pushing or even striking each other, and others clinging onto each other and grappling as they fall to the ground (see below). What they lack in terms of weaponry, they make up for in persistence, which is perhaps a subtle form of aggression.[35]

Weidemeyer's admiral, *Limenitis weidemeyerii*, is a North American species that defends well-defined territories for extended periods: over a month in the breeding season, in southern Alberta, Canada. The same sites are often used in successive generations and are rapidly recolonized by other males if the territory holder is removed. All males of *Limenitis* species (Fig. 7.17) are aggressive, and size and residency status are usually used by contestants to decide contests. The territories of *L. weidemeyerii* are often located along flyways or corridors bounded by vegetation, which females traverse in search of oviposition sites; these sites provide males with vantage points from which they can quickly locate receptive females and challenge any intruders, particularly conspecific males. There was no evidence of any territorial behaviour at sites where the males were just feeding; for example, imbibing honeydew secretions from aphids. Unlike many other butterfly species where size does not seem to matter in territorial contests (see below), the larger *L. weidemeyerii* males were more successful, both in terms of defending territories and ousting other owners. Size and residency are used by contestants to decide contests, and the males gradually assess one another's strengths during a contest. Residents are usually more successful, other things being equal.[36]

The black swallowtail, *Papilio polyxenes*, is a sexually dimorphic North American species, which is a member of a mimicry complex whose model is the aposematic pipevine swallowtail, *Battus philenor* (Fig. 2.18). Only the females are Batesian mimics – in terms of their upper wing surfaces – both sexes look like the model on their underwings and therefore, both should benefit from mimicry when exposing their ventral wing surfaces. The males – described as aggressive – emerge 3–5 days before the females eclose and are significantly smaller and lighter as a result of their more rapid larval and pupal development rates. Males that emerge earlier are more successful in terms of securing and defending a territory; previous success, i.e. experience, seems to be more important in terms of determining success than body size. Dominant males secure superior hilltop territories – where most (80%) successful courtships occur – in which they perch and aggressively chase away other males. The most desired territories are those most visited by females, presumably as a result of their precise location and elevation.[37]

Males of the satyrine butterfly, *Paryphthimoides phronius*, are highly territorial and defend mating areas located in sunny clearings along forest edges, throughout the year in south-east Brazil. Resident males return to the same site on consecutive days and engage in lengthy 'combats' with other males of the same species. These contests were said to include 'linear, spinning and/or spirally flights', which result in one of the contestants leaving the area. At high population densities, new, possibly suboptimal territories are colonized. Territoriality in this species probably coincides with the availability of receptive females, who may take up to 4 days to become sexually receptive, after they have eclosed. For males, their quality and attractiveness to females rests heavily on their ability to secure and hold onto a territory.[38]

Red admirals (*Vanessa atalanta*) (Fig. 3.22) hold territories that they tenaciously defend, chasing any flying insect that ventures into their domain. Their elliptical-shaped territories usually contain one or more ground-based resting spots, such as stones or patches of bare earth (Fig. 3.23). The males spend part of the time perching on their resting spots, and part of the time patrolling their territories. The proportion of perching males to patrolling males gradually shifts over the course of the day, with individuals switching to perching and basking in the afternoon. This enables them to maintain an optimal body temperature in order to challenge intruders, who are confronted and chased off in characteristic helical flights as the butterflies rise upwards to the level of adjacent treetops. The territories are, however, temporary affairs, formed and maintained by male red admirals for a few hours at the end of the day, before dusk. Studies in the USA – by Dr Royce Justin Bitzer at the Iowa University State campus – revealed, somewhat surprisingly, that male owners of territories seldom returned to the same spots on successive evenings. Interactions between males – so-called 'dances' – lasted up to 44 s; the more prolonged interactions occurring between equally sized males.[39]

Other nymphalid butterflies – such as the peacock, *Aglais io* (Fig. 3.27), the small tortoiseshell, *Aglais urticae* (Fig. 3.28) and the comma, *Polygonia c-album* (Fig. 7.10) – also maintain similar breeding territories, which they actively defend against intruders. These nettle-feeding nymphalids sometimes use molehills and bare ground as perching sites in sunlit territories, which usually have a vertical 'edge', created for example, by a row of trees or a wall. Such ground-based territories are typically occupied and defended for a few hours in the afternoons in spring and summer. Females tend to be concentrated along such boundaries and males of *Aglais io* and *Aglais urticae* do not merely investigate the sex of the intruder, they actively defend their territories against other invading males. The Australian painted lady, *Vanessa kershawi*, exhibits behaviour very similar to that of the red admiral, *V. atalanta*, occupying so-called, 'perch' territories on hill-tops or other prominent locations late in the day, chasing intruders and making ascending flights.[40]

The small heath butterfly, *Coenonympha pamphilus* (Figs 3.3 and 3.26), is also highly territorial, but as temperatures rise the males abandon their perches and start searching for females over wider areas. Their territories are used as 'mating stations' – or lekking sites – and males that hold territories are usually larger, and mate more frequently, than those individuals outside the territories. The process is facilitated by the behaviour of virgin females, who search for male territories – using landmark features such as trees or bushes – and engage in lengthy, conspicuous solicitation flights, which increase their probability of finding a male to mate with.[41]

Territory formation and turnover

Certain conditions, or prerequisites, are necessary for the formation and occurrence of territories in butterfly populations. These usually include: a bias – in terms of the ratio of sexually receptive males to sexually receptive females – in favour of males; a certain predictability in terms of the temporal and spatial occurrence of the less abundant sex (usually females); populations of a certain size; and an ability to successfully challenge and expel rivals from a defined site. The establishment of defined encounter sites is one way of coping with low population densities, where the chance of a random encounter between receptive partners is low. A bit like going to parties in humans![42]

Territorial mating systems that involve the defence of resources – such as host plants – are far less common than those where no obvious resources are involved, but they do exist. In several species these involve guarding oviposition sites. For example, males of the white peacock, *Anartia jatrophae* (Fig. 3.16), defend small sites containing their larval host plant, *Bacopa monnieri* (Plantaginaceae), in Ochopee, Florida. Males always appear to return to the same sites containing the host plants after chasing away intruders or moving off to feed on nectar plants elsewhere. As usual, resident males generally win any aggressive encounters, and return to perch on, or near to their previous perching place. They also returned to the same territories with a relatively high degree of fidelity, on subsequent days.[43]

There is often a high degree of consistency, from one day to the next, in terms of the individual male defending a given territory. Nevertheless,

Fig. 3.16. White peacock (*Anartia jatrophae*) female. Krohn Conservatory butterfly show, Cincinnati, Ohio. (Photo by Greg Hume [CC BY-SA 3.0 (https://creativecommons.org/licenses/by-sa/3.0)].)

some residents are displaced, and site fidelity is rarely 100%. Butterflies are, however, highly adaptable, and individuals displaced from one territory can usually shift to another, albeit less desirable in some cases. Territories of lower quality – usually as a result of lower female encounter rates – are more easily abandoned than more desirable ones. Aggressive displacements are more likely to occur where territories are in short supply, as observed for *Callophrys xami* (Lycaenidae) (Fig. 5.3) in an ecological reserve to the south of Mexico City, called Pedregal de San Angel. Individual males were seen to occupy and defend the same territory for an average of 5 h a day, up to a maximum of 23 days. Spontaneous shifts towards more desirable territories also occurred, probably because it takes time to evaluate the quality of a given territory; also, its desirability may change over time. The territories occupied by *C. xami* were located at the intersection of natural and manmade trails, used by females to move through the habitat, and all successful courtships commenced inside such territories.[44]

Territorial contests

As we have seen, some – but by no means all – butterflies are territorial and compete in non-contact interactions over a site or territory. These are the characteristic spiralling – or co-rotating – flights, seen in so many species all over the world, but devilishly difficult to photograph! They have also been called 'spinning wheels', a phrase that graphically describes the pair of butterflies spiralling round and round each other as they move across the ground and usually upwards. Such interactions may be protracted, lasting up to 20 min in some territorial lycaenids, and usually end with the 'loser' being chased away. Researchers vary in how they describe such interactions, using words like 'fights', 'combats', 'duels' and, more frequently, 'contests' to describe them. However, it is often extremely difficult to be certain about the exact nature of these interactions, even though they exhibit a remarkable similarity, across species, genera and families.[45]

True territorial behaviour only occurs when a male repeatedly defends his perching site – and the surrounding territory – by trying to drive away males of the same species; usually by outflying or outmanoeuvring them, via spiralling interactions. The victor in these so-called 'wars of attrition' is usually the resident male, who manages to position himself under the intruder, and thereby drive him up and out of the territory. For example, in the Sara longwing, *Heliconius sara* (Fig. 7.15), a neotropical butterfly found from Mexico to southern Brazil, the dominant male – always the resident – can position himself under his opponent:

> positions himself about 0.3 m below his adversary and, maintaining this geometry, evicts the intruder by flying in tandem slowly (ca. 0.3–0.5 m/s) obliquely upwards to perhaps 15–25 m from the territory. (Hernandez and Benson, 1998, p. 534)

In other words, the resident escorts the intruder up and out of the territory. Strangely enough, in this species, it is *smaller* males who have the territorial advantage: a result the researchers put down to their 'diminished mating prospects' compared with larger males. In other words, it seems that the larger males were generally more successful at mating and tended to avoid combat with their smaller, more aggressive rivals![46]

Contests can occur very frequently in desirable territories and can be protracted. For example, co-rotating flights can last for more than 20 min in *Chrysozephyrus smaragdinus*. Similarly, a resident male thecline lycaenid, *Favonius taxila* – a brilliant green Japanese hairstreak – engaged in 70 separate circling flights, seeing off all intruders, in the space of 47 min. Similarly, a Costa Rican hesperiid – known by the rather uninspiring common name of the approximate flat (*Celaenorrhinus approximatus*) – engaged in nearly 10 'fights' per hour with intruding males, whilst defending its small sunspot territory. These sallies, which consist of circling and ascending flights, typically last less than 1 min and the resident male nearly always wins. *Celaenorrhinus* skippers (Pyrginae), also called sprites or flats, hold their wings flat when resting (Fig. 3.17).[47]

Males of the South American heliconiid, *Actinote pellenea* (Fig. 3.18), defend territories in clearings in low-lying areas near streams in São Paulo state, Brazil. Resident males attack intruders that come within 10–15 m

Fig. 3.17. Common spotted flat (*Celaenorrhinus leucocera*) male, Thailand.

and chase them in horizontal pursuit flights. A small proportion (c. 11%) of these interactions end in a physical fight, with one male grabbing and clinging on to the other, resulting in them both falling to the ground. On the ground, they batter each other with their wings and claw at each other with their legs. The resident always wins – by physically dominating the intruder – perhaps because stronger individuals come to occupy territories over time. It is unclear just how costly the physical interactions are, and most disputes were decided without much physical contact.[48]

Battling swallowtails

Although butterflies lack any obvious weapons with which to threaten or damage rivals, some examples of physical damage have been observed in several swallowtail species. The rare and endangered homerus swallowtail, also called the Jamaican giant swallowtail, *Papilio homerus* (Fig. 3.19)

Fig. 3.18. *Actinote pellenea*, Rio Grande do Sul, Brazil, in April 2010. (Photo by Cláudio Dias Timm – Flickr: Borboleta (*Actinote pellenea*), CC BY-SA 2.0, https://commons.wikimedia.org/w/index.php?curid=14844249)

Fig. 3.19. Homerus swallowtail (*Papilio homerus*), Blue and John Crow mountains, Jamaica. (Photo by V. Turland, from http://whc.unesco.org/en/documents/137172)

– it is featured on several Jamaican postage stamps – appears to engage in what can only be described as fights. Males of this very large papilionid – the largest in the Americas – clash with each other, with audible impacts, in territorial disputes. Territories occur in forest clearings where nectar sources are common. When an intruding male enters the territory of a resident male, they fly towards each other and collide. There then follows a spinning-wheel flight where both males fly in circles around each, until one – usually the intruder – is chased out of the territory. Resident males retained occupancy in 78% (n = 18) of such battles. These interactions can result in considerable damage: in one instance a male lost more than 90% of one wing. Such battles are we must assume, rather rare, and perhaps only occur in this butterfly because of its considerable size: it has a wing span approaching 15 cm. Another aggressive papilionid is the Indra swallowtail, *Papilio indra*: a species complex of at least 12 different subspecies, some of which have been recorded hill-topping. One rare and highly localized subspecies from Colorado, USA, *Papilio indra minori*, engages in physical contests, and males have been observed crashing 'full tilt' into each other with a shower of black scales falling from their wings. The two individuals were said to have emerged 'battered and torn' from their aerial combat.[49]

Crepuscular contests

A few species have evolved to conduct territorial contests during the twilight hours, before dawn or at dusk. These crepuscular butterflies include the evening brown, *Melanitis leda* (Satyrinae) (Fig. 3.20a, b), a common species in tropical regions from Africa to Australia. This is a butterfly with distinct wet and dry season forms, a phenomenon called polyphenism. Male evening browns, studied in Australia, defended perching sites for a short, c. 30-min period after sunset, where they courted females and competed with other males in relatively low light conditions. In these twilight fights, the winners tended to be younger than losers, and residents – i.e. of a perching site – won much more often (77% of contests) than non-residents. Ambient light levels during these evening encounters were at least two orders of magnitude lower than during the day, so the

Fig. 3.20. (a) Common evening brown (*Melanitis leda*) wet season form with bee, Thailand. (b) Common evening brown (*Melanitis leda*), Sulawesi, Indonesia.

brightness contrast of a flying butterfly would be severely diminished. Yet somehow the males managed to cope with the dim light and low temperatures, and successfully court *M. leda* females.[50]

Male evening browns, *M. leda*, engage in highly stereotyped contests, which consist of two discrete phases: circling and diving. At first, they fly together in a circle, one slightly (0.1–0.2 m) above the other, at relatively high speeds. After only a few seconds (4 s on average), they begin the lengthier second phase. Both males take turns in gliding downwards, at an angle of between 20–30°, with wings held wide open, pulling up just before reaching the ground or substrate. The butterfly then flies vertically back upwards and repeats the manoeuvre, slightly further on. The two males were described by researcher Darrell Kemp as looking like little aeroplanes that were continually stalling and recovering. Is it a display of strength or agility? They carry out this diving behaviour for about 45 s on average, rhythmically swooping downwards and flying upwards, in the direction of the other male. The outcome of the contest is decided when one of the males breaks off and flies away. The winner then returns to the place where the contest began.[51]

Another species that likes to conduct its contests and conquests during the twilight hours, is the owl butterfly, *Caligo idomeneus* (Fig. 3.21). This, and others in the genera *Caligo* and *Opsiphanes*, spend short periods of time (c. 15 min) before dawn – just as the stars are disappearing from the sky – and at dusk, defending specific perching sites, often spaced apart along rivers or roads in Espírito Santo, Brazil. They occupy the same sites from one year to the next. It can be quite cool at these times of day and these large butterflies must spend considerable reserves of energy to increase their body temperatures – by thoracic muscular shivering – so that they are ready to court and mate with any receptive females. The advantage of seeking mates in these twilight hours is that they are much less vulnerable to predation, but there is clearly a cost in terms of mating in the cool, and low light levels at the turn of day.[52]

Male *C. idomeneus* were observed to arrive on their territories shortly after 05:50 h and they remained there for only about 15 min, before flying back into the forest. Shivering thermogenesis – via rapid vibration of the wings – enables them to elevate their thoracic temperatures by 10 (*Caligo*) or 15°C (*Opsiphanes*), above ambient. Crepuscular species, such as *Caligo memnon*, have also evolved visual adaptations to dim light – large eyes and enlarged facets (see Chapter 4) – giving them much greater visual sensitivity. Female *Caligo* species visit the display sites of the males and mate with them there. The females choose males at these lek sites, accepting or rejecting partners according to their predilections. Courtships are relatively brief, requiring as little as 20 s, but a pair may remain *in copula* for more than 100 min.[53]

Wars of attrition or mistaken identity?

Circling flights, where two males interact in a territorial dispute, have been called 'wars of attrition'. The resident male almost invariably wins.

Fig. 3.21. Idomeneus giant owl (*Caligo idomeneus*), Jardin Botanica de Quito, Quito, Ecuador. (Photo by D. Gordon E. Robertson [CC BY-SA 3.0 (https://creativecommons.org/licenses/by-sa/3.0)].)

For example, resident male green hairstreaks, *Chrysozephyrus smaragdinus*, in Japan, won 'fights' 98% of the time. The exact means by which such 'persistence duels' are settled, however, remain something of a mystery. Holding on to a territory in this way greatly increases a male's chances of locating a mate, but some researchers have questioned whether such male–male interactions are true wars of attrition. Male butterflies are also notoriously poor at determining the sex, as well as the species, of intruding insects, until they are quite near. Experiments carried out by Professor Ron Rutowski and colleagues on the Empress Leilia, *Asterocampa leilia* (Fig. 4.8), in the Sonoran Desert, Arizona, USA, found that the males did not respond to intruders of the same species if they were more than 3 m away. Males of the chryxus Arctic, *Oeneis chryxus*, were seen flying up from their perch to chase other butterflies, birds, falling leaves or even tossed pebbles! This tendency to investigate any flying object is quite typical, and can result in males spending a lot of their time chasing each other. Such behaviour, seen for example in the anicia checkerspot, *Euphydryas anicia* – a North American species – can produce aggregations of males.[54]

Despite some of the more aggressive territorial interactions described above, contests between most male butterflies are usually non-violent. Violent clashes are probably relatively rare, and it may be that spiralling flights are more of a chance for the resident male to assess the sex of the intruder. It has also been suggested that such male–male interactions are mistaken attempts at courtship – the *erroneous courtship hypothesis* – which break down when it becomes clear to each individual that they are of the same sex! Most butterfly researchers are, however, persuaded that such spiralling interactions are usually genuine territorial contests, rather than just investigations of the sex of an intruder. The slow departure of the non-resident ‘loser’ – usually the individual with the lesser persistence – appears to act as a form of appeasement behaviour in some *Heliconius* species, for example. The decision to desist and back away is precipitated by the aggressiveness of the resident, who – as in *H. sara* – may eject an intruder by flying under it and forcing it upwards and out of its territory. In conclusion, individual species differ in terms of the frequency and duration of such behaviour, but the essential elements of mate-locating strategies – patrolling flights and persistence duels – are broadly similar across a wide range of butterfly species, and most contests are probably settled by persistence, rather than any form of violence. There is, nevertheless, some disagreement amongst researchers about the exact nature of the interactions between males, and whether they are genuine contests, or evaluations of identity. The exact basis on which they are resolved remains somewhat obscure, and it seems to me ripe for some more research using slow motion photography.[55]

Persistence or endurance?

Another way at looking at territorial contests is to consider them as non-contact endurance matches. High-speed video was employed by researchers Paulo Peixoto and Woodruff Benson – from the Universidade Estadual de Campinas, São Paulo, Brazil – to study territorial disputes in two small, tropical satyrine species. They demonstrated that neither species used physical contact to settle their disputes. In one species, *Hermeuptychia fallax*, the younger males – which had a greater fat content – tended to be the ones who became the resident, suggesting that, in this species at least, the contests were settled by the endurance abilities of the butterflies, which might be related to fat reserves and youthful vigour. It is possible that other factors, such as the relative number of parasites they contained, might also have been involved. It has also been shown that butterflies from areas with more stable climatic conditions tend to invest more in terms of territorial defence, perhaps because the returns are greater, in the form of extended opportunities for reproduction.[56]

The outcome of territorial disputes in butterflies will depend on the balance of certain behavioural, morphological or physiological attributes, which, taken together, are called the resource holding potentials (RHPs) of the different organisms. The relative weightings of these RHPs is often in favour of the resident or current holder of a territory. The RHP will be based on a suite of diverse traits, which together determine the fighting ability of an individual, but identifying which of these – size, age, persistence or aggressiveness, for example – are most relevant to a given species, is not straightforward. Some traits, such as experience, may not be obvious. Differences in body size, flight-muscle ratio and age between residents and intruders were not significant in determining the outcome of territorial contests between male green hairstreak (*Chrysozephyrus smaragdinus*) butterflies, for example. Residency, i.e. whether an individual considers itself to be the owner of a territory, as we have seen, is an important consideration, but is not the only factor in determining the outcome of contests between butterflies. Age and fat content may also be important: lipid reserves of resident male green hairstreaks, *C. smaragdinus*, were lower than those of intruders, suggesting that the residents expended more energy during the contests. Those individuals with greater fat reserves – more fuel as it were – may be able to persist for longer during competitions.[57]

Older males – experienced and more persistent?

In one butterfly species, at least, older males are competitively superior to younger ones, irrespective of any differences between them in terms of body size or other biophysical properties. Older male, great eggflies, *Hypolimnas bolina*, are confronted with diminishing opportunities for mating, and consequently, they are prepared to fight harder and show more aggression than their younger rivals. In other words, they are prepared to take more risks in the hope of reaping reproductive rewards. It is rather surprising that, unlike most other animals, size often does not seem to matter much in terms of settling territorial contests in butterflies. This may be because of their lack of weapons – e.g. horns, hooves or teeth – with which to do battle. Not that the possibility of injury does not exist in some interactions between some very large butterflies (see above), but for the most part, butterflies are essentially weaponless. Rather, motivation – and, it is tempting to think, experience as well – appears to be the main deciding factor. Some other, intangible factors, like stamina and memory might also be important. Contests often last longer when the age differences between contestants is small. In an arena where the two contestants are risking relatively little in terms of getting injured, the outcome often comes down to persistence. In some situations, younger males give up more easily, perhaps to fight another day, whilst older males battle on, somehow knowing that this might be their last contest, or last opportunity to win over a mate (see also, section on 'older males' under 'Polyandry' in Chapter 7). Older males are more prepared to take risks, including the risk of damage or being predated, as they engage in territorial contests.[58]

The reason why residents come out on top in most contests over territories is thought to be unrelated to their physical prowess, in many species. Yet in other species – such as red admirals, *V. atalanta* – occupants of territories do tend to be larger. This might be an artefact of sorts, resulting from the fact that 'more intrinsically aggressive' males, over time, come to hold territories as a result of both their fighting prowess and experience. They are, in other words, the 'superior' competitors, more aggressive, or at least more motivated to persist in duels with intruders. Another suggestion is that residents *value* their territory more highly than intruders. In *C. smaragdinus*, resident males may guard their territories for up to

Fig. 3.22. Red admiral (*Vanessa atalanta*) on ground in territory, Galicia, Spain.

Fig. 3.23. Red admiral (*Vanessa atalanta*) perched on rock in own territory, Galicia, Spain.

4 weeks; so, they have invested a lot of time and energy and perhaps they are more motivated to hang on to them, as a result. Additionally, they might also have expended some sort of *start-up cost*, in order to secure the territory in the first place. According to Darrell J. Kemp, understanding the factors that determine residency – i.e. occupying and holding onto a territory – will be the key to fully understanding contest competition in butterflies.[59]

Case study. The speckled wood

In perching species, the males occupy prominent vantage points – typically sitting on top of a leaf high off the ground – surveying the immediate surroundings and waiting for potential mates. The speckled wood butterfly, *Pararge aegeria*, typically engages in perching (Fig. 3.24a, b), although males also patrol a wider area, carrying out what has been called a 'fly-and-search strategy'. So, the two strategies co-occur in this species. It is said that patrolling males are more efficient at locating females than perching males, but they must maintain a higher body temperature.

Speckled wood butterflies are variable in colour, and darker males – which warm up faster – are therefore, more likely to patrol than paler individuals. When a perching male spies a passing female, he takes off rapidly and attempts to intercept her. If she is interested, there is a rapid flight chase, after which she lands, and he alights close by. There then follows a sequence of courtship behaviour, followed by copulation. Where the male chases after a passing female who has already mated, she quickly ends the affair by dropping down into the vegetation.[60]

In temperate, forest-dwelling species such *P. aegeria*, males are typically seen defending sunlit patches of the forest, waiting to rendezvous with passing females. Paler males are more likely to sit in sunlit patches – making short investigatory flights – than darker males, who fly more frequently *between* sunlit patches. Resident males make occasional scouting flights, but generally spend most of their time within the same large sunspot territory. The sunlit spots are chosen by the males to facilitate their perception of passing females and, simply as places to warm up. Although females are not attracted to these sunlit territories per se, they are much more visible, or apparent, to resident males as they fly though these natural spotlights and hence more easily discovered (Fig. 3.25). Therefore,

Fig. 3.24. (a) Speckled wood (*Pararge aegeria*) perching on leaf, Yorkshire, UK. (b) Speckled wood (*Pararge aegeria*) perching, Galicia, Spain.

Fig. 3.25. Speckled wood (*Pararge aegeria*) in sunspot, Galicia, Spain.

Fig 3.26. Small heath (*Coenonympha pamphilus*) resting on hare's tail grass head, Galicia, Spain.

resident males are much more successful at mating than non-resident males; not because they are any more attractive, but rather because the territory itself gives them an advantage in terms of detecting and pursuing females. Interestingly, perching males have larger thoraxes than patrolling males, which gives them a higher acceleration capacity, and presumably they are better at intercepting itinerant females.[61]

Sunspots are valuable pieces of real estate, and hence worth fighting for, especially in shadier, closed-in forests, where sunlit patches are something of a rarity. When there are many sunspots and a large area of more open forest is bathed in sunlit patches, their value diminishes. The chances of intercepting a female in each sunspot diminishes as the total number of sunspots increases, and males often then switch tactics; i.e. from perching to patrolling. The value of owning a nice sunspot in which to warm up also decreases as ambient temperatures rise, so, not surprisingly, the intensity of aggressive male interactions generally decreases as the summer progresses and males spend more time patrolling rather than perching.[62]

Male speckled wood butterflies aggressively defend their sunlit territories, fighting with each other in, often protracted, 'spinning-wheel' flights where they 'bounce into each other' and try to prevent one another from landing. These contests are often repeated numerous times – up to 20 consecutive spinning-wheel flights were recorded between the same pair of males in one study in Sweden – before the outcome of the contest is finally decided, and one male gives up and flies off. Such persistence is impressive! Although individuals differ in terms of their capacity, or propensity for fighting.[63]

A classic study carried out in Wytham Woods, near Oxford, in England, showed that when patrolling speckled wood males came down from the canopy and challenged resident males for occupancy of their sunspot territories, the resident always won, regardless of how long the territory had been occupied. The intruding males – who are unable, from a distance, to ascertain whether a sunspot is occupied – were always driven back and forced to find their own, unoccupied sunspots, or to carry on patrolling for females. In other words, the contests were always decided based on who owned the territory – a setup that is eloquently called a '*bourgeois strategy*' as it favours the property owner. It has been suggested that prior occupancy is the cue that settles such disputes, in this and other butterflies. In which case, the arrangement that the resident always wins, could be the manifestation of an evolutionary stable strategy (ESS), based on the premise that the individual who has landed in the territory first – no matter for how long – considers himself to be the resident. The spiralling flights, therefore, might not be real fights, but could be simple displays, which allow individuals to settle the fact that the territory is already occupied.[64]

Subsequent research by other scientists suggested that the reason why resident speckled wood males have an advantage over intruders is because they are hotter! They warm up nicely in their little sunspots. This argument was turned on its head however, by another study which showed that warmer individuals can fly for longer and this was why they won

Fig. 3.27. Peacock (*Aglais io*) basking with wings open, Galicia, Spain.

Fig. 3.28. Small tortoiseshell (*Aglais urticae*) on thistle, Yorkshire, UK.

contests; but yet another study later ruled out differences in body temperatures between residents and non-residents as being of any significance in terms of settling contests. Nevertheless, it seems that males can win back a territory that they previously held – in other words usurping a resident – but only after a 'longer than average fight'. The male may have left his territory unguarded in order to go off and mate, for example.[65]

Notes

[1] Scott (1974a), Benson *et al.* (1989), and Vane-Wright and Boppré (1993).
[2] Scott, (1973a, 1974a), Shapiro (1977), Rutowski *et al.* (1997) and Wiklund (2003). Habitual patrolling is an ancestral behaviour according to Dennis and Shreeve (1988).
[3] Based on Kokko and Rankin (2006), and Wickman and Rutowski (1999).
[4] Rutowski (1991) and Rutowski (2019, pers. comm.). Scott (2006, p. 72) used the terms: 'perch to await females' and 'patrol to seek females'.
[5] Rutowski (1994), Odendaal *et al.* (1989). Variations in mate-locating behaviour are discussed by Dennis and Shreeve (1988).
[6] Scott (2006, 2010). Quotation from Scott (2006).
[7] Scott (1974a), Fischer and Fiedler (2001). Wickman and Wiklund (1983) describe three types of male–male territorial interactions: the horizontal flight-pursuit; horizontal-spiralling flight-pursuit; and the true spinning-wheel flight.
[8] Fric and Konvička (2000).
[9] Bergman *et al.* (2015) and Rutowski (2000a). The body posture of male butterflies can vary with perch location (see Rutowski (1991)).
[10] Douwes (1975), Rutowski (1994), Schurian and Fiedler (1996) and Cordero (2000). N.B. Scott (2006) criticized the use of the word territorial in studies that had not rigorously determined that individuals return to the same spot.
[11] Dennis and Williams (1987), Dennis and Shreeve (1988), Dreisig (1995), Ravenscroft (1994), Rutowski *et al.* (1994), Rutowski (2000a), and Kemp and Krockenberger (2004).
[12] Dennis (1982b, 1983), Matsumoto (1987), Baguette *et al.* (1998) and Rutowski (2000a). *Colias eurytheme* males have larger eye surface areas compared with females (Merry *et al.*, 2006).
[13] Douwes (1975), Dennis (1982a), Wickman (1988), Fischer and Fiedler (2001), Turlure and Van Dyck (2009), Ide (2010) and Takeuchi (2010).
[14] Brussard and Ehrlich (1970) and Adamski and Witkowski (2006).
[15] Dennis and Williams (1987).
[16] Based on Shields (1967), Scott (1970), Alcock (1987), Brown and Alcock (1990), Alcock and Dodson, 2008).
[17] Shields (1967) studied hill-topping on Dictionary hill (324 m high), in San Diego County, southern California. Hill-topping studies were carried on Mt Katsuragi (960 m alt.) and Mt Nijo (517 m), between Osaka and Nara Prefectures, in central Japan (Navez and Ishii, 2007a, b). Sáfián (2013) studied hill-topping behaviour by the giant African Swallowtail, *Papilio antimachus*.
[18] Mueller *et al.* (2010) and Corbera *et al.* (2011). Stronger-flying species are better able to establish territories on summits (Dennis and Dennis, 2008).
[19] Prieto and Dahners (2006, 2009).
[20] Scott (1973a).
[21] The motivation to persist may be more important than physical differences (Kemp, 2000b). Martínez-Lendech *et al.* (2007) and Da *et al.* (2016).
[22] Chaves (2004) and Chaves *et al.* (2006).
[23] Chaves (2004) and Chaves *et al.* (2006).

[24] Chaves (2004) is a PhD dissertation, written in Portuguese, supervised by Prof. Woodruff Whitman Benson, of the Universidade Estadual de Campinas, Brazil. Chaves *et al.* (2006) is in English.
[25] Knapton (1985), and Clayton and Petr (1992). Landmark mating strategies may be favoured where females are scarce, and resources are widely dispersed (Daily *et al.*, 1991).
[26] Scott (1973b), Alcock (2010) and Sáfián (2014). N.B. *Apallaga safiani* Libert, 2014, was previously known as *Celaenorrhinus rutilans* (Mabille, 1877).
[27] Alcock and O'Neill (1986) and Ferreira *et al.* (2016).
[28] Scott (1974c), Alcock (1984, 1994), Baughman *et al.* (1988), Wickman (1988), and Alcock and Dodson (2008).
[29] Lawrence and Samways (2001, 2002) and Bonte and Van Dyck (2009).
[30] Hall and Harvey (2002) and Callaghan (1983).
[31] Sáfián (2014) describes hill-topping as a mate-location strategy in the giant African swallowtail *Papilio antimachus* Drury, 1782 in Liberia. (See 'Tracking African Giant Swallowtail – *Papilio antimachus*', https://www.youtube.com/watch?v=_CI4Fs3CJB4). *Papilio antimachus* has recently been found up to 1650 m in the Guinean part of the Nimba Mountains, where males ascend far above the treeline and perch on grasses whilst displaying (Szabolcs Sáfián, pers. comm.).
[32] Rutowski *et al.* (1989) and Rutowski (1997a).
[33] Braby and Nishida (2011).
[34] Dennis and Shreeve (1988), Bennett *et al.* (2012), and Cordero and Soberón (1990).
[35] Scott (1974a, 2006) argued that territoriality is rare or absent in butterflies because they do not meet the main criteria for territoriality, i.e. site tenacity, the ability to visually recognize conspecific males and intentionally aggressive interactions. Other researchers disagree and consider that territoriality does exist: 'this behaviour is now confirmed' (Dennis and Shreeve, 1988). Territorial males of a few butterfly species engage in intense physical fighting (Carvalho *et al.*, 2016).
[36] Rosenberg (1989), Rosenberg and Enquist (1991), Lederhouse (1993) and Anon. (2017).
[37] Lederhouse *et al.* (1982), Codella and Lederhouse, 1989), Lederhouse (1982), and Lederhouse and Scriber (1996).
[38] Peixoto and Benson (2009a, b). Quotation from Peixoto and Benson (2009b).
[39] Territorial activity for red admirals, *V. atalanta*, in Iowa, USA, lasted for c. 1.5–3.2 h, from mid- to late afternoon until near sunset (Bitzer, 1995). Butterflies chase each other in ascending helical spirals above the territories (Bitzer and Shaw, 1979).
[40] Baker (1972), Dennis (1982b, 2004), and Dennis and Sparks (2005). Male *V. kershawi* defended concrete slabs as territories functioning as mate encounter sites in Perth, Australia (Alcock and Gwynne, 1987).
[41] Wickman (1985a, b, 1986, 1992).
[42] Based on Peixoto and Mendoza-Cuenca (2013).
[43] Males of the white peacock butterfly, *A. jatrophae*, defended sites containing their larval host plant (Lederhouse *et al.*, 1992).
[44] Cordero (1997), Cordero and Soberón (1990), Cordero (1997) and Cordero *et al.* (2000).
[45] Kemp and Wiklund (2001).
[46] Hernandez and Benson (1998).
[47] Alcock (1988), Takeuchi and Imafuku (2005a, b). Spiralling flights in the chryxus Arctic, *Oeneis chryxus*, lasted more than 45 s and only ended when the butterflies were well above the treetops (Clayton and Petr, 1992).
[48] Carvalho *et al.* (2016).
[49] Emmel and Garroway (1990), Lehnert *et al.* (2013) and Eff (1962).
[50] Kemp (2002c, 2003).
[51] Kemp (2003).
[52] Srygley and Penz (1999).

[53] Srygley (1994), Freitas *et al.* (1997), Srygley and Penz (1999), and Frederiksen and Warrant (2008).
[54] Rutowski *et al.* (2001), Takeuchi and Imafuku (2005a). Male *Hipparchia* (grayling) species are relatively poor at identifying conspecific partners from a distance (Pinzari *et al.*, 2018). *Euphydryas* males spend much of their time chasing each other (Odendaal *et al.*, 1988). Males of the white peacock, *Anartia jatrophae*, chased both other males as well as a variety of other insects, including dragonflies, that flew near them (Lederhouse *et al.*, 1992).
[55] Benson *et al.* (1989), Kemp and Wiklund (2001), Kemp (2006b), Takeuchi *et al.* (2016) and Takeuchi (2017). For a rebuttal of the issue, of whether territoriality in butterflies is no more than the investigation of the sex of the intruder by the incumbent, see Dennis and Williams (1987).
[56] Peixoto and Benson (2011) and Peixoto *et al.* (2014).
[57] Based on Parker (1974), Takeuchi (2006), Martínez-Lendech *et al.* (2007), Peixoto and Benson (2011, 2012), and Vieira and Peixoto (2013).
[58] Kemp (2000a, b), Kemp and Wiklund (2001) and Kemp (2002a, b). Contests lasted longer when age differences were small (Kemp, 2002b).
[59] Bitzer (1995), Kemp and Wiklund (2004), Takeuchi (2006), Kemp (2013, 2018) .
[60] Perching and patrolling strategies co-occur in the speckled wood butterfly (Van Dyck, 2003). Speckled wood males adopt *either* a territorial, wait-and-fight strategy *or* a fly-and-search strategy (Velde *et al.*, 2012). Darker males warm up faster than pale males (Van Dyck *et al.*, 1997a, b; Van Dyck and Matthysen, 1998) so they can spend more time flying from one sunlit patch to another. Patrolling males locate females at a faster rate than perched males (Shreeve, 1984), but Van Dyck and Matthysen (1998) considered that there is no evidence that one mate-location strategy is *always* superior to the other. Mated female speckled woods often drop down into the vegetation when pursued by males (Bergmann, 2011).
[61] Bergman *et al.* (2007). Van Dyck *et al.* (1997a, b), Berwaerts *et al.* (2002), and Bergman and Wiklund (2009a).
[62] Based on Wickman and Wiklund (1983), Shreeve (1984) and Van Dyck (2003).
[63] Quotations from Wickman and Wiklund (1983).
[64] The rule for settling contests being: 'resident wins, intruder retreats' (Davies, 1978). Morphological and 'biophysical' factors have little relevance in settling contests, but age, and how long an individual has left in which to reproduce, may influence their motivation and aggression (Kemp, 2013). Prior occupancy is a cue to settling disputes (Smith and Parker, 1976). What appears to be a 'territorial defence' may simply be an evolutionarily stable strategy due to the asymmetry between the contestants in time of arrival in the sunspot (Davies, 1978). Patrolling males are unable, from a distance, to ascertain whether a sunspot is occupied (Hardy, 1998).
[65] Wickman and Wiklund (1983), Rutowski (1991) and Bergman (2011). Austad *et al.* (1979) thought that resident males persisted for longer in contests because they maintained higher body temperatures in sunspots. Similarly, Stutt and Willmer (1998) concluded that warmer individuals could fly for longer and were therefore more likely to win a contest. However, Kemp and Wiklund (2004) found no differences in body temperature between residents and non-residents, which appears to rule out this idea. After mating, females spend their time selecting oviposition sites and ignore other males (Velde, 2011). Territory selection by males is based on properties of the sunspot that facilitate *both* 'visual perception and thermoregulation' (Velde *et al.*, 2011).

4 Seeing and Being Seen

Butterfly Colour Vision

Butterflies are such colourful, sun-loving creatures, it would be surprising if they did not see their world in full colour. Yet, although it was long assumed that they did so, it is only relatively recently that it has been unequivocally proven that they have colour vision. Butterflies have sophisticated colour vision, with a very broad range of light perception – in some species ranging from ultraviolet through to red (300–700 nm) – amongst the broadest of all animals. The ability of honeybees to differentiate between colours was demonstrated by the Nobel Prize-winning Austrian ethologist Karl von Frisch over 100 years ago. The eyes of butterflies are however, much more diverse than those of bees – in terms of both their anatomy and physiology – due to the number and diversity of photoreceptors as well as the distribution of coloured accessory pigments lining the rhabdom (see below), which act as filters, shielding the photopigments from stray light.[1]

Colour vision is the ability to perceive differences between light composed of different wavelengths. The brains of animals with colour vision can compare the information received by special photoreceptors (see below) sensitive to different wavelengths of light, and thereby construct colours between these points on the spectrum. One advantage of colour vision is that it increases the contrast between objects – like a bright red apple in a green tree, for example – which gives the viewer better object recognition. The ability to see the world in colour – to detect naturally occurring visual signals of different wavelengths – gives an animal a way of detecting and recognizing a wide range of objects, including food and host plants, nectar sources (flowers!), potential mates, rivals and would-be predators. A set of photoreceptors attuned to different parts of the colour spectrum – i.e. with different *spectral sensitivities* – is what is required to enable colour vision.

© Raymond J.C. Cannon 2020. *Courtship and Mating in Butterflies: Reproduction, Mating Behaviour and Sexual Conflicts* (R.J.C. Cannon)

Such spectral sensitivity also opens possibilities for subtle co-evolutionary struggles and alliances, e.g. between butterflies and plants, mimics and models and between the sexes (e.g. based on wing colours).[2]

To demonstrate that butterflies have true colour vision and are not using some other method – like just going by the brightness of an object – scientists needed to carry out some ingenious experiments. A group working in Japan, used an experimental set-up to show that Japanese yellow swallowtail butterflies, *Papilio xuthus* (Fig. 2.2a, b), can tell the difference between different coloured discs – i.e. on the basis of their chromatic content – regardless of how bright they are, even though they do have an innate preference for bright colours. The butterflies were trained to associate a certain colour with the presence of food – sugar solution – but they took longer to learn some colours than others. Once they knew how to distinguish the colour associated with a reward, they were tested to see if they could still find the right colour when the light intensity was varied. Red and yellow were the colours most easily learnt by the Japanese yellow swallowtails, perhaps because they had an innate preference for flowers of these colours.[3]

Like all sighted insects, butterflies have compound eyes (Fig. 4.1). Eyes probably first evolved in the mists of the pre-Cambrian period, but the first well-preserved, definitive eyes are to be found on trilobite fossils in the early Cambrian epoch, some 540 million years ago (mya). This unprecedented innovation – sight – could have been the stimulus for the remarkable explosion of life forms that evolved at that time. Vision would have given animals a massive evolutionary advantage in terms of finding food and avoiding predation. Trilobites and other arthropods from this period already possessed compound eyes of a kind, at least superficially like those of species alive today. The first insect fossils with well-developed compound eyes have been found in rocks dating back to the Devonian period and the basic form of the insect compound eye has in effect been conserved and utilized by insects for at least 360 million years. Compound eyes have been said to represent a paradigm of evolutionary conservation.[4]

Fig. 4.1. Small tortoiseshell (*Aglais urticae*) eye close-up, UK.

The Structure of the Compound Eye

Insect compound eyes are composed of numerous individual light-gathering elements called *ommatidia*. The eye of a *Papilio* butterfly (Fig. 4.2) typically contains about 12,000 – superficially identical – such ommatidia; the compound eyes of the more primitive glacial apollo butterfly – which has been called a living fossil – consist of about 5,000 ommatidia, but the number of ommatidia is principally determined by the size of the compound eye. At the top of each ommatidium sits a facet lens and immediately below the lens is a crystalline cone (Fig. 4.3). Together they function as the optical apparatus, which focuses incoming light onto the photoreceptors below. The facets are tiny – approximately 15–30 µm in diameter – yet clearly visible in a close-up photograph of a butterfly eye (Fig. 4.4). On the surface of the facet lens, there is in some butterflies an array of tiny protuberances – about 250 nm high – which reduce the reflectance of the compound eye. This so-called corneal nipple array is probably a throw-back to ancestral moths and has been lost from some butterfly groups, such as the papilionids, during their evolution. The height and breadth of the tiny protuberances varies between different groups and species of butterflies, with the nipples being most pronounced in nymphalid butterflies, but completely lost in papilionids. The function of the nipple array is probably glare reduction, which is less important in day-flying as opposed to night-flying Lepidoptera, and has therefore been lost in the more highly evolved butterflies.[5]

Below the lens, the ommatidia each contain nine photoreceptor cells, which together form a long, cylindrical, optical-sensing structure called a *rhabdom*. The rhabdom is formed by the fusion of nine light-sensitive organelles called rhabdomeres and is essentially a stack of microvilli – tiny little invaginations of the photoreceptor cell membranes – that contain the visual pigments. The technical term for the fused rhabdom is an *optical waveguide* because it channels the light downwards through the embedded visual pigments towards a reflecting surface – called a *tapetum* – at the base of the ommatidium. The tapetum is composed of air-filled cells that act as mirrors and bounce any unabsorbed light back up through the

Fig. 4.2. Red Helen (*Papilio helenus*), Doi Inthanon, Thailand.

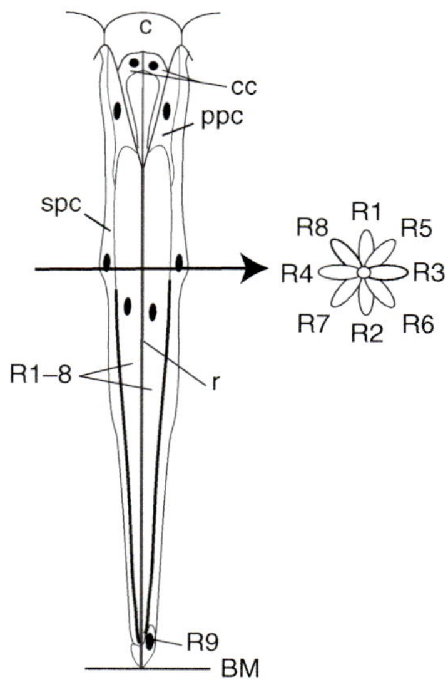

Fig. 4.3. Drawing of a butterfly ommatidium showing the cornea (c), crystalline cone (cc), rhabdom (r), photoreceptor cells (R1-9), primary pigment cells (ppc), secondary pigment cells (spc), and basement membrane (BM) modified from Macias-Muñoz *et al.* (2019). Cross-section shows photoreceptor R1-8 cell bodies. (Reproduced with the kind permission of Professor Adriana Briscoe, Professor of Ecology and Evolutionary Biology at the School of Biological Sciences at the University of California, Irvine, California.)

Fig. 4.4. Scanning electron microscope image of a pyralid moth showing typical lepidopteran eye facets. Side view of head. (http://remf.dartmouth.edu/images/insectPart3SEM/source/3.html).

rhabdom. Like the corneal nipple array, the tapetum is an archaic structure that evolved in nocturnal moths and is probably of little adaptational significance for day-flying butterflies. It has been lost completely in papilionids and in the pierid genus, *Anthocharis*, which includes the common orange tip butterflies (Fig. 3.5).[6]

Although the ommatidia in a butterfly eye look very similar on the surface, there are usually three different types (I, II and III) in terms of both the photoreceptors – spectral photoreceptor classes – contained within their cells, and the arrangements of pigments surrounding the rhabdom. Recent research has revealed that there are at least six ommatidial types in *Heliconius* butterflies – each containing different combinations of photoreceptor cells expressing blue, green, magenta and long-wavelength opsins. The lattice-like structure of the butterfly compound eye is in effect a patchwork – or retinal mosaic – of these different ommatidia, with different photoreceptor sensitivities. Clusters of pigments (yellow, red and orange) line the rhabdom of most, but not all, butterflies and function as filters of light propagating along the rhabdom; the different pigments are distributed and combined with a fluorescing material in different ways in the three types of ommatidia in *Papilio* swallowtail butterflies. These three different types of ommatidia have been identified in many different species and are randomly arranged within the compound eye.[7]

Eye Shine and Differences Between the Sexes

Any light that is not absorbed by the visual pigments in the rhabdomeres on the way down is reflected upwards, towards the lens. Light that emerges from the eye again is called *eye shine*, or eye glow. Studies have shown that in some species – such as the North American ruddy copper butterfly, *Lycaena rubidus* (Lycaenidae) (Fig. 4.5) – the distribution of visual pigments differs between the two sexes, such that the eye shine of

Fig. 4.5. Ruddy copper butterfly (*Lycaena rubidus*), male. (Photo by Alan Schmierer [CC0].)

the males is mainly green-orange, whilst that of the female is blue-green. There are also differences between the wing colours of the two sexes in this species: the males have bright iridescent colours whilst the females are more muted. The dorsal, or upward-facing, parts of the eyes in the ruddy copper also differ between the sexes, with the males having a sort of two-colour vision, based on UV- and blue-detecting visual pigments, which it is thought help it to detect flickering moving objects, such as airborne males of the same species. Butterflies are very good at resolving fast-moving objects in their visual field, probably better than us humans. Butterflies often have subdivisions within their compound eyes, with specialized zones for different types of activity. For example, the downward looking areas of the eye may be specialized for foraging, whilst the forward-looking zones may be specifically tuned for detecting mates, silhouetted against the bright sky.[8]

The Light-Gathering Molecules

The process of vision in butterflies begins with the absorption of light by visual pigments called *rhodopsin*. Rhodopsins are composed of proteins called *opsins*, combined with a light-sensitive molecule (11-cis-3-hydroxyretinal) – derived from vitamin A3 – called a *chromophore*, which sits in a special binding pocket in the protein. The interaction of the chromophore with critical amino acid residues within the opsin molecule determines the absorbance spectrum of the combined visual pigment, a phenomenon known as *spectral tuning*. Opsins are essentially molecules that facilitate the conversion of energy from photons into electrical signals that are sent to the brain. In simple terms, they do this via a change in the shape of the protein: *photoisomerization*. In other words, light results in a change from one structural form of the protein, or isomer, to another. A photoreceptor usually only contains one type of opsin molecule, for example absorbing UV, violet, blue, green or red light. A novel type of photoreceptor has, however, been found in the papilionid butterfly, *P. xuthus*, which contains two opsin molecules within a single ommatidia, which absorbs both green and orange-red light.[9]

The spectral sensitivity, meaning the wavelengths of light that are detected by rhodopsin pigments, is determined by a small number of amino acids in the opsin molecule. In other words, small changes in the building blocks of the optically sensitive proteins can change the wavelength at which peak absorbance takes place, thereby changing the way an animal sees colour. This shows how evolution might work to produce adaptations in colour vision, via small changes in the DNA that codes for the proteins which enable vision to occur. Duplications of the gene that codes for the opsin molecule are thought to have produced some of the different classes of photoreceptors found in many butterflies, such as the blue-sensitive pigments of lycaenid eyes. In pierids for example, blue opsins have diversified into violet and blue; in lycaenids, blue opsins have diversified into blue and blue-green pigments, and so on.[10]

All butterfly eyes contain UV-, blue-, and green-sensitive rhodopsins, with peak sensitivities in the following regions of the spectrum: 300–400 nm (UV light), 400–500 nm (blue light) and 500–600 nm (long-wavelength light). We humans use a red–green–blue trichromatic system and can discriminate colours across a range of the electromagnetic spectrum from 400 to 700 nm. Butterflies use their longer wavelength red receptors for recognizing host plants and flower colours. As an example, the small white butterfly, *Pieris rapae crucivora* (Fig. 7.11), has three types of long-wavelength photoreceptors, two of which have spectral sensitivities peaking in the red (620 nm) and deep red (640 nm) part of the spectrum. The evolution of flower colours and butterfly vision probably went hand in hand, a phenomenon called coevolution. In species that are part of a mimicry group, such as the *Heliconius* butterflies (Figs 2.6 and 2.8), distinguishing subtle differences in red colour patterns on the wings of individuals of the same, and mimetic species, is critical. In species like this, which have evolved to resemble one another in terms of wing colour patterns, there may have been additional selection pressures to evolve more specialized visual pigments to better detect these subtle differences in colour. The compound eyes of *Heliconius* species are unique in that they contain one more UV opsin than in other butterflies; the duplicate UV opsin genes may have evolved to facilitate species recognition in *Heliconius*. Additionally, *Heliconius* butterflies have also evolved unique – 3-hydroxy-DL-kynurenine (3-OHK)-based – yellow wing pigments (see Chapter 2, this volume), which provide them with an expanded 'palette of discriminable colours' on their wings compared to other non-*Heliconius* butterflies.[11]

Recent research has, however, revealed that butterfly eyes are extremely diverse – every family is different in terms of their visual pigments. This diversity encompasses the number of opsin genes, the eye pigmentation and the spectral sensitivities of the photoreceptors. These differences in spectral organization are thought to reflect both the evolutionary history of the different species and their contrasting lifestyles. The spectral tuning of the visual pigments may also have been driven by evolutionary pressures to do with effective communication between members of the same species, as well as developing adaptations to do with finding flowers and the best sites for laying eggs. In other words, they are not just adaptations to lighting conditions, but are more to do with the social or biological needs of the butterfly.[12]

Examples of the diversity of butterfly eyes include: the monarch, *Danaus plexippus* (Fig. 5.19), with three photopigments (UV, violet and green); the ruddy copper, *Lycaena rubidus* (Fig. 4.5), with four photopigments (UV, violet, blue and yellow); and the small white, *Pieris rapae*, with three photopigments (UV, violet and blue); but the male *P. rapae* has a special double-peaked blue photoreceptor, which helps it to discriminate the different wing colours of the male and female. As research into butterfly vision has progressed over recent years, examples of species with ever greater numbers of photoreceptors have been discovered. The Japanese yellow swallowtail, *Papilio xuthus* (Fig. 2.2a, b) was found to contain at least six types of photoreceptors: peaking in the UV (at 360 nm), violet (400 nm), blue (460 nm), green (520 nm) and red (600 nm) zones of the colour spectrum, as well as having a sixth type with a very large bandwidth at about 270 nm. The golden birdwing, *Troides aeacus formosanus* (very similar to *Troides aeacus thomsoni* shown in Fig. 10.5), from Taiwan, was found to have nine spectral types of photoreceptor, with sensitivity peaks in the UV (360 nm), violet (390 nm), blue (440 nm), blue-green (510 nm), single-peaked green (540 nm), dual-peaked green (550 nm), orange (580 nm), red (610 nm) and deep red (630 nm) parts of the spectrum.[13]

More recently, the eyes of the common bluebottle butterfly, *Graphium sarpedon* (Fig. 10.35) – an Asian swallowtail with blue-green iridescent wings – were found to contain 15 distinct photoreceptors with spectral sensitivities divided into six classes: UV (one), violet (one), blue (three), blue-green (one), green (four) and red (five). The researchers were surprised to find so many: 'We have studied colour vision in many insects for many years, and we knew that the number of photoreceptors varies greatly from species to species. But this discovery of 15 classes in one eye was really stunning' said Professor Kentaro Arikawa, Professor of Biology, at Sokendai-Hayama, Graduate School of Integrated Science, Yokohama City University, Japan. The 15 photoreceptors were not, however, distributed evenly across the eye: four were confined to the upper part of the eye, whilst eight were in the underside of the eye, suggesting specialized functions other than colour vision.[14]

These butterflies with such a large array of photoreceptors must surely have amongst the richest colour vision in the butterfly world, if not the animal kingdom. Only the mantis shrimp has been found to have more receptors (16), but it could not differentiate closely similar colours anything like as well as a butterfly. Researchers have asked what are butterflies doing with all these photoreceptors? Four is probably more than enough to achieve good colour vision over a wide range of the spectrum from 300–700 nm. The reason, probably, is that not all these photoreceptors are utilized for colour vision. For example, only four of the six photoreceptors – UV, blue, green and red receptors – in the swallowtail, *P. xuthus*, were used for colour vision, making it tetrachromatic. The other, *achromatic* photoreceptors not used for colour vision – all found in type II ommatidia – are thought to be used for specialized activities such as for detecting signals – e.g. high-frequency signals (motion), UV wing colours and food plants – or light of a low intensity. There are also slight differences in the wavelengths of peak absorbance by individual photoreceptors of the same type, between different butterflies. We can only imagine how this might affect their perception of colours, but it is fascinating to think that two species of butterflies, perhaps occupying different niches in the same habitat, might see their surroundings in slightly different ways.[15]

Gathering the Light in Different Ways

In butterflies, each individual ommatidium effectively samples a small portion – between 1 and 3° – of the overall visual field and corresponds to a single pixel of visual information. The light sampled from these tiny adjacent areas of the visual field passes through the facet lens and crystalline cone and is channelled down – protected by a surrounding screen of pigment cells – onto the rhabdom below. These are called *apposition* eyes and are different from the eyes of moths – which have *superposition* eyes – where individual rhabdoms receive light from adjacent facet lenses and crystalline cones, not just the one in its own ommatidium. It is because rhabdoms in night-flying moths gather light from more than one lens and cone that they are so sensitive. The individual rhabdoms in moth eyes are also fatter than those in butterflies, giving them a better chance of absorbing any light rays. Skipper butterflies, which are the most primitive type of butterfly, also have superposition eyes, like moths. Butterflies are unique in having a sort of apposition eye called an *afocal apposition eye*, because incident light is focused to an image at a point inside the crystalline cone and not on the rhabdom. The light is then refocused into a 'parallel pencil of rays' that enters the rhabdom. The set-up is the same as an astronomical telescope and, apparently, it is a system that is well adapted to daylight-flying insects. So perhaps butterflies can be thought of as having an array of thousands of tiny little telescopes in each of their compound eyes![16]

Within a given type or group of butterflies, smaller individuals will have proportionally larger eye surface areas and ommatidial facet diameters than larger individuals. This is because these dimensions – eye surface area and facet diameter – are generally inversely proportional to body size in most insects. In other words, smaller insects have proportionally better vision for their size. Males also have larger eyes than females in many species – up to 30% bigger – presumably because they take a more active role in searching for potential mates (see Chapter 3, this volume). Visual acuity, or sharpness, also varies across different parts of the eye surface. For example, visual acuity is greatest in the anterior (front) and ventral (underside) regions of the eyes of orange sulphur butterflies, *Colias eurytheme* (Pieridae) (Fig. 2.15), where the facets are wider and with a narrow interommatidial angle. These are the regions of the eye that the butterfly probably uses most when searching for mates, flowers and – in the case of females – places to lay eggs; see the large green eye of the congeneric, clouded yellow, *Colias croceus* (Fig. 4.6). The eye facets are also larger in the lateral areas – i.e. the area facing forwards and out to the sides – of the eyes of males of two fritillary butterflies – the bog fritillary, *Boloria eunomia* (Fig. 7.1a, b) and the cranberry fritillary, *Boloria aquilonaris* – an adaptation for spotting and flying after females. The ventral two-thirds of the eyes of male northeast-Asian wood whites (Pieridae) have a roughness – compared with the smooth dorsal one-third of the

Fig. 4.6. Clouded yellow (*Colias croceus*) with green eye, Asturias, Spain.

compound eye – due to an irregular distribution of facets, of two distinct sizes. It is hypothesized that this sexually dimorphic feature may have evolved in males of *L. amurensis* to facilitate mate finding.[17]

The visual field of butterflies is remarkably large (Fig. 4.7). For example, with one eye or the other, orange sulphur butterflies, *C. eurytheme*, can see a remarkable 93% of the spherical space around them; they only have one small blind spot: behind their heads. The Empress Leilia butterfly, *Asterocampa leilia* (Fig. 4.8) – a nymphalid from south-western USA and Mexico – has a huge visual field, with each eye covering separate hemispheres with little more than a 10° overlap; in other words, their vision is mostly monocular. They can, however, use their small area of binocular vision – which is along the frontal midline of the body – to help them when needed, e.g. when landing on a leaf. This butterfly also has zones at the front, and around the equator of the eye, with enlarged facets, especially in males. These so-called *equatorial acute zones* are the parts of the eyes with the greatest sharpness and are probably used by the males when perched on a prominent spot looking out for females. The acute zones may also be used to monitor the flow of objects through the optical field as the butterfly flies along.[18]

Three-Colour and Four-Colour Vision

Butterflies evolved from moths, or moth-like ancestors, which have a trichromatic (three-colour) visual system based on the ancestral set of UV-, blue- and green-sensitive photoreceptors. The same system is found in bees and wasps. We humans also have a trichromatic system, but based on blue-, green- and red-sensitive photoreceptors. The three basic dimensions of colour that we trichromatic humans can recognize are: hue, saturation and brightness. The basic insect trichromatic system (UV, B and G) diversified over time with the evolution of red receptors, which are much more common in lepidopterans than, say, bees and wasps. Red receptors

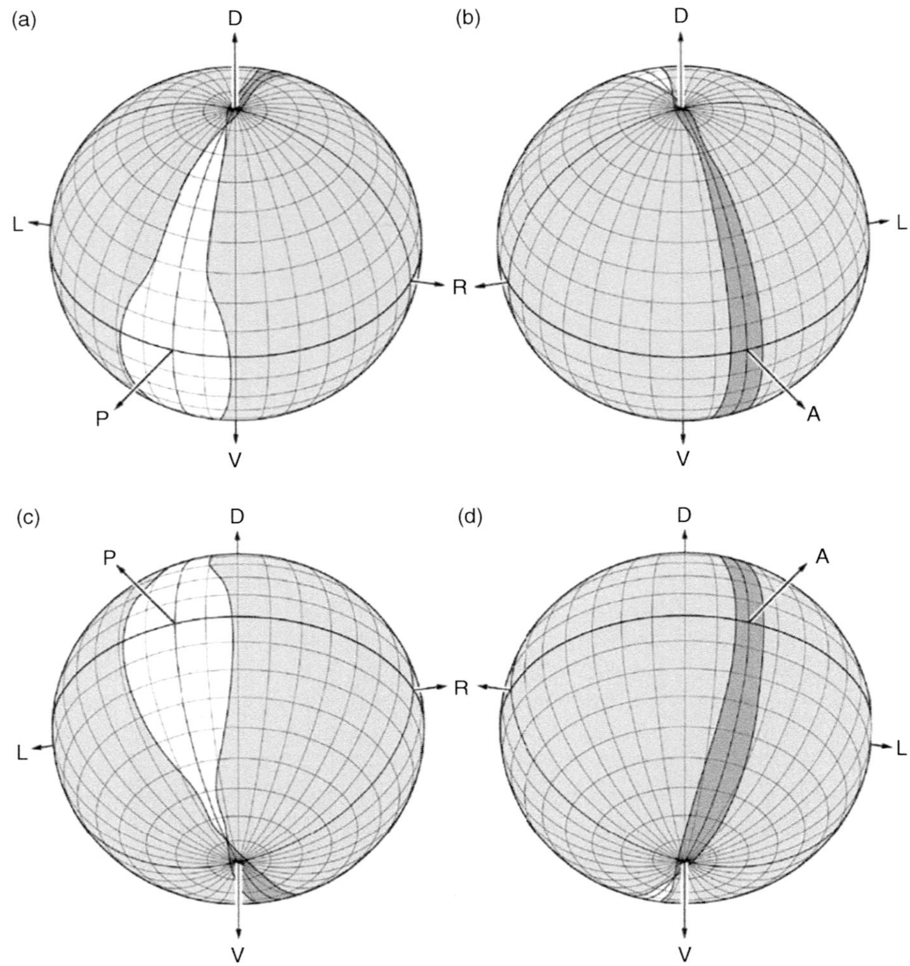

Fig. 4.7. Visual field of *Colias eurytheme* (Lepidoptera, Pieridae) males shown from four perspectives: (a) dorso-posterior view, (b) dorso-anterior view, (c) ventro-posterior view and (d) ventro-anterior view. The white area is that part of the space around the butterfly that is not detected by either eye. The light shading is the region of monocular vision, while the darker shading is where vision is binocular. Abbreviations: D, dorsal; V, ventral; A, anterior; P, posterior; R, right; L, left.. (From Merry *et al.*, 2006.)

are said to have evolved at least four times during the evolutionary development of the Lepidoptera (i.e. moths and butterflies) and then lost again in some groups, such as the Nymphalidae. It could be argued that trichromacy is ideal for colour vision – allowing separate dimensions of hue, saturation and brightness – and there is no obvious explanation for why some animals have evolved more than three types of spectral receptors.

Fig. 4.8. Empress Leilia (*Asterocampa leilia*). (Photo by Alan Schmierer [CC0].)

One explanation is that different sets of receptors are used for different behaviours, and another evolutionary force or driver – namely sexual selection – has driven the evolution of additional photoreceptors. For example, there may have been a coevolution of photoreceptors and adaptations in wing coloration associated with sexual signalling; new receptors were needed to see the new colours![19]

Seeing Ultraviolet and Polarized Light

Butterflies also have an area of the eye specialized for detecting polarized light. The ommatidia in a region on the upper edge of the eye – called the dorsal rim area – contain rhabdoms with UV opsins tuned to detect polarized skylight. Monarch butterflies, *D. plexippus*, use this narrow strip of about 100 upward-facing ommatidia on the dorsal or hind margin of the eye for orientation during their long migrations. This narrow ribbon at the back of the eye of the monarch butterfly is said to be monochromatic, with highly polarization-sensitive photoreceptors. Butterflies of the genus *Papilio* – such as the orchard swallowtail, *Papilio aegeus*, and the Asian swallowtail, *Papilio xuthus* – have polarization-sensitive photoreceptors in all regions of the eye, not just the dorsal rim. They have *false colour* vision based on their polarization-sensitive photoreceptors, which they use to help them find food and oviposition sites. The false colours perceived by butterflies as a result of their sensitivity to polarized light may amplify, or at least augment, the colours of iridescent and UV-reflecting wings on butterflies of their own, or other species. Butterflies with polarized reflectance patterns are more typical of forest habitats, such as the tropical forest understorey. Waxy or shiny materials like leaves are biological light polarizers, and female *Papilio* butterflies are thought to be able to distinguish between horizontal and vertical leaf surfaces via their angle of polarization; so, to them, horizontal leaves look green, whilst vertical leaves appear blue-green or reddish. Plants with shiny leaves probably appear more colourful, so butterflies enhance their perception of their environment in this way. Some butterflies, including *Heliconius* species, also use polarized light for mate recognition, and it has been suggested that it gives them a private communication channel, undetected by vertebrate predators.[20]

Hindsight: Genital Photoreceptors

One remarkable feature of butterflies, which was only discovered – by accident – in 1980, is that they have light-sensitive genitalia! A Japanese scientist working on the nervous system of the Asian swallowtail butterfly, *P. xuthus*, discovered that some neurons were responding directly to light. It turned out that this butterfly possessed genital photoreceptors – known as *extraocular photoreceptors* – which occurred in both the male and female of the species. A great deal of work was carried out over the following decades by Professor Kentaro Arikawa and co-workers to more fully understand this system. They demonstrated that males use their genital photoreceptors during mating to ensure that their coupling with the female is correct, i.e. is a perfect fit. In other words, if the complex reproductive parts are not properly in place during the act of mating (or coupling) – for example, if there is a slight gap between the two pairs of genitalia – this is registered by the photoreceptors.

Each sex has two pairs of photoreceptors on their genitalia. The precise position of these photoreceptors has been mapped on the hind ends of both males and females. The females use their photoreceptors in a different way: they help her determine whether the ovipositor is sufficiently extended, and unobstructed during egg laying. It is thought that they also allow the female to detect whether the ovipositor is clean and unobstructed, in a way that only a light-sensitive receptor could do. It turns out that genital photoreceptors may exist in all butterflies, at least based on a test sample of 16 species from seven families, including skippers (Hesperiidae), but were not found in moths. This suggests that they are an adaptation evolved by diurnal butterflies, taking advantage of the possibilities offered by mating in daylight. All of the photoreceptors are sensitive to light in the ultraviolet to blue part of the spectrum, which the researchers think allows them to detect contrast – differences in *luminance* or colour – in bright sunlight.[21]

Pseudopupils

When taking pictures of butterflies, especially when using a flashgun or pop-up flash, a regular pattern of dark spots often appears in the compound eye (Figs 4.9 and 4.10). The dark spot near the centre of the eye, as

Fig. 4.9. Common sailor (*Neptis hylas*), Thailand. Pseudopupils in compound eye visible using flash.

Fig. 4.10. Yerbury's sailor (*Neptis yerburii pandoces*), Thailand. Pseudopupils in compound eye visible.

well as the less well defined *accessory* spots scattered around about it, are called pseudopupils. The largest and darkest spot – the *principal* pseudopupil – is that part of the compound eye which is facing directly towards the camera (see Fig. 4.1). These pseudopupils are not fixed but move with the angle of view of the observer, or lens. The principal pseudopupil appears as a dark spot because of the black – usually jet-black – pigments clustered closely around the cones. So, in the principal pseudopupil, one is in effect looking right down a small group of ommatidia – down their vertical axes – and there is little, or no light scattered back. We are looking right down into the photoreceptors and the light is completely absorbed, so this region looks black. In the principal pseudopupil, the black pigment is seen directly through the facet lens of the ommatidium, whereas in the accessory pseudopupils, the pigment is observed through the facet lens of the neighbouring ommatidia. Butterflies have the richest patterns of pseudopupils in the invertebrate world, but their shapes and patterns vary to some extent in different groups. Pseudopupils are said by Professor D. Stavenga to be 'optical revelations of the interior of compound eyes', providing deep insights into its structure and function.[22]

Butterflies Using Their Eyes

Butterflies have clearly evolved their unique eyesight for a reason. You only need spend a short while observing butterflies in the field to realize what superbly visual animals they are: often flying surprisingly fast through their three-dimensional habitat, interacting spontaneously with other butterflies and other insects, skipping rapidly from flower to flower. It comes as some surprise then, to learn that the resolution of the insect compound eye is quite poor, and they can only see the details of the patterns on flowers for example, from relatively short distances. Nevertheless, the speed with which butterflies and other pollinators, such as bees for example, can move from flower to flower, does not suggest that they are in any way constrained by the limitations of their compound eyes.[23]

As well as having superb colour vision, which we can only suppose helps them to find the right kind of flower for their needs, butterflies are also able to distinguish colours under different types of light, technically speaking under different spectral illuminations. This ability is called *colour constancy* and it is something they share with us humans. In laboratory experiments, Japanese yellow swallowtail butterflies, *P. xuthus*, were able to pick out the correct coloured patch they had been trained to associate with food, even when it was illuminated with different coloured lights (green, yellow, red and blue). Somehow, they were able to convert the colours seen under coloured light to that seen under white light. Clearly, this ability to recognize the correct colour, regardless of the light conditions – which vary greatly between sunny and overcast days – and the location of the flower, is a huge advantage to a day-flying and feeding insect.[24]

Foraging butterflies may, however, use a variety of different signals – such as size, shape and odour – as well as colour, to select an appropriate flower, but colour remains the most important cue for most nectar-foraging butterflies, as studies on the monarch, *D. plexippus* have shown. Butterflies have varying degrees of preference for certain types of flower, but apart from some species that specialize on one species of flower, they are not usually restricted to visiting one sort of flower when others are available. In other words, they exhibit preferences for certain types of flower – partly based on innate colour preferences – but do not exhibit 'flower constancy' when foraging. For example, adult female cabbage white butterflies, *P. rapae* (Fig. 4.11), have an innate bias towards the colour green, but readily alter this bias, via learning, towards another colour, e.g. red, after several hours of experience. In temperate regions, butterflies are usually generalists in terms of their choice of flowers whilst nectaring, but in tropical regions, some species specialize on one or a few plant species. This phenomenon of flower specialization can, however, be seasonal and limited to times of abundancy when it is easy to be picky![25]

In conclusion, butterflies have a remarkable array of photoreceptors, which give them a spectral sensitivity that is almost unrivalled in the

Fig. 4.11. Small white (*Pieris rapae*), close-up of eye. (Photo by Richard Bartz a.k.a. Makro Freak [CC BY-SA 2.5 (https://creativecommons.org/licenses/by-sa/2.5)].)

animal kingdom. Unlike birds, bees and us primates – where there is little variation in the visual photoreceptors within each group – butterflies have evolved a diversity of photoreceptor sensitivities, with between three and 15 different types of spectral receptors, and tetrachromatic colour vision. Why butterflies have evolved such a diversity of colour vision is not altogether clear but it is almost certainly to help them find food sources, mates and places to lay their eggs. Eyes are metabolically costly organs to maintain, so we can infer that the visual adaptations that butterflies have evolved are useful, if not essential to them in terms of their survival and reproduction. In the words of one leading researcher, Professor Adriana Briscoe (2008, p. 1812) of the University of California, Irvine, USA: 'the butterfly eye is one of the most beautifully honed instruments of evolution'. There is, however, still a great deal more to learn about butterfly eyes, especially concerning the differences between species in terms of the number of photoreceptors, their spectral sensitivities and how they are utilized by the butterflies in their unique life histories. Additionally, very little is known about how all the visual information received by photoreceptors is processed in the brain of a butterfly.[26]

Notes

[1] von Frisch (1914), Blackiston *et al.* (2011), Frentiu *et al.* (2007) and Briscoe (2008). Pigments in the primary and secondary pigment cells are thought to shield the photopigments from stray light; the filtering pigments that coat the rhabdom selectively absorb light and tune the photoreceptor cell, there is also a pupillary pigment near the distal end of the retina within the photoreceptor cells that reacts to light like the human pupil, migrating towards the rhabdomeres in bright light to shut down the light into the ommatidium (Professor Adriana Briscoe, pers. comm, 13 May 2018). The pigments in the pigment cells surrounding the photoreceptor set of each ommatidium function as shields for stray light and therefore are often called screening pigments. They are usually blackish, absorbing throughout the whole wavelength range. The pigments lining the rhabdom are located within the photoreceptor cells and are usually distinctly coloured, yellow, red, purplish, depending on species and cell type (Doekele Stavenga, The Netherlands, 2019, pers. comm.).

[2] Pichaud *et al.* (1999) and Marshall and Arikawa (2014).

[3] Kinoshita *et al.* (1999).
[4] Schwab (2012) and Land and Nilsson (2012). 'Pervasive taxonomic presence and high design similarity suggest that insect compound eyes represent a paradigm of evolutionary conservation' (Buschbeck and Friedrich, 2008, p. 448).
[5] Arikawa *et al.* (2004), Matsushita *et al.* (2012) and Stavenga *et al.* (2006a).
[6] Frentiu *et al.* (2007), Takemura *et al.* (2007), and Sison-Mangus and Briscoe (2009).
[7] Arikawa *et al.* (2003), Qiu and Arikawa (2003), Takemura *et al.* (2007), Takemura and Arikawa (2006), Uchiyama *et al.* (2013) and McCulloch *et al.* (2016). Butterfly eyes have a strikingly heterogeneous organization, with specific sets of spectral detectors combined in different types of ommatidia (Arikawa *et al.*, 2003). 'Although the overall lattice of ommatidia is of crystalline regularity, the different types are randomly arranged within that regular lattice' (Arikawa *et al.*, 2003, p. 4532). 'The ommatidia of types I, II and III bear four (UV, narrow-band blue, double-peaked green, red), three (violet, single-peaked green, broad-band) and two (wide-band blue, double-peaked green,) receptor classes, respectively' (Koshitaka *et al.*, 2008).
[8] Sison-Mangus *et al.* (2006) and Marshall and Arikawa (2014). The lycaenid *Lycaena rubidus* eye contains four photopigments: one UV (P360), two blue (P437 and P500) and one long-wavelength (P568) (Briscoe, 2008; quoting Bernard and Remington, 1991).
[9] Briscoe (2008), Arikawa *et al.* (2003), and Buschbeck and Friedrich (2008). Photon absorption causes a cis–trans isomerization of retinal, which induces a conformational change of the opsin protein (Rister and Desplan, 2011).
[10] Sison-Mangus *et al.* (2006), Osorio and Vorobyev (2008) and Brandley *et al.* (2013).
[11] Briscoe (2008), Briscoe *et al.* (2003), Wakakuwa *et al.* (2004), Zaccardi *et al.* (2006), Frentiu and Briscoe (2008), Yuan *et al.* (2010) and McCulloch *et al.* (2016). Quotation from Briscoe *et al.* (2010).
[12] Although recent studies have indicated that the colour vision of butterflies is almost species specific (Awata *et al.*, 2010), not enough studies have been carried out to work out the amount of variation within families (Professor A. Briscoe, UK, 2019, pers. comm.). Frentiu and Briscoe (2008) and Matsushita *et al.* (2012).
[13] Bernard and Remington (1991), Arikawa *et al.* (2003, 2004, 2005) and Briscoe (2008).
[14] The 15 different photoreceptors in the eyes of common bluebottles are described in Chen *et al.* (2016). Quote taken from Frontiers blog: https://blog.frontiersin.org/2016/03/28/fifteen-shades-of-photoreceptor-in-a-butterflys-eye
[15] Koshitaka *et al.* (2008) and Chen *et al.* (2016).
[16] The spatial resolution of insect compound eyes is much coarser than that of humans: a single pixel of the human visual system covers about 0.008° whereas that of diurnal insects is typically about 1.0° (Takeuchi *et al.*, 2006). Quotation from Nilsson (1989).
[17] Rutowski (2000b), Merry *et al.* (2006) and Turlure *et al.* (2016). The interommatidial angle is the angle that separates the optical axes of adjacent ommatidia (Bergman and Rutowski, 2016). See fig. 1 of Uchiyama *et al.* (2013).
[18] Rutowski and Warrant (2002).
[19] Briscoe and Chittka (2001), Kelber and Osorio (2010) and Koshitaka *et al.* (2008). Red receptors in nymphalid butterflies are typically due to heterogeneously expressed deep orange or red filtering pigments (Blackiston *et al.*, 2011; Zaccardi *et al.*, 2006), which coat the rhabdoms, together with a single green opsin. However, in some butterflies of the tribe Nymphalini the filtering pigments appear to be absent, so they have lost their red receptors (Briscoe and Bernard, 2005; Professor Adriana Briscoe, UK, 2019, pers. comm.).
[20] Kelber *et al.* (2001), Sweeney *et al.* (2003), Sauman *et al.* (2005), Douglas *et al.* (2007), Labhart *et al.* (2009) and Blackiston *et al.* (2011). See polarized light photographs of butterfly wings in Douglas *et al.* (2007).
[21] Arikawa *et al.* (1980), Arikawa and Aoki (1982) and Arikawa (2001).

[22] Stavenga (1979 and pers. comm.) and Glaeser and Paulus (2015). 'The pigments in the pigment cells surrounding the photoreceptor set of each ommatidium function as shields for stray light … they are usually blackish, absorbing throughout the whole wavelength range' (Stavenga, 2 January 2019, pers. comm.).

[23] de Ibarra *et al*. (2015).

[24] Kinoshita and Arikawa (2000).

[25] Snell-Rood and Papaj (2009), Pohl *et al*. (2011), Cepero *et al*. (2015) and Jain *et al*. (2016).

[26] Rutowski (2003), Arikawa *et al*. (2004), Briscoe (2008) and Macias-Munoz *et al*. (2016).

5 Courtship Behaviour

What is Courtship?

It may seem obvious that courtship is a preliminary stage in the mating sequence, occurring prior to copulation, but it is worth considering exactly what it comprises. Definitions usually emphasize one individual attempting to reduce the resistance of another to copulation, either by inducing a change in their motivational state, or by advertising qualities that may influence the choice of the other towards the courter. The function of courtship is clearly both to attract and persuade, via a sequence of behaviours and the exchange of information, which may include visual, olfactory, behavioural or acoustic signals. These cues, especially colours, can produce mutual stimulation, although it is usually just one sex, typically the male, who is bright and flashy and emits pheromones. Acceptance of a mate during courtship may depend on triggering internal activation thresholds, often via multimodal stimulation, i.e. involving more than one cue.[1]

We can think of courtship as having two main functions. First, as a means by which each sex can recognize the other as being a member of the same species, i.e. conspecific mate recognition; and, second, as a way of deciding whether each individual wishes to proceed to copulation. In everyday language, it is to see whether they are sufficiently attracted by each other to want to mate. As we shall see, sex is largely consensual in butterflies – with a few notable exceptions – and there are frequently many repeated courtship attempts, sometimes only reaching the early stages, before the courtship sequence is eventually concluded, and the individuals form a copulating pair. Courtship provides an opportunity to determine whether an individual is ready to mate and, as a way of facilitating the process, bringing together the partners in time and space.[2]

© Raymond J.C. Cannon 2020. *Courtship and Mating in Butterflies: Reproduction, Mating Behaviour and Sexual Conflicts* (R.J.C. Cannon)

Courtship has evolved as a way of facilitating mate choices. It provides a means by which one sex, usually the male, carries out a repertoire of behaviours with the purpose of communicating information on identity and quality. In the process, advertising his genetic inheritance, or at least an indication of this, e.g. via ornamentation and/or pheromones. The nature of the courtship ritual varies enormously from species to species, but usually follows a general pattern that may be simple or elaborate. The sequence of behaviour reveals signals or cues – e.g. visual and chemical – which, if successful, elicit responses in the receiver, enhancing their receptivity and allowing the interaction to proceed towards the final stages of mating. Recent research has, however, identified the existence of two-way lines of communication between the sexes. Females make choices about mating and fertilization and may communicate their decisions with males via changes in their behaviour. Finally, it is worth noting that courtship displays involve a cost as well as a potential benefit. Elaborate displays – particularly in exposed locations – draw attention to the performer and expose them to the risk of predation. Males may have to trade off these costs against the rewards of copulation and decide for how long, or how vigorously, they wish to court a female, especially when she appears to be unreceptive. Females may also be vulnerable to predation of course, although in general, they tend to be less showy, less mobile and less apparent than males.[3]

The word courtship is used here to refer to behaviour occurring before mating commences, i.e. prior to intromission of the aedeagus. It has become clear however, that there are some patterns of behaviour that occur after intromission, referred to as 'copulatory courtship'. The purpose of these actions is variable but can include, for example: (i) preventing female movement, and (ii) inducing oviposition. In butterflies, such movements are often obscured during copulation, e.g. by folded wings, but may involve subtle movements of the legs or abdomen. It remains unclear how important copulatory courtship is in butterflies as there are very few published examples, but there are anecdotal accounts of seemingly trivial behaviour – such as leg tapping during copulation – which have unknown effects. As always in attempting to understand butterfly courtship, there is scope for much more research and careful observations.[4]

It is worth emphasizing at the outset that there are many species for which courtship behaviour and mating has rarely, if ever, been recorded. For many others there are only anecdotal accounts scattered in the literature – sometimes in papers reporting on unrelated topics – and often not systematically described in a publication. In attempting to review the subject therefore, one must rely on a rather eclectic range of sources, with many gaps and sketchy details. Some species lend themselves more easily to being studied, because their behaviour is more apparent, complex and obvious. In others, courtship is brief, non-existent or difficult to observe. So, we are left with an assortment of different studies from which to try and draw conclusions. Nevertheless, it seems likely that the courtship sequences of butterflies are mostly not species-specific, and the same or very similar patterns occur in species within the same genus, or even within the same tribe or family. There remains, however, a considerable amount of behavioural diversity within the Papilionoidea, as a whole.[5]

Courtship Patterns and Diversity

Researchers have categorized male courtship patterns into a number of separate phases. The precise details vary greatly from species to species, but typically consist of a series of distinct modular phases – or behavioural acts – which can be terminated at any stage by either party. Progression from one stage to the next is elicited by both visual and olfactory cues – i.e. stimuli – which can function over different ranges, or distances. Potential mates are usually identified by sight and confirmed by smell. In other words, prospective mates are usually identified visually, but as the butterflies get close to each other, pheromones – chemicals particular to each species – become the most important form of communication. For example, visual cues play a major role in attracting males at the beginning of the courtship in the leopard lacewing, *Cethosia cyane cyane* (Fig. 2.5), but olfactory cues – in this case the chemical 'cedrol' – are used by males to identify females at close range. Similarly, males of two lycaenid species – *Chrysozephyrus smaragdinus* and *Neozephyrus japonicus* – first respond to moving objects and then perceive their colour when close. There is in effect a spectrum of courtship behaviours, from the simple to the complex, depending on species and, to some extent, individual receptiveness. Nevertheless, there is a basic pattern of courtship behaviour upon which evolution has moulded different manifestations. Brief and unelaborate courtship routines may, for example, be adaptations to reduce the risk of predation during this vulnerable phase, when individuals are preoccupied with mating.[6]

It is not always easy to determine which sex initiates courtship. In general, males initiate proceedings, but there are many examples of females approaching perching males and leading them in pursuit flights that quickly result in copulation. Despite such female solicitation (see below), males are usually the seeking sex. For example, in the common blue butterfly (*Polyommatus icarus*) this involves a display of fluttering wings by the male – but the sequence is very often terminated prematurely in the face of female indifference or refusal, and rarely proceeds as far as copulation. The male butterfly does not usually force himself upon a female (but see monarchs, below), and there is generally an element of choice for both sexes – the major exception being pupal mating (see Chapter 7, this volume). Males respond to moving objects but do not perceive detailed form or colours until they are close, and consequently are usually unable to distinguish between males and females of their species, in brief aerial interactions.[7]

Tabulating butterfly behaviour

There are relatively few really detailed studies of butterfly courtship and mating behaviour in which all of the individual 'acts' or behaviours are analysed and recorded. Studies using video cameras can be carried out to produce so-called ethograms, which are catalogues of all the individual behavioural components (acts) – carried out by both sexes – in terms of their sequence and duration. For example, the courtship and mating of the Julia butterfly, *Dryas iulia alcionea*, was broken down into 24 behavioural acts, pre- and post-coupling, involving both males and females. Breaking down the complex repertoire of behaviours in this way, it is possible to analyse the mating behaviour of butterflies scientifically and to carry out experiments to investigate the effects of different factors – environmental, temporal and behavioural – in the field and laboratory. That said, there are dangers of coming to erroneous conclusions if the experimental set-up is not ecologically relevant to conditions experienced in the wild.[8]

One of the most comprehensive examples of cataloguing butterfly behaviour on record, must be that of Takeshi Ohtani (University of Hyogo, Sanda, Hyogo, Japan), who, together with his colleague Michiya Yamamato, continuously followed individual Japanese cabbage white, *Pieris rapae crucivora*, butterflies 'all day long'! This remarkably dedicated study, employing a technique the researchers termed 'single-individual trailing method', generated a total repertoire of 35 discrete – i.e. distinctly different – behaviours for this species under natural field conditions. Movements of the head, legs, wings, proboscis and abdomen were meticulously recorded, in an attempt to break down behaviour into essential, non-divisible elements. The researchers also invented a shorthand method of documenting these behaviours, which were recorded at 5-s intervals, until the butterfly died or was lost. A heroic dedication to the cause of science! In a related study, two observers followed the movements of adult *P. rapae crucivora* continuously throughout the day 'as far as possible'. Fifteen males and 26 females were 'chased continuously' and detailed maps of flight patterns were obtained, showing how most females emigrated from the area following copulation.[9]

Lycaenids – Good Vibrations!

The blues

Courtship behaviour in many species of blues, such as the common blue, *Polyommatus icarus* (Fig. 2.22), is rarely observed in the field, and it is sometimes said that there is little or no courtship in this species. However, courtship behaviour leading to copulation is often very difficult to observe in lycaenids because most females will have already mated, very soon after they eclosed, and are thereafter, largely unreceptive. So, courtship does occur, but it is often brief and variable, i.e. with or without initial courtship. Pairs *in copula*, however, are frequently encountered (Fig. 7.8). Male–male competition for receptive females is also intense, and mate refusal behaviour by females – wing fluttering – is commonly observed.[10]

Courtship behaviour of the common blue butterfly, *Polyommatus icarus* (Polyommatini), was described by Helge Knüttel and Konrad Fiedler, based on field observations in northern Bavaria, Germany. They noted that males frequently approach females and attempt to mate, but the courtship sequence is rarely completed, and the male usually gives up at some point – presumably after determining that the female is unreceptive – and flies off. The male courtship behaviour itself consists of fluttering above the female, then alighting next to her – sometimes still fluttering – and attempting to copulate. The female may refuse copulation when the male attempts to grasp her with his genitalia. The male is unable to force a copulation because the female has to actively evert a sclerotized genital opening (gonoporus) for copulation; thus, both sexes exhibit mate choice in this species. Newly eclosed, polyommatine (members of the subfamily Polyommatinae) females remain motionless, close to the place where they emerged, but when patrolling males approach them, they reveal themselves by making a short flight, or 'dance call'. Females probably detect fast-approaching males by the colourful signals generated by their rapidly beating wings.[11]

The courtship sequence of Reverdin's blue, *Plebejus argyrognomon* (Polyommatini), does not differ from that of other lycaenids, such as the common blue, *P. icarus*, the small blue, *Cupido minimus* (Polyommatini), (Fig. 5.1) and the silver-studded blue, *Plebejus argus*. The 'most sustained' courtship sequence in *P. argyrognomon* consists of an initial rejection – i.e. a mate refusal posture – followed by a 'sexual chase', which terminates with the female alighting and adopting a receptive posture. Mating then occurs, with the male landing parallel to the female, bending his abdomen round and attaching to her, before moving into the typical, reversed copulatory position: i.e. back to back. The refusal posture – vibrating half-opened wings, with a raised abdomen – of the females is induced by

Fig. 5.1. Little blue (*Cupido minimus*), Somiedo, Asturias, Spain.

pheromones released from the wings of the males as they flutter around her. If the female is highly receptive, there is no refusal posture and the courtship simply consists of the male 'fluttering around' followed by mating. Pheromones are also thought to be produced by the females, stimulating the final stages of the male courtship sequence.[12]

David Measures (1937–2011) was an artist and meticulous observer of butterfly behaviour. He had the amazing ability and talent to be able to paint butterflies in the field and record their behaviour and natural history in annotated sketches and watercolour paintings. He was able to observe and record details of butterfly courtship behaviour: for example, that of the Chalkhill blue, *Polyommatus coridon* (Fig. 5.2). He describes how the male 'hurls himself up at her' the moment he sees a female, later alighting alongside her, perching 'head to head'. He describes how, once joined in copulation, they walk about or change their position, flying down into the grass and then back onto a white umbel head, whilst remaining *in copula* for at least a half hour. He describes the courtship in quite lyrical terms:

> the cock's [male's] wings are put forward to shelter and enclose the female, so her antennae brush his scent glands. They bob and bow, sidling gradually against each other, until finally they are back to back, and mating can begin.[13]

Three coppers

The courtship behaviour of the bronze copper, *Lycaena hyllus* (Lycaeninae) – a North American lycaenid – was described by David and Bonnie MacLean, based on observations made in the summers of 1988 and 1989, in Minnesota, USA. Competition was intense in a dense population consisting of up to 25 adults per 10 m^2 and groups of males – 10 or more – were seen chasing each other. Despite the high numbers, there were few observations of mating pairs, and most females were probably mated soon after they emerged from their pupae. Consequently, rejections by already

Fig. 5.2. Chalkhill blue (*Polyommatus coridon asturiensis*) female on vetch, Somiedo, Asturias, Spain.

mated females were a regular occurrence. Courting males landed next to resting females and both sexes engaged in rapid wing fluttering. The male typically moved around to a position behind the female and placed his head and antennae beneath her wings: a behaviour described as 'nudging'. Females continued their wing fluttering and blocked attempts to mate by the males, who usually gave up and flew off after a minute or so of these 'rejection dances'. Rejection behaviour was also a feature common of courtships observed in the Turkish fiery copper, *Lycaena ochimus*, in Adana Province, south-central Turkey. Courting males vibrated, or flickered their wings, but in almost all cases, were rejected by the females, either by rapid wing movements or flying off.[14]

The mating behaviour of the tailed copper, *Lycaena arota*, another North American species, was described by James A. Scott, based on observations made in 1969 and 1971, at different locations in Colorado, USA. Male *L. arota* are territorial and typically perch on shrubs and trees, 1–2 m above the ground. When a female passes by, the male gives chase, flying closely (c. 14 cm) behind her. When the female lands, the male alights closely behind her and vibrates his wings in a semi-open, V-shaped position, in bursts of either 2 or 10 vibrations/s. Some males, rather ungallantly it seems to me, did not vibrate their wings at all, but just walked up to the female and attempted to copulate! In either case, if the female remains still – a sign of receptivity – the male moves alongside and curls his abdomen round towards her in an attempt to copulate. Unreceptive females, however, flap their wings vigorously – at 'almost full stroke' – 10 or more times per second. Female *L. arota* rarely mate more than once, so 'rejection dances' were frequently observed, and only two observations of copulation were made.[15]

Hairstreaks

In more complex courtship sequences, there are four main stages with a number of specific acts within each one: i.e. pre-courtship, pre-coupling, coupling and post-coupling behavioural sequences. The following is based on the courtship behaviour of the xami hairstreak, *Callophrys xami* (Theclinae, Eumaeini) (Fig. 5.3), a green-tinted lycaenid found in Mexico and Texas.

1. *Interception.* When a female flies close (<1 m) to a flying or, more likely, perching male.
2. *Courtship flight.* The male flies closely behind the female, parallel to the ground, for 30 s or so.
3. *Alighting.* The pair alight on vegetation close to each other.
4. *Fluttering.* The male walks in front of the female, facing her head on whilst fluttering vigorously and, it is thought, emitting pheromones.
5. *Mirroring.* The male takes up a parallel position beside the female, still fluttering his wings.

Fig. 5.3. Xami hairstreak (*Callophrys xami* subsp. *xami*) adult in Ciudad de México, Mexico. (Photo by Mario Castañeda Sánchez.)

6. *Copulation.* The male moves the tip of his abdomen to make genital contact with the female, and stops fluttering. The pair commence copulation, adopting the 'tail to tail' *in copula* position.
7. *Finalization.* The female starts to walk, and the pair eventually separate and fly away.[16]

Glen A. Gorelick observed the courtship and mating behaviour of two species of North American *Callophrys* hairstreaks – *Callophrys dumetorum* and *Callophrys viridis*, as they are now called – in California in 1967 and 1968. Males approach females in flight, reportedly 'forcing' them to alight. As the female perches motionless on a leaf, the male vigorously vibrates his wings, then moves around in a clockwise direction, 180°, until he is facing outwards; the female faces one direction with her antennae oriented upwards, whilst the male faces the opposite direction, with his antennae pointing downwards. The entire courtship sequence lasts no longer than 15 s. The courtship behaviour of *C. viridis* was described as being identical.[17]

The Colorado hairstreak, *Hypaurotis crysalus* (Theclinae), is a beautiful violet, montane species (Fig. 5.4), which occurs in the south-west USA and northern Mexico. The mate-location and mating behaviour were studied by James A. Scott, in Colorado, in 1970. Males patrol Gambel oak (*Quercus gambelii*) canopies searching for females in the late afternoon and early evening. Scott showed that the males are attracted by the violet – probably UV – wing colours of both basking and flying females. They land on the same leaf and move alongside the female, who responds by closing her wings. The male flicks his wings open and closed for several seconds – through a decreasing amplitude – whilst bending his abdomen laterally, either left or right, towards the female. Receptive females 'passively accept' the male's advances, whilst unreceptive, i.e. recently mated ones, fly or turn away, elevating their abdomen if the male makes contact. Courtship is brief, usually lasting less than a minute, rarely for as long as 5 min, and may be facilitated by a female pheromone (unidentified).[18]

Fig. 5.4. Colorado hairstreak (*Hypaurotis crysalus*) dorsal side of male. (Photo by Megan McCarty – own work, CC BY-SA 3.0, https://commons.wikimedia.org/w/index.php?curid=11526410)

A brief encounter between mating black hairstreaks, *Satyrium pruni* (Theclinae), was observed by Vince Lea and other members of the Cambridgeshire and Essex Butterfly Conservation Trust, in 2006, in Nottinghamshire, UK. After a brief, spiralling flight, ascending upwards for c. 50 cm, the pair landed on a bush and then repeated the courtship dance. They then retreated to the centre of a blackthorn bush, where they stood side by side, before coupling in the typical back-to-back position. The pair remained *in copula* for 25 min; on separation, one butterfly (sex unrecorded) flew off, but the other one remained *in situ* for 5 min, cleaning its antennae, before it too departed the scene.[19]

The apefly

Courtship flights are often an important prelude to mating. In the apefly, *Spalgis epius* (Miletinae) (Fig. 5.5) – so-called because of the supposed resemblance of its pupa to the face of an ape – a courting pair may engage in up to four such courtship flights within a period of about 4–5 min, before the female alights on a perch in the male's territory. During these short flights, the male follows closely behind (c. 12–18 cm), and slightly above, the female, who may alight a few times before finally settling down and raising her abdomen. The male alights a short distance (4–8 cm) behind the female, then moves alongside her and curls the tip of his abdomen around – usually to the right side – to make genital contact. After coupling, the male moves around so that the pair are tail to tail, and they remain like this for slightly less than an hour, only moving if forced to do so by disturbances like rain or predators. Apefly (*S. epius*) males are highly territorial and perch on several places in the tree canopy, relatively close (20–25 cm) together, within a defined territory. The males mark these sites in some way, chemical unknown, by dragging the tip of their abdomens

Fig. 5.5. Apefly (*Spalgis epius*). (Photo by Vengolis [CC BY-SA 4.0 (https://creativecommons.org/licenses/by-sa/4.0)].)

across the foliage. As well as chasing off intruders, males spend short periods of time (2–3 min) basking, with their wings spread out. The arrival of a female initiates the courtship flight by the male, and the pair fly horizontally through the canopy.[20]

Pierids – Wing Buffeting

The courtship of many pierid species is usually very brief, lacking some of the elaborate displays seen in other butterflies. For example, there is 'essentially no courtship' in the Mexican pierid, the Madrone butterfly, *Eucheira socialis* (Pierinae, Aporiina), according to Professor Arthur M. Shapiro: males simply approach, and attempt to copulate with the females, who either accept them or walk away! Male pierids usually approach perching females and briefly buffet them with their wings; the females either exhibit mate refusal behaviours or respond positively by extending their abdomens. For example, in the cloudless sulphur, *Phoebis sennae* (Coliadinae, Coliadini), females either remain motionless and accept the male's advances, or assume a brief, 'open wing-raised abdomen' rejection posture. Receptive little sulphur, *Eurema lisa* (Coliadinae, Eureminі), females also assume rigid postures and extend their abdomens out from between their hindwings (see below). Likewise, courtship behaviour in the checkered white, *Pontia* (=*Pieris*) *protodice* (Pierini), is 'rapid and highly stereotyped', without an aerial phase and with no abdominal extension by the female. Similarly, the courtship behaviour in the sulphur butterflies, *Colias eurytheme* and *Colias philodice* (Coliadinae, Coliadini), is also extremely brief without any elaborate displays. In the Japanese sulphur, *Colias erate*, newly emerged, virgin females simply accept courting males – by exposing their abdomens – and then copulate for about 45 min. Mated females are, however, mostly unreceptive to male courtship advances, so most of the females encountered by patrolling males in the daytime demonstrate this by adopting mate refusal postures at their perching sites; e.g. with wings spread, and abdomen upturned. It is possible that these rapid courtships are evolved responses to predation pressures, as pierids are relatively palatable and therefore vulnerable to predation.[21]

There are, however, some interesting variations in pierid courtship, with lengthy and relatively elaborate behaviour in some species. For example, mate-avoiding behaviour lasts longer in some species (e.g. *Pieris melete* and *Pieris canidia*) than others (*Pieris napi* and *Pieris rapae*) and is frequently entirely absent, e.g. in *P. rapae*. Male pierids generally identify females using visual cues, and wing colours appear to initiate mating in many species; differences in UV reflectivity between sexes vary according to species, and even subspecies (see Chapter 2). Only a stationary visual stimulus is required to release the entire sequence of male courtship behaviour in *Colias* species. When *Colias erate* males encounter a mature female who has returned to a state of receptivity, they hover around her if she is perching, or fly near the female attempting to inhibit her flight course. A receptive female alights on the vegetation and remains motionless, extending her abdomen towards the male from beneath her hindwings. After alighting on the edge of the female's wings, and then moving down next to her, the male bends his abdomen around towards the tip of the female's abdomen, for copulation.[22]

Three little sulphurs

Ronald L. Rutowski carried out an intensive investigation of the courtship behaviour of the little yellow, *Eurema lisa* (Coliadinae, Euremini), over 3 years (1972–1974) in Florida, USA. Eight distinct behavioural events (written in *italics*, below) were identified and associated with successful courtships. Remarkably, these all usually occur in less than 4 s, and the pair then remain together, *in copula*, for between 30 min and 1 h. In the field, most *E. lisa* females are courted and mated soon after they emerge, close to their pupation site. In the experiments using outdoor cages, virgin females were placed on a perch, where they remained whilst a male was introduced to the cage to initiate courtship. Interactions were recorded on videotape or film.[23]

The males approach the females and make contact (*initial contact*), using either their wings or legs, sometimes landing on the female's wings or thorax before taking off again. The female responds to this buffering by fluttering her wings (*first wing flick*) continuously as the male flies around her, repeatedly making contacts with her using his wings or legs. The female ends her flutter response (*last wing flick*) by spreading her wings and curving her abdomen out from between her wings (*extends abdomen*). Some successful courtships did, however, occur without any wing fluttering, and these were of even shorter duration (1.5 s or less). After the female extends her abdomen and remains still, probably in response to a pheromonal cue, the male typically *alights* on her wing tips and moves – still flapping his wings – into a position where he is holding on to her thorax. The male then *begins probing* the female's abdomen with his genitals, curling the tip of his abdomen down towards the inner margin of her hindwings. The male uses his claspers to grab the tip of the female's abdomen (*genital contact*) and eventually they both become quiescent and the male *closes his wings*.[24]

Males of the barred sulphur, *Eurema daira* (Coliadinae, Euremini) (Fig. 5.6) – which occurs in Argentina, extending north as far as the southern USA – have a unique 'wing-waving' display, which they perform during courtship: the male courts the female by landing next to her and waving one of his forewings in front of her. The purpose of this wing-waving behaviour may be to expose the UV-reflecting area on the dorsal forewing surface; this probably acts as both a behavioural 'releaser' for the female, and as a way for her to assess the identity and quality of the male. Professor Rutowski described how, with each upsweep of his forewing, the male butterfly rubs the trailing edge along the length of the female's antenna. This stroking behaviour may occur several times until the female is ready to accept the male for copulation. Antennae are frequently held wide open and/or held backwards, at some stage during courtship, which may be a courtship feature common to females of most species (see Chapter 6, this volume). Virgin female, *Eurema daira*, actively solicit male courtship via aggressive contacts with males and elaborate posturing (see Chapter 6, 'Female Solicitation', this volume).[25]

The dainty sulphur, *Nathalis iole* (Coliadinae) (Fig. 5.7) – the smallest pierid found in North America – also has an unusual wing-spreading behaviour, with males placing themselves in front of the female, facing away, with wings spread, quivering slightly. During this display – which was only observed in about a quarter of all courtships – the fore- and hindwings are held slightly apart for about 5–8 s, exposing a red-orange coloured sex brand, on the dorsal surface of the male hindwing. The spread-wing display is performed in response to an initial rejection by the female – displayed by wing fluttering or by adopting a full mate-refusal posture – and presumably conveys information about the identity, sex and quality of the male.[26]

Fig. 5.6. Barred yellow (*Eurema daira*), Lake June, Winter Scrub State Park, Lake Placid, Florida. (Photo by Judy Gallagher [CC BY 2.0 (https://creativecommons.org/licenses/by/2.0)].)

Fig. 5.7. Dainty sulphur (*Nathalis iole*) Lake Houston Nature Park, New Caney, Texas. (Photo by Judy Gallagher [CC BY 2.0 (https://creativecommons.org/licenses/by/2.0)].)

Fig. 5.8. Large marble (*Euchloe ausonides*). (Photo by Ken-ichi Ueda [CC BY-SA 4.0 (https://creativecommons.org/licenses/by-sa/4.0)].)

In the large marble, *Euchloe ausonides* (Pierinae, Anthocharini) (Fig. 5.8), a North American species, if a patrolling male encounters a flying female, they hover near to each other. The female then lands, and the male follows, landing behind her and bending his abdomen towards her to initiate copulation. Resting females may reject courting males by spreading their wings and raising their abdomen up to a position near the vertical (see Chapter 6, 'Refusal and Rejection'). Occasionally, they fly upwards, to over 7 m above the ground – a similar tactic to that displayed by other species (see Chapter 6, this volume) – to escape the attentions of the male. Occasionally, it is the male however, who gives up and flies away, after having hovered together for a short while.[27]

See images of courting Iberian marbled whites (*Melanargia lachesis*) (Box 5.1) where the male, shown approaching from the right (1 and 4), flies and hovers over the female (2), touching her with his legs (3).

Japanese small white

The courtship behaviours of the Japanese small white, *Pieris rapae crucivora* (Pierinae, Pierini) (Fig. 7.11), and other *Pieris* species, were studied

Box 5.1. Courting behaviour of Iberian marbled whites (*Malanargia lachesis*).

Fig. B5.1a. Iberian marbled white (*Melanargia lachesis*) courting sequence 1.

Fig. B5.1b. Iberian marbled white (*Melanargia lachesis*) courting sequence 2.

Fig. B5.1c. Iberian marbled white (*Melanargia lachesis*) courting sequence 3.

Fig. B5.1d. Iberian marbled white (*Melanargia lachesis*) courting sequence 4.

by Yoshito Suzuki and colleagues at Kyushu University, Fukuoka, Japan, based on observations made in the outdoor cages. Briefly, the behavioural sequence consists of the following:

- *Mating approach*: the male lands next to a resting female and faces in the same direction, continuously fluttering his wings and catching the female's forewing with his legs.
- *Copulation attempt*: the male curves his abdomen around in an attempt to couple with the female genitalia.

Successful copulations last for about 2 h. Males also approach females in flight, flying in a zigzag pattern whilst positioned slightly below and behind her. Male *Pieris melete* and *Pieris canidia* may exhibit more elaborate behaviours, described as *hovering* and *caressing*, but these are not essential for successful copulation. During the preliminary, aerial stages of courtship, where males are pursing flying females, they are presumably displaying the colours on their dorsal wing surfaces. Male *P. rapae crucivora* recognize females using a mixture of near-UV and visible light reflected off their wings.[28]

Steven Peter Courtney – who carried out a PhD study on the orange tip, *Anthocharis cardamines* (Pierinae, Pierini) and *P. napi* (Pierinae, Pierini) at Durham University, in the UK – described the general mating pattern as follows. Males respond to a visual stimulus of approximate Pierinae size and colour, followed by recognition of the female, courtship, acceptance (or rejection) of the male and, finally, copulation. Courtney used a cine camera to record courtship behaviour of *P. rapae* and showed that the male approaches the female in flight, from below, and flies into her from underneath, in a zigzag flight path, as also described by Suzuki in Japan (above). However, he concluded that the peculiar zigzag flight of the male occurs *as a result* of his contact with the female. In other words, the male loses and regains height as a result of the collision. This was interpreted as an aggressive behaviour by the male, attempting to induce the female to land, and hopefully copulate.[29]

The behaviour of *P. rapae crucivora* was studied in a remarkable undertaking carried out on the campus of Hokkaido University, Japan, in 1976. Focusing just on the courtship and mating aspects of this study – what the investigators called temporally integrated behaviours – seven component patterns were identified. Mating activity usually commences with a *prenuptial flight*, involving a virgin female and one or more males. The female then lands and remains motionless, with her wings closed (*keeping closed*). The male alights soon afterwards and stands side by side (*paralleling*) with her, then flutters his wings and catches the leading edge of the female's forewing with his fore- and midlegs (*catching wings*). The male then curves his abdomen round and inserts it between the hindwings of the female, making genital contact (*grasping genitalia*). Once joined, the pair engage in a *nuptial flight*, the male carrying the female, who is upside-down with her wings closed. *Copulation* lasts about an hour, or 5090 s, on average![30]

Fig. 5.9. African pierid (*Pseudopontia mabira*) (Pieridae, Pseudopontiinae). (Photo by Szabolcs Sáfián.)

A forest 'ghost'

The courtship behaviour of a delicate African pierid, *Pseudopontia gola* (Pieridae, Pseudopontiinae) – very similar to the closely related *Pseudopontia mabira* (Fig. 5.9a, b), was described by Hungarian lepidopterist, Szabolcs Sáfián, based on observations made in wet lowland rainforests of western Liberia and eastern Sierra Leone. Adults fly very slowly, close to the ground in these shaded African forests. Males compete for access to newly emerged females, which may get mated before their wings have hardened and they can fly. Courtship of more mature females is, however, a more elaborate affair. The pair engage in a 'joint flight', and if the female is receptive, she alights on a plant low down in the dense undergrowth with her wings held horizontally, wide open. The male then begins a lengthy 'flapping flight' directly above the female, which often continues for over 10 min. Mating eventually occurs, but sometimes some males lose their way during the slow-motion hovering flight or are driven off by other males. In such cases, the females continue to hold their wings wide open – in what is presumably, an acceptance posture – for a couple of minutes, and if the male does not return, they simply fly away! *Pseudopontia* species are completely white, with no visible wing or body markings and some species, such as *P. gola*, have only recently been described. They are known as ghost butterflies because of their clear, translucent, white wings and their very slow, erratic flight. Very little is known about larval host plants of *Pseudopontia* species or other aspects of their biology.[31]

N.B. The courtship behaviour of wood white butterflies is described in Chapter 6, this volume.

Hesperiids

Hesperia skippers

James A. Scott studied and compared the courtship and mating behaviour of two skippers – the Pahaska skipper, *Hesperia pahaska*, and Snow's skipper, *Ochlodes snowi* (both Hesperiinae, Hesperiini) – in different locations in Colorado, USA, in 1969 and 1970. In *H. pahaska*, the perching male flies up and follows closely (5 cm) behind a passing female. The female lands and the male 'flutters about' her for a short time before also alighting, and then curving his abdomen towards her, in the usual manner. Some more wing fluttering may occur in both sexes prior to copulation, although many such approaches were unsuccessful. Unreceptive females vigorously fluttered their wings, 'from vertical to horizontal', about 10 times per second.[32]

The mating behaviour of *O. snowi* is reportedly identical to that of *H. pahaska*, and very similar to the cobweb skipper, *Hesperia metea*. Scott describes how the female sometimes hovers, just above the ground (3 cm), with the male also hovering and moving around behind her; both sexes beating their wings through wide arcs. It is interesting to note that of the 55 courtships observed by Scott, only two copulating pairs were seen, demonstrating just how elusive such behaviour can be. The mating behaviour of another hesperiid – the Simius roadside skipper, *Notamblyscirtes simius* (Hesperiinae) – was also described by Scott, at the same location in Chaffee county, Colorado. When a flying *N. simius* female is approached by a male, she carries out a rapid zigzag dance – darting rapidly back and forth, or up and down, 10 or 20 times – just above the ground. The male mirrors the movements of the female whilst flying very closely (5–8 cm) behind. The females appear to seek out males by flying to hill-topping sites on nearby ridgetops and small hills, where perching males congregate.[33]

Arthur M. Shapiro studied the cobweb skipper, *H. metea*, in Pennsylvania, USA, in 1964, and described the mating behaviour. Males pursue females who cross their 'transient' territories, flying around her for a few seconds until she lands, usually on a blade of grass. The male lands close behind and moves closer, fluttering his wings. The female may take flight again or, if receptive, allow the male to move up to a position alongside her. The male then curves the tip of his abdomen around so that the extruded genitalia contact the caudal tip of the female. This sequence may be repeated a few times before the female acquiesces and exposes her genitalia; copulation then takes place with the male and female facing in opposite directions. The entire mating process in *H. metea* lasts about 3 min.[34]

The fiery skipper

Irene Shapiro studied courtship and mating behaviour in the fiery skipper, *Hylephila phyleus* (Hesperiinae, Hesperiini) (Fig. 5.10a, b), on the University of California, Davis campus in 1974, using a modification of the fishing rod technique pioneered by Tinbergen (see Chapter 1, this volume). Wild-caught males and females were glued onto fishing lines and dangled in front of perching males. A fluttering movement of the wings was found to be essential for eliciting an investigatory response by the male. Courtship

Fig. 5.10. (a) Fiery skipper (*Hylephila phyleus*). (Photo by Regular Daddy – own work, CC BY-SA 3.0, https://commons.wikimedia.org/w/index.php?curid=7531125) (b) Fiery skipper (*Hylephila phyleus*) Copulating pair, female on top. (Photo by Eugene Zelenko – own work, CC BY-SA 4.0, https://commons.wikimedia.org/w/index.php?curid=2485752)

and successful matings were achieved in experiments using virgin females and the courtship sequence was described in terms of seven behaviours:

1. *Investigative behaviour.* The male flies towards a passing female.
2. *Settling behaviour.* The male lands behind the female.
3. *Head–wing behaviour.* The male thrusts his head between the female's hindwings.
4. *Fluttering behaviour.* The male rapidly flutters his wings.
5. *Curving behaviour.* The male curves his abdomen round towards the female's genitalia.
6. *Coupling behaviour.* The male clasps the female and inserts his aedeagus.
7. *Uncoupling.* After copulating for between 40 and 72 min, the pair separate.

Females were relatively still during courtship, although some extruded their genitalia, perhaps as a rejection posture. The rejection signal in a related species, *Hesperia dacotae* (Hesperiinae, Hesperiini), was described as the female holding her abdomen up between closed wings and periodically jerking the wings forwards. The courtship sequence in *H. phyleus* can be terminated by either sex at any stage; for example, if disrupted by another male. Many of the courtships ended however, as a result of rejection by the female, in the form of 'shuddering': rapid fluttering of both fore- and hindwings whilst held out at a 45° angle. The key element of the courtship appears to occur in phase three, where the male thrusts his head between the female's hindwings, 'touching her genital area'. The extent to which visual, tactile and chemical stimuli play a role in inducing successful matings can only be guessed at, but pheromones are likely to be important at several stages.[35]

The courtship behaviour of a related, South American species, *Hylephila fasciolata* (Hesperiinae, Hesperiini), was described from observations made in Chile. The sequence started with the pair flying low over grass; the female landed, closely followed by the male, a few centimetres behind her. The male then approached the female 'in small jumps', each time fluttering his wings in a sort of courtship 'dance', probably disseminating pheromones. The female remained still until the male got closer, and then vibrated both of her hind legs in an up and down motion, which was interpreted as rejection behaviour. When the male approached to within a few millimetres of the female, she flew away.[36]

Giant skippers

Observations of the courtship behaviour of several species of *Agathymus* (Hesperiinae, Erionotini) were made by Don and Viola Stallings and colleagues, in Coahuila state, Mexico, in 1976. These so-called giant skippers feed and develop on different Agave species. Female Estelle's giant skipper, *Agathymus estelleae*, emerged in the early morning and climbed up the leaves of their host plants – *Agave lecheguilla* – as their wings expanded and hardened. Males soon appeared, and the first on the scene appeared to establish a territory, remaining as a 'sentry' until the female had finished expanding her wings. Other males took up positions downwind of the female, all awaiting her maiden flight. After 2–3 h, the female took off and 'in every instance' flew directly over the first male to have established a 'sentry position'. This male then took off and followed her, appearing to touch her in flight. The pair would then land a short distance (c. 40 m) away and were *in copula* within 3–4 s. Sometimes, mating did not occur, and the female would take off again, pursued by other males. Copulation lasted for between 1 and 5 h.[37]

Observations of another, closely related species, Mary's giant skipper, *Agathymus mariae*, revealed slightly different courtship behaviour. Instead of establishing a 'sentry position' close to the newly emerged female, male *A. mariae* perched in an open area and waited for the females to find them, on her maiden flight (circling close, c. 70 cm from the ground). The males appeared to change their 'territories' on occasion, interspersing perching with extensive patrolling, 1 m above the ground. When a perching male spied a female flying overhead, he took off in pursuit; the female then dropped down onto the ground, or onto the host plant. Copulation then followed, after a very short (c. 5-s) period, as in *A. estelleae*. These examples provide an interesting comparison of the subtle variations in courtship patterns that have evolved between congeneric species. It would be fascinating to discover how the male sex pheromones (MSPs) varied between these two species.[38]

The Essex skipper

Observations of about 300 courtship sequences, some recorded on video, revealed that male Essex skippers, *Thymelicus lineola* (Thymelini) – known as the European skipper in North America, because it was introduced – would briefly hover over the female, moving back and forth or side to side, sometimes flying off and returning to repeat the same behaviour two or three times, before eventually landing and attempting to mate. The male typically landed behind and to the side of the female, then advanced diagonally and touched the underside of her wing with his antennae, whilst curving his abdomen and extending his claspers. Receptive females moved forward slightly, and the male followed, flexing his abdomen to bring his claspers into contact with the female's genitalia. This courtship sequence was brief and rarely took more than 5 s, but the majority (c. 60%) of approaches were unsuccessful. Males frequently returned up to three times before a successful coupling occurred. The pairs remained *in copula* for 59 min, on average (n = 21). No physical contact occurred between the wings of the males and the antennae of the female, and significantly fewer matings occurred – between males and virgin females – when the androconia were scraped off the wings of the males, or the antennae of the females were painted with clear nail polish! The authors of this Canadian study, Kenneth Pivinik and colleagues, concluded that osmophores – the terminal particles of the androconia (see Chapter 9, this volume) – are broken off, and pheromones are released, as the males hover over the females during courtship.[39]

Nymphalids

Courtship appears to vary, from the simple or straightforward, to the highly elaborate in this family. For example, there are no distinctive courtship displays by either sex in the American snout butterfly, *Libytheana carinenta* (Libytheinae) (Fig. 5.11), whereas those in the buckeye butterfly, *Junonia coenia* (Nymphalinae) (below), are a little more complex, but still highly variable. In *L. carinenta*, males approach perched or flying females; those in flight are chased by the males until they alight on vegetation. Receptive females remain motionless with folded wings, as the

Fig. 5.11. American snout or common snout butterfly (*Libytheana carinenta*), Merrimac Farm Wildlife Management Area, Nokesville, Virginia, USA. (Photo by Judy Gallagher [CC BY 2.0 (https://creativecommons.org/licenses/by/2.0)].)

male moves alongside and starts to probe between her hindwings with his curled abdomen. After making genital contact, the male turns to face away from the female in the usual back-to-back position. These snout butterflies remained *in copula* for between 45 and 140 min, and the female carried the male if they were disturbed and flew off.[40]

The buckeye

Perching male buckeyes, *Junonia coenia* (Fig. 5.12), investigate passing objects and pursue females, whom they recognize via visual cues, i.e. colours and flight patterns. Receptive females alight nearby after a short flight and the males land a few centimetres behind them. The following behaviours can occur, but some courtships only last for 30 s or so, from sighting to copulation.

1. *Male hovering.* Male beating his wings over a small amplitude above, or slightly downwind of the female.
2. *Male fluttering.* Male on the ground moving wings with greater amplitude and slower frequency.
3. *Male nudging.* Male, with antennae pointing backwards, pushing his head underneath the hindwings of the female.

Receptive females raise their wings vertically, whilst unreceptive ones keep their wings spread. Female wing *flapping* behaviour whilst on the ground is a 'rejection dance' in this species, and is particularly associated with already mated females; being able to communicate unreceptivity is valuable for both sexes, as it avoids them wasting their time and energy. A later study on *J. coenia*, carried out in the laboratory, identified four distinctive behaviours: nudging, chasing, head dipping and abdomen curling. Confusingly, the same terms were given to different behaviours by different researchers! Nudging is said to occur when a hovering male brushes

Fig. 5.12. Buckeye butterfly (*Junonia coenia*) in the Jamaica Bay Wildlife Refuge, Broad Channel, Queens, New York. (Photo by Rhododendrites [CC BY-SA 4.0 (https://creativecommons.org/licenses/by-sa/4.0)].)

his legs against the wings of a female who is perched on the ground. Head dipping (previously called nudging by James A. Scott) is when the male lowers his head under the abdomen of the female and tries to move it into a position for copulation.[41]

The courtship sequence of the pearl crescent, *Phyciodes tharos* (Nymphalidae, Melitaeini), was also described by James A. Scott. The main ritualized behaviours are summarized as follows:

1. *Male wing display.* The male holds his wings open at c. 50° from the vertical, motionless or 'imperceptibly' vibrating, whilst standing in front of the female.
2. *Male wing fluttering.* In response to wing fluttering or movement by the female, the male spreads his wings and flutters them for a few seconds.
3. *Male bending.* The male holds his wings slightly apart and crawls under one of the spread wings (left or right) of the female, whilst bending his abdomen right round, in a 180° arc.
4. *Female mating posture.* Receptive females hold their wings up at 50–60° from the vertical and elevate their abdomens by about 20° from the horizontal.

Unreceptive females flutter their wings in a 'rejection dance' when males approach them and try to mate, sometimes also turning or crawling away, or even dropping down into the vegetation. These behaviours are variable and subject to repetition, depending on the degree of receptivity of the female, so that the duration of successful courtships varies enormously: from 5 s to 8 min.[42]

Common and danaid eggflies – all of a quiver

Ronald L. Rutowski observed and described courtship and mating in the common eggfly, *Hypolimnas bolina* (Fig. 2.1), in north Queensland, Australia, in 1989. Virgin females were released into large flight cages containing perching males. The males flew up and approached the flying females from behind; they then positioned themselves below the females, flying with shallow wing beats and wings spread. Following this display, a receptive female will alight, with the male landing nearby and facing her. The male then moves round, positioning himself so that his head is about halfway along her body. He then curls his abdomen around and inserts it between the inner margins of the female's hindwings. This probing behaviour may end in courtship, or the female may turn her abdomen away and fly off. It is now known that the courtship behaviour of the male in *H. bolina*, where he positions himself beneath the females, effectively maximizes the UV brightness, the visible area and the flash effect emanating from the UV markings on the dorsal surfaces of his wings. The ritualized flight of the male during this courtship display is characterized by a higher wing beat frequency – at 11 beats per second – and a shallower wing beat amplitude than that seen in normal, or regular flight. This behaviour maximizes the so-called 'flash effect' from the wings, creating a short, strobe-like effect, aimed at mesmerizing the female (Fig. 2.17).[43]

George Stride was probably the first person to record details of courtship and mating in this genus – eggflies and diadems – based on observations of the danaid eggfly, *Hypolimnas misippus* (Nymphalinae, Kallimini) (Fig. 1.15a, b), in flight cages and in the wild, in Ghana, c. 1954 (see Chapter 1, this volume). Although the complete sequences of courtship were never witnessed for a given pair, the behavioural interactions generally followed a set pattern, depending on the receptivity of the female. Males generally intercept females in flight, and often make physical contact with them. Virgin females often exhibit an escape reaction in response, by falling out of the air, with wings closed, onto the vegetation, some 1–2 m below. If the female continues flying, the male positions himself just (7–15 cm) below her, and commences a characteristic, slow, quivering flight: the wings being rapidly vibrated, with the body held more closely to the vertical than in normal flight. Females usually reciprocate in a similar manner; i.e. quivering flight, with their bodies being held almost completely vertical. The pair fly together like this, the male below the female, but if the female is unreceptive, i.e. already mated, she starts to ascend, slowly but steeply, up to a height of 9–12 m or more. She does not stop rising until the male breaks away; he then stops quivering, and returns to the ground, terminating the interaction. More receptive females return to the ground without ascending, holding their wings loosely open. N.B. unreceptive females remain still, with their wings tightly held together above their backs. The male may sit quietly alongside her for 10 minutes or so, his response seemingly reliant on the activity of the female. Females who open their wings stimulate the males to push their way forward, underneath their wings, whilst curling their abdomens around in the characteristic manner. Once joined, the pair move into a back-to-back position, and the male becomes quiescent ('cataleptic' according to Stride).[44]

Banded and scarlet peacocks

The banded peacock, *Anartia fatima* (Nymphalinae, Victorinini) (Fig. 2.23a, b), is both a perching and patrolling species: when patrolling, the males fly in slow zigzag patterns close to the ground/vegetation searching for females. Flying *A. fatima* females are chased and courted by males who position themselves slightly above and beyond her. When a patrolling male finds another member of the species, he dives down to inspect the individual, who will either fly up in response – if it is another male or an already mated female – or remain stationary if it is a virgin female. In the case of the latter, the male starts courting by fluttering above the female for 30 s or more. Receptive females expose their abdomens by folding their wings over their thorax; courting males alight next to the female and bend the tips of their abdomen forward in the characteristic manner. Mating

Fig. 5.13. Scarlet peacock (*Anartia amathea roselia*) male, Argentina.

often occurs shortly after a female has emerged, before her wings have fully hardened. In the scarlet peacock, *A. amathea* (Fig. 5.13), males also swoop down and flutter above females, who perch on vegetation with outspread wings. This species has a characteristic rejection behaviour: as the male flutters a few inches above the female, she closes her wings above the body and rocks from side to side![45]

The squinting bushbrown, *Bicyclus anynana*

Another courtship sequence involving six consecutive steps was described for the African butterfly, *Bicyclus anynana* (Satyrinae, Elymniini) (Fig. 5.14), by Professor Caroline Nieberding at the Institute of Biology, Leiden University, Leiden, The Netherlands, and co-workers. In summary, these courtship acts consist of the following in this crepuscular species:

1. *Localization*. The male first locates and approaches a female.
2. *Orientation*. The male orientates his body towards the posterior part of the female.
3. *Flickering*. The male initiates a rapid sequence of opening and closing his wings (c. 5–17 flicks/s) and exposing their dorsal surfaces.
4. *Thrusting*. The male touches the edge of the female wings with his head; his antenna or legs may also make contact with the side of the female, and androconial hairs fan out above the wing surface when the wings are open.
5. *Attempting*. The male curls his abdomen round in order to grasp the tip of the female abdomen with his claspers.
6. *Copulation*. Following genital contact, the male moves until the copula position is reached.

This courtship sequence occurs very rapidly in *B. anynana* – taking only 14 s on average – after which the pair remain joined *in copula* for about 30 min. Rejection by the female, if and when it occurs, is most likely to happen after the so-called *thrusting* phase, when presumably she has had a chance to sample and assess the quality of the male via his aphrodisiac pheromones

Fig. 5.14. Squinting bushbrown (*Bicyclus anynana*) female, dry season form from a lab strain (Louvain-La-Neuve, Belgium). (Photo by Gilles San Martin – own work, CC BY-SA 3.0, https://commons.wikimedia.org/w/index.php?curid=17649097)

and sexual adornments; namely the white pupils in the centre of the dorsal forewing eyespots (see Chapter 2, this volume). The combination of eyespots of a certain size, pheromones of an attractive odour and wing flickering at a certain frequency – described by Gilles San Martin and colleagues as a visual 'Morse code' – during courtship, appear to act as sort of behavioural 'releaser' for the female, triggering her receptivity to the male. The transfer of pheromones from male to female probably occurs during the flickering and thrust phases of courtship. Remarkably, females can also perform all these courtship behaviours towards males, depending on the temperature at which the larvae and/or pupae develop.[46]

The squinting bushbrown, *B. anynana*, is remarkable for having two distinct seasonal forms – wet (WS) and dry season (DS) phenotypes – which are produced by differences in temperature during the larval stages. As well as morphological differences in eyespots and background coloration of the wings (Figs 2.13 and 2.14), there are also differences in terms of courtship behaviour. Essentially, the females are more 'choosy' in the WS, and the WS males engage in vigorous courtship behaviour, as described above. Females choose males based on their dorsal forewing eyespots: in particular, the UV reflectivity of these so-called, sexual ornaments. In the DS, the sex roles are reversed, and the females do most of the courting, displaying their sexual ornaments this time, also in the form of the central white pupil in the dorsal eyespots. DS males become much choosier in the DS as they become the ones driving sexual selection (see Chapter 2, this volume).[47]

It is worth emphasizing that it is the *dorsal* eyespots – visible from above when these butterflies spread their wings – in particular the small, white, UV-reflecting wing scales at the centres of the eyespots – that are the features used as sexual ornaments in the alternating male/female sexual selection of this butterfly. The multiple eyespots on the *ventral* sides of WS *Bicyclus* butterflies – which become reduced and tiny in the DS female (Figs 2.12 and 5.14) – have probably evolved to serve a different function: namely predator deflection. The large eyespots on the ventral surfaces of the forewings are often hidden by the hindwings in *B. anynana*, but become apparent as soon as the butterfly lands and closes its wings and are sometimes displayed when the butterfly senses danger. Moths also do this: flashing a pair of large eyespots that look – to us – like the eyes of an owl. Enough perhaps to frighten any small bird contemplating giving them a peck![48]

The behavioural plasticity of the two forms – WS and DS phenotypes – of *B. anynana* is quite remarkable, but even more surprising is the fact that both the patterns on the wings and the courtship behaviour of the males are determined by temperatures experienced during certain critical periods in their development. For example, high average temperatures (e.g. 27°C) – such as those experienced during a typical WS – during pupal development, and extending into adulthood, determine the rate at which males engage in courtship behaviour. The critical period for determining which type of wing morphology the butterflies develop is also temperature dependent, but occurs slightly earlier, during the late larval stages. Although the exact mechanisms involved in inducing these different phenotypes are not fully understood, recent experiments suggest that environmentally induced changes in steroid hormone levels during early pupal development have a latent effect on the behaviour of the adult. The ability to respond to ambient conditions during development allows this species to adjust to the changing costs and benefits of engaging in courtship and mating during the different seasons.[49]

Heliconius butterflies

Heliconius species are slightly unusual in showing no obvious differences in wing colours or patterns between the sexes. Their appearance 'is primarily adapted for mimicry and defence' according to Professor Chris Jiggins in his 2016 book, *The Ecology and Evolution of Heliconius Butterflies*. Nevertheless, males of *Heliconius melpomene* exhibit strong mating preferences for females with their own type of wing pattern. Although these striking aposematic wing colorations may have evolved as warnings to predators, they also function as mating cues. Some *Heliconius* butterflies, like *Heliconius erato* (Fig. 2.6), have two very different mating strategies: pupal mating (see Chapter 7, this volume) and active courtship (below). *Heliconius* females usually mate only once; most, soon after pupal eclosion. The courtship behaviour of *H. erato phyllis* is a complex and highly variable affair, not at all the stereotypical behaviour seen in some other species. Nevertheless, one behaviour, called *androconial exposition*, occurs in all successful courtships: the male separates the forewings from the hindwings to expose a silver-coloured area on the anterior margins of the hindwings, the location of the androconial scales. The male showers the female with pheromones during a distinctive 'hovering flight', above and behind her, which is essential in terms of her recognizing and accepting the male.[50]

Fourteen so-called *behavioral units* were identified by Andre Klein and Aldo de Araújo, based on observations of *H. erato phyllis* in open-air insectaries in Porto Alegre, Brazil. Seven behaviours were exclusively carried out by males, five by females, and two in which both sexes are active simultaneously. I have summarized these as follows:

Male

- *Inspection in flight:* male approaches alighted female with body directed towards her.
- *Hovering*: male remains in flight over the alighted female (5–15 cm above).
- *Alighting on wings*: male in flight, briefly puts his legs on her anterior wings (<2 s).
- *Slow flapping*: male slowly opens and closes wings.
- *Androconia exposition*: male flaps wings quickly, without opening much, and separates forewings from hindwings, exposing the silver-coloured androconial patch.
- *Abdomen bending*: male bends abdomen towards the female.
- *Touching with proboscis*: male touches several regions of the female with proboscis.

Male and female

- *Flight interaction:* changes in trajectory by at least one of the individuals.
- *Flight pursuit:* the male pursues female in rapid flight, often touching her on the posterior and ventral regions.

Female

- *Posterior vibration:* female separates forewings from hindwings slightly, vibrates hindwings and elevates abdomen, exposing a yellow gland.
- *Contraction:* Alighted and with wings closed, the female approximates both forewings and hindwings, increasing their overlapping area.
- *Rotation:* female turns, changing her orientation to the male.
- *Dislocation*: female departs from the male by walking or brief flights (<2 s).
- *Opening and separation*: female opens wings and separates forewings from hindwings, exposing the anterior dorsal margins of the hindwings.

Female *Heliconius* butterflies may not respond to the courtship attempts of the males, and can adopt a rejection posture, with her abdomen raised vertically and her scent glands exposed. If she has already mated, this behaviour releases an antiaphrodisiac (see Chapter 9, this volume), which deters all but the most ardent males. In some cases, courtship may be brief, or lacking altogether in *Heliconius* butterflies, like *H. erato* and *Heliconius himera*. For example, in most successful courtships in *H. erato phyllis*, the male usually only executes two 'units' of behaviour: *androconia exposition* and *abdomen bending*. The persistent occurrence of androconia exposition in courtships that resulted in copulation suggested to the researchers that a pheromone signal is used and is essential to female acceptance.[51]

Separate observations of mating in this species, in outdoor insectaries in Ecuador were described as follows:

> in our mating experiments with newly eclosed females, mating was often rapid and lacked obvious courtship elements … males flew aggressively toward females, sometimes hovering briefly, before landing and attempting to mate. Females could and often did reject males by spreading their wings and raising their abdomens out of reach of male claspers.[52]

Fritillaries and checkerspot butterflies

Surprisingly little has been written about courtship in this very well-studied butterfly, Edith's checkerspot, *Euphydryas editha* (Nymphalinae, Melitaeini), presumably because it is so brief, or non-existent. There is only one generation per year in this short-lived species, and both sexes live for about 10 days on average. However, the males emerge much earlier (4–8 days) than the females. Fertilizable, i.e. unmated females are thus only available for short periods, and they only mate once. Newly eclosed females remain motionless on the ground in low vegetation for about an hour after eclosion whilst their wings harden. During this time therefore – before they can fly – they are unable to reject courting males, so the first mate to find a teneral female is usually the one who mates with her.[53]

In *Euphydryas* species, the males have little time in which to evaluate the quality of their potential mates, and often have difficulties in distinguishing males from females. Consequently, males spend much of their time chasing each other! One male–male chase was observed to last for 22 min in *Euphydryas anicia*, studied in Colorado, USA, with the roles of pursuer and pursued switching between the two individuals and some mistaken attempts at copulation occurring! Males engage in 'scramble competition' with many individuals – up to eight or nine – scrambling after an unmated female (as observed in *E. anicia*). Males typically only respond to flying objects, and chase females in aerial pursuits, once they have taken off from the ground. Virgin female *E. anicia* are chased for longer than mated females. Courtship is very brief. Once the pair lands, the male moves behind the female, bends his abdomen and they mate, moving into a back-to-back position within a minute of first landing. The small male investment in reproduction – i.e. a spermatophore of less than 2% of bodyweight – may be the reason for the absence of courtship in this species.[54]

Courtship in the Chalcedon checkerspot, *Euphydryas chalcedona*, is fairly similar to that of other nymphalids, including *H. erato*, *Limenitis camilla* and *Junonia coenia*, two of which are described in this chapter. It is a lengthy affair – lasting up to a minute! – compared with other *Euphydryas* species. Males often encounter females in flight; the pair alight and the female walks a short distance with the male following behind her. The male then moves towards, and slightly alongside the female, whilst curling his abdomen around and attempting to insert it between the female's hindwings. Successful copulations occur when the male clasps the female, and she remains still, allowing copulation to continue. Unreceptive females flutter their wings for more than a second, and then crawl or fly away. After mating, which lasts about 77 min on average, males seal the female's copulatory opening with a mating plug (see Chapter 7). Pheromones are not thought to be involved in courtship in *E. chalcedona*, and females may evaluate males on some aspect of their behaviour during the pursuit and courtship phases. However, little seems to be known about mate choice in checkerspots.[55]

Courtship displays can be very brief and difficult to observe in the field. In 1981, Professor Ronald Rutowski and John Schaefer used the then technology – a super-8 movie camera – to record the courtship behaviour of the Gulf fritillary, *Agraulis vanillae* (Heliconiinae, Heliconiini) (Fig. 5.15a, b). The most striking aspect of courtship in this species is a *wing clap* display by the male, as he stands head to head with the female. During these displays, which last for about 11 s on average, the male repeatedly claps his wings open and shut, catching the female's antennae between them. This presumably transfers his pheromone, which stimulates coupling. Prior to alighting next to the female, the male may hover above her for 15 s or so, and she may – or may not – flutter her wings in response. The wing clapping behaviour did not occur in all successful courtships – it

Fig. 5.15. (a) Gulf fritillary or passion butterfly (*Agraulis vanillae*) dorsal view. (Photo by Korall – own work, CC BY-SA 3.0, https://commons.wikimedia.org/w/index.php?curid=7334198) (b) Gulf fritillary or passion butterfly (*Agraulis vanillae*) ventral side, Isla de Margarita, Venezuela. (Photo by Wilfredo R. Rodriguez H. – own work; source materials for this file are available from Commons Archive: Lepidoptera in Isla Margarita.NEF, CC0, https://commons.wikimedia.org/w/index.php?curid=23477372)

was omitted altogether in seven out of the total of 24 – illustrating the spectrum of courtship behavioural sequences that may occur between different pairs. The wing clapping seems very similar to the 'bowing' behaviour performed by the male grayling, *Hipparchia semele* (see Chapter 6, this volume) and to that performed by the closely related, *Argynnis paphia* (Argynnini) (Fig. 5.16).[56]

The courtship of the silver-washed fritillary, *Argynnis paphia* (Heliconiinae, Argynnini), is relatively well known and has been described by several authors. Males seek females using visual cues: bright orange/yellow colours and motion flickering (see experiments by Magnus, described in Chapter 1, this volume). The males rarely make mistakes, but can sometimes be fooled into approaching yellow, fluttering leaves, or even yellow-orange car sidelights! (N.B. yellow pierid butterflies are not of the right colour and do not attract the attention of male *A. paphia*). After approaching the female, the male flies around her in tight circles and, if she is receptive, she takes off and flies in a straight line, unlike the usual undulating flight exhibited by this species. The male then performs a unique pursuit flight, swooping down beneath the female, then darting up in front of her, until he is positioned above, and slightly behind her. The male repeats this behaviour, swooping down and darting up in front of her several times, presumably stimulating her visually and chemically, via the release of his male pheromone. If the female accepts the male, the pair perform a 'wheeling' flight, both circling each other until the female eventually alights on a leaf or flower. The female's antennae are brought into contact with the alar androconial organs in the later stages of this complex courtship sequence. Virgin female *A. paphia* are also said to expose and direct a pair of lateral clubs from her abdomen to chemically stimulate an approaching male. Unreceptive, female silver-washed fritillaries fly away from the males, whilst mature virgin females are said to flutter their wings in a whirring movement.[57]

James A. Scott studied the courtship and mating biology of the dotted checkerspot, *Poladryas minuta* (Nymphalinae, Melitaeini), in Colorado, USA. Courtship behaviour is quite variable in this species, depending on

Fig. 5.16. Silver-washed fritillary (*Argynnis paphia*) Galicia, Spain, worn male.

the age and receptivity of the female. Some courtships are very brief: the male alights behind the female, curves his abdomen laterally, and copulation occurs within 2 s. Other courtships may be substantially longer, up to 2 min, with a number of additional behavioural components. Males may *hover* for a few seconds, a few centimetres above and downwind of the females, beating their wings though a small amplitude. Depending on the response of the female, the male may land and *flutter* his wings more slowly. Females also flutter their wings, whilst the male is on the ground behind her. One behaviour that occurred in many successful courtships, particularly when the female had spread her wings early in the courtship sequence, was called *male nudging*. The male holds his wings about 40° above the horizontal with antennae bent backwards and pushes his head under the female's spread hindwings. This enables the male to position himself alongside the female so that his laterally curving abdomen is in place for copulation. Remarkably, males may repeat the courtship attempts 20 or 30 times before the female becomes quiescent and adopts a receptive body pasture, so that copulation can occur. However, most courtships typically lasted for only 10–15 s.[58]

When male Seminole, or Texan crescents, *Anthanassa texana seminole* (Nymphalinae, Melitaeini), encounter a female on the wing, they briefly spiral around each other, before the female alights on a nearby leaf. The male then 'loops' about 10 cm behind and above her, a behaviour described as being similar to that of the silvery checkerspot, *Chlosyne nyeteis* (Nymphalinae, Melitaeini). Unreceptive females flutter their wings for a few seconds before departing upwards with the male following in pursuit. The pair may descend again and repeat this sequence and, if the female is receptive, she spreads her wings, vibrates them and arches her abdomen upwards. The male then alights alongside her, closes his wings and nudges her by contorting his abdomen. He then moves around for a rear-to-rear coupling, after which they both close their wings, and remain *in copula* for 1–2 h. If they are disturbed, the female flies off, carrying the male.[59]

Recent research by Manuela Pinzari and colleagues in Italy has demonstrated that the marsh fritillary, *Euphydryas aurinia* (Fig. 6.8), has a very simple courtship behaviour. The males immediately attempt to clasp onto the female genitalia in a manner very similar to that reported for some other *Euphydrya* species, including *E. chalcedona*, *E. editha* and *E. anicia* (described above), thus revealing a common courtship behaviour throughout this genus.[60]

Satyrids

Perhaps the best studied butterfly in this group is the speckled wood, *Pararge aegeria* (Satyrinae, Satyrini, Parargina), which is discussed in detail in Chapter 6, this volume. There are however, perhaps not as many detailed observations on the courtship and mating of satyrid species in general. Their predilection for dark or semi-shaded habitats makes studying them quite a challenge in some situations, like tropical rainforests. The mating behaviour of the Ridings' satyr, *Neominois ridingsii* (Satyrinae, Satyrini), was studied by James A. Scott, in Colorado. Virgin females flying over perching males are quickly pursued and the pair soon land nearby. Either sex may flick their wings open (30° to 70° from the vertical), then the male approaches the female and 'touches the end of her abdomen' with his head. The male then spreads his wings and faces the female, head to head. These flicking and nudging behaviours are not, however, essential, as some copulations occurred without them, suggesting the possibility of chemical communication as well. Unreceptive females simply fly away from courting males.[61]

Danaiids

The courtship and mating behaviour of the queen butterfly, *Danaus gilippus* (Fig. 5.17), (Danainae, Danaini, Danaina) was comprehensively investigated by Lincoln Brower and colleagues (see Chapter 1, this volume). The essential element of the courtship in *D. gilippus*, and probably all members of the subfamily Danainae (apart from the monarch) is the act of hairpencilling. The courtship starts with an *aerial pursuit* of the female by the male. He catches up with her, and in a rapid bobbing motion, sweeps his extruded abdominal hairpencils over her head and antennae. This *aerial hairpencilling* phase disseminates a so-called, 'shower of gold': pheromone transfer particles (see Chapter 9, this volume). The female responds by alighting on some herbage, and the male continues to buffet her, in this *ground hairpencilling* stage. The male 'dances' rapidly in front of the perching female, who may flutter her wings in response. The male then retracts his hairpencils and performs a *hovering and striking* behaviour, where he hovers above the female, intermittently dropping down onto her back (or dorsum). Receptive females fold their wings in response to hairpencilling, which probably

Fig. 5.17. Queen (*Danaus gilippus berenice*), underside, Grand Cayman. (Photo by Charles J. Sharp – own work, from Sharp Photography, CC BY-SA 4.0, https://commons.wikimedia.org/w/index.php?curid=47521550)

stimulates the male to *attempt copulation*. The male drops down beside the female and grasps the undersides of her fore- and hindwings with his legs. He then bends his abdomen round and attempts to bring it into contact with the tip of hers, whilst also touching her head and antennae via alternative movements of his own, left and right antennae: *antennal palpation*.[62]

After joining together in copulation, the male opens his wings to about 60°, then quickly closes them – *wing-snapping* – before commencing a *post-nuptial flight*, carrying the female. These flights sometimes took a pair up to a height of 24 m or so, traversing up to 300 m. An identical, post-nuptial flight is carried out by the plain tiger, *Danaus chrysippus* (Fig. 7.5). Courtships do not always proceed as far as copulation however, and either sex can break off during any of the initial phases. For example, females can evade and avoid males both in the air and on the ground (in foliage). Females also *reject* males by not folding their wings dorsally, in response to hairpencilling. Choices can clearly be made by females, and they can drive off males by a variety of evasive movements, using their wings and abdomen. Males can also *desert* females, sometimes departing after spending as much as 20 s courting, before hovering in front of her and flying off. As we have seen in other species, both sexes have the power of rejection, and can break off the courtship at any stage prior to copulation. The density of the local population is said by Brower *et al.* to be an important factor in determining the outcome of a given courtship. Perhaps because there are more individuals to choose from, but also as a result of harassment of the courting couple by other males.[63]

The courtship behaviour of the plain or African tiger, *Danaus chrysippus* (Danainae, Danaini, Danaina), is very similar to that of the queen, *D. gilippus*, but has not been investigated to the same extent. George Stride describes how courting female African tigers, *D. chrysippus*, fly in a jerky manner, consisting of a series of short dashes. The males fly above the female and drop down to brush her head with their hairpencils (also called anal brushes). The male flexes his abdomen forward during each of these 'dips', extruding the hairpencils.[64]

During courtship, the male chestnut tiger, *Parantica sita* (Danainae, Danaini, Danaina), spins vertically around the female with his hairpencils extended. If she responds, he flies slowly in front of her, 'wiggling his wings and displaying fully splayed hairpencils', according to researcher Professor Keiichi Honda and colleagues. There follows a pursuit phase where the male chases the female for a short while before she lands and displays an abdomen-curling acceptance posture. The male lands next to her and copulation takes place, often lasting for more than 12 h. The males have already loaded up their hairpencils with pheromone (danaidone), by rubbing their protruded, half-expanded hairpencils against the alar patches on their wings, prior to courtship. In the related, blue tiger, *Tirumala limniace* (Danainae, Danaini, Danaina) (Fig. 5.18), males also chase females – sometimes in groups forming a chain – but only some 6–7 days after they have emerged from their pupal cases. Courtship in *T. limniace* is reportedly most intense between 7 and 11 days after eclosion and is supplemented by continual flower visiting and nectaring.[65]

Fig. 5.18. Broad blue tiger (*Tirumala limniace*) female, Chiang Dao, Thailand.

Monarch butterflies: hell's angels?

Monarch butterflies, *Danaus plexippus* (Danainae, Danaini, Danaina) (Fig. 5.19) – as was mentioned previously – are rather unusual, in that the males are often able to force themselves on the females during courtship. They engage in 'aerial takedowns' where they knock the females out of the air and onto the ground; there they engage in a sometimes-lengthy struggle. The male tries to grasp the tip of the female abdomen with his abdominal claspers, but the females struggle vigorously and try to avoid mating by making their abdomens unavailable. Struggling females elevate their abdomens upwards by 90°, or curl them between their legs whilst resisting males. Hairpencilling behaviour – see Chapter 9, this volume – is not usually observed in these forced matings, but these abdominal organs are, nevertheless, often extruded during the struggles. Chemical communication is, however, not thought to play a role in these interactions, but cannot be completely ruled out. The females do their best to resist by trying to make their genitalia unavailable, adopting a range of body positions to resist the mating attempts of males. This resistance behaviour is relatively successful, but the males are extremely persistent. The mating drive of the male appears to be so strong that other male monarchs – and even different species – are sometimes tackled in misdirected attempts at mating. For example, one male monarch was seen knocking an American painted lady, *Vanessa virginiensis*, onto the ground and pinning it to the pavement, in one such misidentification.[66]

Fig. 5.19. The monarch (*Danaus plexippus*). (Photo by PublicDomainPictures from Pixabay.)

This coercive mating behaviour occurs on the overwintering grounds of the monarch butterflies, in Mexico, and is thought to be an adaptation – perhaps a recently evolved one – associated with their mass migration. The males have used up their fat reserves during their southerly autumn migration and are unlikely to re-migrate to new pastures in the spring. So, this unusual behaviour could be an act of desperation, their only chance to mate before dying. It is, however, a classic example of a conflict of interest between the sexes, with females resisting the persistent advances of males. Despite being generally stronger than the female, male monarchs are only able to force a mating in about a quarter (26%) of attempts if the female puts up some form of resistance. If the females offer no resistance, a much higher number (62%) of attempts result in successful coupling. Researchers were unable to determine, however, with any degree of confidence, whether the females were resisting the males right up to the end, or whether they did finally acquiesce. In other words, were some males better at forcing females into mating, or were they more attractive to the females – who eventually yielded to them – in some way? The ratio of males to females has been changing over recent decades, at the Mexican overwintering sites, with a trend towards male-biased populations: c. 53% female in the 1970s, but only c. 43% female in 2008.[67]

This apparently aggressive behaviour by male monarch butterflies has resulted in quite a lot of emotional or anthropomorphic language, as researches have tried to understand and describe the forced matings. Miriam Rothschild famously described the males as 'thugs' and 'hells-angels' and other researchers have also wondered about the nature of such forced matings. David Smith questions whether female monarchs lack choice to the extent reported, and reports that refusal is routine and exercised by females c. 90% of the time, in his experience. The male's penis (aedeagus) is long and sharp and the female's orifice (ostium bursae) is a toughened funnel, seemingly well adapted to receiving such a 'weapon', so they both appear to be relatively well prepared for this type of rough and tumble, even though forced copulations can produce damage and lead to premature deaths. Monarchs also have tougher wings than other danaiids, so they seem to have evolved several adaptations along with their supposed violent behaviour. It has been suggested that male adaptations that inflict harm on females may arise as negative side-effects from other adaptations, which are otherwise beneficial. In conclusion, it is not altogether clear just how harmful this coercive behaviour is, in *D. plexippus*.[68]

Other examples of forced matings and aerial 'takedowns'

There are other species that have a similar behaviour to that of the monarch, i.e. with apparent forced matings and no observed courtship. These include: the leopard lacewing, *Cethosia cyane*; the tawny coster, *Acraea terpsicore* and many *Acraea* species (Fig. 5.20) including *Acraea andromacha*, *Acraea encedon*, *Acraea quirinalis* and A. *epaea*; the clearwing swallowtail, *Cressida cressida*; and some *Actinote* species. In the heliconine species, *Actinote pellenea* (Heliconiinae, Acraeini), for example, males grab the females in mid-air, and both fall to the ground. Copulation lasts for about an hour and the pair fly off, out of the male's territory, to the cover of dense vegetation. In these species, however, it is not altogether clear to what extent the female may be acquiescing in the forced mating. Researchers have speculated that the female genital apparatus (*sinus virginals*) must be everted to affect a mating and this cannot be forcibly induced by the male, suggesting a degree of cooperation. This illustrates the difficulties of attempting to discern the extent of coercion and cooperation in such interactions. It has also been suggested that the presence of female pheromones obviates the need for courtship in *Acraea* species, such as *Acraea andromacha andromacha*. Mating in the rare Brazilian heliconiid, *Actinote zikani*, also occurs without any obvious courtship occurring. A patrolling male follows a female, flying for a short distance (2–5 m) in a straight line, about 2 m above the ground, and then spiralling down to the ground. The male grasps the abdomen of the female with his valves, apparently forcing a copulation.[69]

In the east African butterfly, *A. quirinalis*, the male lands on top of a resting female and knocks her to the ground. He grasps the female with

Fig. 5.20. Elegant acraea (*Acraea egina*), The Gambia.

his legs, and bends his abdomen *downwards*, from a position on top of the female. It is very unusual, if not unique, for male butterflies to commence courtship from a position *on top* of the female, like this. In *Acraea* species, the male hangs downwards from the female during copulation, with his abdomen twisted round 180° from the vertical. Male *Acraea* butterflies also attach a mating plug or sphragis to prevent the female from remating. Males of *A. andromacha andromacha*, observed in New Caledonia, were seen to pounce on females in mid-air, fall to the ground and commence copulating – the male covering the female with his wings outstretched – without any courtship occurring. Although multiple matings sometimes occur in *Acraea* species, they are usually monogamous, as a result of the attachment of sphraga (see Chapter 7, this volume). In the Japanese luehdorfia, *Luehdorfia japonica* (Fig. 8.6), males also capture unmated females in mid-air, fall to the ground and copulate. No courtship or mate refusal behaviour was observed in this species, and copulations lasted from 32–154 min (n = 7).[70]

In the clearwing swallowtail, *Cresdida cressida* (Papilioninae), also known in Australia as the big greasy, females are captured on the wing in an 'aerial takedown'. Described by Dr Albert George Orr of Griffith University, Brisbane:

> mating occurs when the male approaches the female from above and stoops [swoops?] like a hawk, diving at great speed and striking her with considerable force, carrying her off in a single motion.

If the female acquiesces in the mating, copulation may occur in mid-air, but more frequently she struggles and tries to reject the male, and they both fall to the ground. What happens next usually depends on the relative sizes of the two individuals. Large females may be able to escape from smaller males and fly away, but if the male manages to force a mating, the pair fly off *in copula* on a post-nuptial flight. They can remain together for between 17 h and 3 days, a remarkably long period of time, during which the male both forms a spermatophore and secretes a large, horny sphragis. The latter is a considerable investment, representing up to 7% of the male's body weight, a much greater investment than that of the spermatophore in this species. Once the sphragis has hardened, it acts as both a physical barrier to the female copulatory opening, and a visual deterrent to other males; a signal that she has already mated.[71]

The fact that there are at least three possible outcomes in these encounters between male and female clearwing swallowtails, *C. cressida* – i.e. acceptance, forced mating or escape – illustrates the diversity of behaviour that can exist in a species. That some encounters are marked by a lack of conflict suggests that there are circumstances when it is in the interests of both sexes to mate. Where force is used, we must conclude that it is not in the interests of the female to mate, but she may nevertheless have some post-copulatory, cryptic ways and means of levelling up the contest. Even in situations where it is in the interests of both sexes to mate, there are still considerations of exactly how much in terms of resources to invest in a given encounter. Such decisions may be based on whether each individual has mated before, and how much time has elapsed. Males of the North American snow apollo, *Parnassius clodius* (Papilionidae, Parnassiinae, Parnassiini), also engage in what has been described as a 'rape' type of mating, in which the males engage the females and copulate by 'brute force'. American researchers David McCorkle and Paul Hammond described this behaviour, thus:

> *P. clodius* males, upon sighting a female, chase rapidly after her and literally attack her from behind. Upon grasping the female in mid-air, the male and female drop abruptly to the ground. The female then lies limply on the ground, often with wings crumpled in the vegetation, while the male sits on top of her in copulation. Parnassius have extremely tough, tear-resistant wings, which may be an adaptation to this rough mating behavior.

The authors considered that this so-called, 'rape' mating behaviour was a 'primitive' trait.[72]

Papilionids

Alexandra Freeman investigated mate choice in the mocker swallowtail, *Papilio dardanus* (Papilionini), and observed courtship behaviour of butterflies placed in flight cages, within a temperature-controlled greenhouse. Both males and females approached each other in flight, flying up underneath the other butterfly and brushing them with the upper surfaces of their wings. When males did this, and the female was receptive, the pair would land, and mating would occur. Both males and females curved their abdomens around in order to make genital contact. After the male had clasped onto the female, he became motionless, and was carried by the female in flight if the pair were disturbed. They remained *in copula* for c. 1 h before separating. Unreceptive females exhibited a fast, fluttering flight, or, rarely, landed and flapped their wings vigorously to prevent the male from approaching. If continually harassed by several males, the female would drop to the ground and remain motionless. Males were generally uninterested in mating on the day that they emerged, but females would sometimes mate on the day they emerged, despite being generally less active. Both sexes were, however, ready to mate on the second day after eclosion, and by the third day, both sexes pestered each other vigorously.[73]

The reproductive behaviour of *Papilio peranthus* (Papilionini) was studied at the Gita Persada Butterfly Park, Lampung, Indonesia, by Nismah Nukmal and colleagues, from the University of Lampung. The four sequential stages involved in mating – locating, recognition, courtship and copulation – were broken down into discrete behavioural acts (*italics*), the durations of which were recorded in experiments involving different combinations of males and females. Male *P. peranthus* encounter females whilst patrolling in an *unorientated* manner, but then pursue the female in an *oriented* flight. The female lands, and the male carries out a *circling flight*, before *hovering* above her, *touching* her lightly with his wings, or the tip of the abdomen. This courtship phase is relatively brief, lasting 1–4 min. The male then *mounts* and *grasps* the female, making genital contact with his claspers, and the pair remain *in copula* for about an hour; the male hanging below the female. Males of the common Mormon, *Papilio polytes* (Fig. 5.21), hover in front of females during courtship, displaying only their dorsal wing surfaces. In *Papilio glaucus*, there is a high incidence of courtship solicitation by the females, although only a relatively small proportion (17%) of courtships involving virgin females led to copulations. There was also a reported lack of antennal contacts between males and females in this species.[74]

Birdwing butterflies

The courtship patterns of male *Troides* butterflies (Papilioninae, Troidini) fall into two categories, depending on whether the female is a virgin or has already mated. When a newly eclosed virgin female is encountered, the male performs a brief (c. 30-s) display, rapidly fluttering his wings before alighting next to her and engaging the female with his claspers. If the female is receptive – according to Michael Parsons, who studied the courtship behaviour of the oblong-spotted birdwing, *Troides oblongomaculatus papuensis* (Fig. 5.22), in Papua New Guinea – she opens her genital aperture by raising her abdomen. Coupling then occurs almost immediately, and the pair remain *in copula* for about 5 h. Female birdwing butterflies are much larger than the males, and the female carries the male. A different sort of behaviour occurs when a male *T. oblongomaculatus papuensis* encounters a female who has already mated. In these circumstances, the courtship display by the male is sustained and repeated, but it seldom results in mating. The courtship behaviour of *T. oblongomaculatus papuensis* is probably typical of all *Ornithoptera* and *Triodes* species (both genera are in the tribe Troidini). If a male encounters a female in flight,

Fig. 5.21. Common Mormon (*Papilio polytes* f. *cyrus*), female non-mimetic form, Thailand.

Fig. 5.22. Oblong-spotted birdwing (*Troides oblogomaculatus papuensis*) male, Manki Range, Bulolo, Morobe Province, Papua New Guinea. (Photo by Didier Descouens – own work, CC BY-SA 4.0, https://commons.wikimedia.org/w/index.php?curid=16187799)

he may trace out an elliptical trajectory as he flies around and in front of her, brushing her antennae with his androconia – which are located in a pocket along the inner margins of his hindwings – and hovering above her. The androconial scales may detach and stick to the female's antenna. Female *Ornithoptera* species accept a second mating attempt if the mating plug received from the first mate is still soft, which is why the males of some species guard the female for a while, once they have mated (see 'Mate Guarding' in Chapter 6, this volume). A small proportion (10–20%) of Richmond birdwing, *Ornithoptera richmondia* (Fig. 5.23) females in Queensland, Australia, contained more than one spermatophore, showing that mating plugs can be overcome (or undermined!).[75]

Multimodal Displays – the Components of Courtship

Males of many species, rather than waiting for females to come to them, i.e. fly past, actively search for mates amongst vegetation, where females may be resting in certain characteristic positions. Most female lycaenids sit facing downwards, with their wings folded (Fig. 5.24), which makes them more attractive to males, than ones facing upwards. Males respond to motionless females that are adopting this position, but sexual behaviour, as shown for the pale grass blue, *Zizeeria maha argia*, is released by the colours and patterns on the ventral wing surfaces. Indeed, male *Z. maha argia* exhibit little or no sexual discrimination and will adopt a courtship posture in response to the ventral surfaces of either sex. Colours, such as blue – in blue-winged species such as *Graphium sarpedon* (Fig. 10.35) – are important visual cues in terms of eliciting courtship behaviour. Sometimes the specific colours and patterns of the hindwing underside of

Fig. 5.23. Richmond birdwing (*Ornithoptera richmondia*) plate from *Icones Ornithopterum* (1898–1906). (By Robert Henry Fernando Rippon (died January 1917) [Public domain].)

the females serve to attract the male. Or to put it another way, they match his search image.[76]

Courtship can often be broken down into separate aerial and ground phases, with an intermediate *air-ground* phase in which the male performs certain aerial behaviours – such as hovering or alighting on the female's wings – over or above the female, who has already landed on a substrate

Fig. 5.24. Lesser grass blue (*Zizina otis*) perching facing downwards, Thailand.

of some sort. The air-ground phase of courtship in *Dryas iulia alcionea* is the most complex part of the courtship process. As stated above, the initial attraction is probably visual in both sexes, but chemical signals become more important as the pair get closer together. Nevertheless, mating displays usually contain a multitude of signals that are sensed by the receivers in a variety of different ways, including light, sounds, smells and tastes. Intriguingly, recent research suggests that courtship provides an opportunity for both sexes to effectively taste each other, in order to determine the epicuticular composition of different body parts. In other words, using their legs and antennae to interrogate, via chemosensory contact, the chemical composition of their suitor. Different, closely related species can coexist together as a result of the diversity of such signals – a sort of multimodal niche separation – amongst species occupying the same habitat.[77]

A fascinating element of butterfly courtship is that, in most species, females can resist and reject attempts by the male to mate. In other words, males cannot force a mating if the female is unreceptive. For example, copulatory success in the Julia butterfly, *Dryas iulia* (Fig. 6.2) depends on acceptance by the female: males that engaged in prolonged courtship activity were no more successful than less persistent males. There is, however, one well-known exception, the monarch butterfly (see above). This ability to decide whether to mate with a potential suitor gives the female the ability to select mates in a manner that maximizes her own individual fitness and, therefore, her choices are a key driver in the evolutionary process. The male also has a choice about whether to terminate a copulation prior to passing on an ejaculate, and there are also mechanisms by which choices can in effect be made by the female, during and after copulation (see Chapter 8, this volume). In summary, females decide *whether* mating occurs, and males decide *when* to terminate it (once copulation has commenced).[78]

Determinants of Success or Failure

Unsuccessful courtships are frequently observed in most species, and the process can be terminated at any stage of mating, including whilst *in copula*. For example, *Callophrys xami*, hairstreaks may choose to terminate mating after just a few minutes of coupling. It is not easy for observers to decide which sex initiates termination in many cases, and it is likely that both males and females can evaluate their partner and make choices about whether to continue, at any stage of the courtship. For example, some male cloudless sulphurs, *Phoebis sennae*, were observed to terminate their courtship even though the female appeared to be receptive. In other cases, the female may have been unreceptive, or the male may have been lacking in terms of persistence! Males may vary the amount of time they spend in courtship activities, depending on the receptivity of the females, as shown for *Colias philodice* and *Colias eurytheme*: males spent less time involved in courtship, early in the day, when – highly receptive – virgin females were emerging.[79]

In some species, courtship is a lengthy affair, with the male following the female for hours before mating occurs. This includes butterflies such as the peacock, *Inachis io*, and red admiral, *Vanessa atalanta*, for example, where copulation is seldom observed. Interestingly, *successful* courtships took longer in the orange sulphur, *Colias eurytheme* (Fig. 2.15) than unsuccessful ones. Perhaps because the unsuccessful males got the 'brush off' from the female prior to finishing their courting, which requires a certain minimum amount of time, or effort. There is some evidence to suggest that the courtship performance of a male is a reliable indication of his quality; for example, in terms of physiological factors such as his metabolic efficiency. In other words, there are 'condition-dependent' characters that are genuinely indicative of quality, and on which the female bases her choice of a mate. Certain genotypes have fitness-related advantages in terms of mating success. Courtship flights are also a way in which females can view, and presumably assess such factors, including the coloration on the ventral surfaces of the male's wings.[80]

The release of pheromones by the male is a crucial element in courtship and their effect is to increase the receptivity of the female. Key aspects of the courtship are carried out by the male with the aim of bringing his pheromone-emitting structures into contact with the female, so that she can receive the chemical messages they convey. The final stage of courtship has been described as 'chemical contact' when species recognition occurs via contact between the antennae of the male and the body of the female. The whole courtship process has been described as involving four main phases: random flights; visual recognition; short-range olfaction; and chemical contact (RVOC for short). See Box 5.1 on courting Iberian marbled whites (*Melanargia lachesis*). Whatever the complexity, or otherwise, of courtship behaviour, it is always contingent upon local environmental conditions as well as the availability of receptive females. Courtship activity relies, especially, on adequate light levels and sufficiently high temperatures for both flight and mating; although the latter may proceed at a slower rate at lower temperatures. The optimal temperature for mating by male buckeye butterflies, *J. coenia*, for example, is reportedly, 32–34°C, i.e. very high, but light levels are also critical, and little mating occurs at low light levels, even if temperatures are optimal.[81]

Butterflies can use bright sunlight to increase their thoracic temperature, enabling them to fly even when ambient temperatures are suboptimal. The more active a butterfly, the more likely it is to mate, so light levels – the intensity of solar radiation – can affect mating rates, e.g. via increased courtship activity, and perhaps even female choice (for more active males). Mate-locating activities by male *P. rapae crucivora* butterflies in Japan correlated best with sunlight intensity when air temperatures were relatively constant. The timing of mate-locating activities is, however, usually determined by the availability of receptive females, which frequently eclose in the early mornings. Newly emerged female *P. rapae crucivora* remained close to their cast-off skin until they were ready to take off, c. 2.5 h after eclosion, depending on thermal conditions. The rate of emergence of females, and the rate at which they become available and apparent to males via an initial flight, also depend on ambient temperatures.[82]

Notes

[1] Sappington and Taylor (1990a) suggested that acceptance or rejection may depend on 'internal thresholds', perhaps involving several 'hierarchically arranged cues', including the quality or quantity of the courtship pheromone(s). Courtship displays are a form of advertising (Quigley and Fusani, 2018).

[2] Wiklund (1977), Alexander *et al.* (1997), and Quigley and Fusani (2018).

[3] Rutowski (1981a) and Burk (1982). Male–female interactions may in fact involve two-way avenues of communication, stimulation, persuasion and influence (Rodríguez, 2015).

[4] Eberhard (1991, 1994).

[5] Scott (1973a, 2014) and Henning (1990).

[6] Shapiro (1977), Silberglied (1977), Rutowski *et al.* (1984c), Imafuku and Kitamura (2015) and Li *et al.* (2017). There are both 'differences and similarities' between the mating systems of unrelated butterflies (Alcock, 1988).

[7] Scott (1973a). Both sexes exhibit mate choice and rape is impossible, according to Knüttel and Fiedler (2001). Male small coppers, *Lycaena phlaeas daimio*, are unable to discriminate between the sexes in aerial interactions (Suzuki, 1976).

[8] Mega and Araújo (2010), and Nieberding and Holveck (2017).

[9] Ohtani (1985) and Yamamoto (1981).

[10] Lundgren (1977), and MacLean and MacLean (1990).

[11] Häuser (1993), Knüttel and Fiedler (2001) and Bálint *et al.* (2012).

[12] Lundgren and Bergström (1975) described the courtship sequence in *Lycaeides argyrognomon*, now known as *Plebejus argyrognomon* (Lycaenidae).

[13] Measures (1976).

[14] MacLean and MacLean (1990) and Schurian and Fiedler (1996).

[15] Scott (1974b): *Lycaena arota* males vibrated their wings whilst holding them c. 60° from the vertical (c. 120°).

[16] Cordero (1993).

[17] The two species studied by Gorelick (1971) were *Callophrys dumetorum* (Boisduval) and *Callophrys viridis* (Edwards). Scott *et al.* (2012) describes the modern nomenclature for these species.

[18] Scott (1974d) carried out some 'preliminary experiments' on *Hypaurotis crysalus*, which suggested that males are attracted to the violet colour of the wings, rather than the orange spots (see Fig. 5.4). In this short-lived species, males live for c. 5 days, females c. 9 days.
[19] Lea (2006).
[20] Dinesh *et al.* (2010) and Dinesh and Venkatesha (2013).
[21] Silberglied and Taylor (1978), Rutowski (1977a, 1979, 1981a, 1983a, 1984a), Silberglied (1984), Shapiro (1988), Watanabe *et al.* (1997), Hasegawa and Watanabe (2008), and Watanabe and Irie (2011).
[22] Watanabe and Irie (2011).
[23] Rutowski (1977a).
[24] Obara and Hidaka (1968), Rutowski (1977a, 1989) and Scott (1986a).
[25] Rutowski (1983b), Imafuku *et al.* (2000) and Daniels (2007).
[26] Rutowski (1981a).
[27] Scott (1975c).
[28] Suzuki *et al.* (1977), Tigreros *et al.* (2014) and McQueen and Morehouse (2018).
[29] Courtney (1980)
[30] Ohtani (1985) and Ohtani and Yamamoto (1985).
[31] Mitter *et al.* (2011) and Sáfián (2015). See African butterfly database: https://www.abdb-africa.org
[32] Scott (1973b).
[33] Shapiro (1965), Scott (1973b, 1973c). N.B. *Notamblyscirtes simius* was originally called *Amblyscirtes simius.*
[34] Shapiro (1965).
[35] Shapiro (1977) and Dana (1991).
[36] Benyamini (1992).
[37] Stallings *et al.* (1985).
[38] Stallings *et al.* (1985).
[39] Three hundred courtship sequences, including 25 where matings took place, were observed by Pivnick *et al.* (1992). Matings were recorded by Pivnick and McNeil (1985). The European skipper was first introduced into Canada, near London, Ontario, c. 1910, as overwintering eggs on imported seeds of Timothy grass (Layberry *et al.*, 1998).
[40] Rutowski *et al.* (1997). N.B. *Libytheana bachmanii* is synonymous with *Libytheana carinenta*.
[41] Scott (1975b), Hafernik (1982), and McDonald and Nijhout (2000).
[42] Scott (1986b).
[43] Rutowski (1992). Male courtship behaviour in *H. bolina* enhances the perceived brightness of the UV signal (White *et al.*, 2014) consistent with the preference of the female for brighter UV males (Kemp, 2007a).
[44] Stride (1956, 1958) and Owen (1971).
[45] Emmel (1972), Fosdick (1973), Taylor (1973a), Silberglied *et al.* (1979), and Preston-Mafham and Preston-Mafham (1988). Adrian Hoskins also described mate rejection behaviour in *A. amathea*, in which an already-mated female 'rocks her erect wings continually from side to side, to indicate her lack of interest' in the male (https://www.learnaboutbutterflies.com).
[46] Breuker and Brakefield (2002), Robertson and Montiero (2005), Nieberding *et al.* (2008), and San Martin *et al.* (2011).
[47] Ng *et al.* (2017) found that males were reluctant to mate with females that did not have visible dorsal eyespot centres, i.e. the wing ornaments were essential for mating.
[48] Monteiro *et al.* (2015) and Bhardwaj *et al.* (2017).
[49] Nieberding *et al.* (2008), Bear and Montiero (2013) and Bear *et al.* (2017). Mating costs are probably higher in the dry season, when the males give females a higher-quality spermatophore and, as a result, live shorter lives (Prudic *et al.*, 2011).
[50] Klein and de Araújo (2010).
[51] Klein and de Araújo (2010).

[52] McMillan *et al.* (1997).
[53] Moore (1987) and Odendaal *et al.* (1985, 1989, 1990).
[54] Male *Euphydryas anicia* tend to chase any object of roughly the same size as a *Euphydryas* butterfly (Odendaal *et al.*, 1990), which can lead to the 'incidental' formation of male aggregations (Odendaal *et al.*, 1988).
[55] Rutowski and Gilchrist (1987), Dickinson and Rutowski (1989), and Ehrlich and Hanski (2004).
[56] Rutowski (1980a, 1984b) and Rauser and Rutowski (2003). Rutowski and Schaefer (1984) filmed both successful and unsuccessful courtships in *Agraulis vanillae*, using a Beaulieu 4008 ZM 2 super-8 movie camera.
[57] Magnus (1950, 1958), Bastock (1967), Treusch (1967), Freeman (1998) and Ockenfels *et al.* (1998).
[58] Scott (1974c).
[59] Based on Ross (2002).
[60] Pinzari *et al.* (2018), Rutowski and Gilchrist (1987), Scott (1986a) and Odendaal *et al.* (1990).
[61] Scott (1973c).
[62] Brower *et al.* (1965).
[63] Brower *et al.* (1965).
[64] Stride (1958).
[65] Honda *et al.* (2016) and Li *et al.* (2015).
[66] Pliske (1975a), Hill (1976), Winter (1985), Frey *et al.* (1998; *pers. observ.*), quoted in Solensky (2004), Boppré (1993), Frey (1999) and Smith (2014).
[67] Hill *et al.* (1976), Frey (1999), Solensky and Oberhauser (2004) and Davis and Rendón-Salinas (2010).
[68] Rothschild (1978), Rothschild and Marsh (1978) and Smith (2014). Brower *et al.* (2007) wondered whether forced matings were a newly evolved 'wrong' or a long established 'rite'. Harmful male traits may arise as 'negative pleiotropic side effects' of adaptations that produce other advantages (Morrow *et al.*, 2003).
[69] Epstein (1987), Orr (1995, 1999a), Francini *et al.* (2005) and Carvalho *et al.* (2016).
[70] Epstein (1987), Matsumoto (1987), Tsubaki and Matsumoto (1998), and Sourakov and Emmel (1997).
[71] Orr and Rutowski (1991) and Orr (1999a).
[72] McCorkle and Hammond (1985).
[73] Freeman (1998).
[74] Krebs (1988), Nukmal *et al.* (2017), and Low and Montiero (2018).
[75] Parsons (1983), Orr (1999b), and Orr and Kitching (2010).
[76] Wago *et al.* (1976), Wago (1977), Pelmyr (1982) and Kato and Yoshioka (2003). The blue pigment in *G. sarpedon* wings has been identified as sarpedobilin, a kind of bile pigment (Kato and Yamada, 2001).
[77] Mega and Araújo (2010). Mating displays are composed of 'multiple signals in multiple sensory modalities' (Westerman *et al.*, 2014; Heuskin *et al.*, 2014).
[78] Wiklund *et al.* (1993), Mega and Araújo (2010), and Wiklund and Kaitala (1995).
[79] Rutowski (1979), Marshall (1982a) and Cordero (1993).
[80] In *Polygonia c-album, Aglais urticae, Inachis io* and *Vanessa atalanta*, males can follow females for hours before mating (Wiklund, pers comm., in Bergman, 2011), Watt *et al.* (1986) and Papke *et al.* (2006).
[81] Andersson *et al.* (2007), McDonald and Nijhout (2000) and Westerman *et al.* (2014). Hidaka (1973) used the acronym 'RVOC' to describe the courtship process.
[82] Rutowski *et al.* (1996), Hirota and Obara (2000), Hirota *et al.* (2001) and Liao *et al.* (2017).
[83] Sappington and Taylor (1990a) suggested that acceptance or rejection may depend on 'internal thresholds', perhaps involving several 'hierarchically arranged cues', including the quality or quantity of the courtship pheromone(s). Courtship displays are a form of advertising (Quigley and Fusani, 2018).

6 Body Language

A variety of stereotypical behaviour has evolved in butterflies, which enables non-receptive females to curtail the attentions of courting males; typically in virgin females just after emergence (eclosion), and for a period after mating. The advantages of such signals are that they protect the female from the negative effects of male harassment, and they also save the male from wasting time and energy on fruitless courtship attempts. Refusal postures are, however, not universal, and are completely lacking in some species. Acceptance postures are also widespread, but – like refusal behaviours – vary in nature, from species to species, although certain elements – like lowering the abdomen – are commonly observed. The same posture may mean different things in different species however, or in different circumstances. Males in some species, are also able to signal the fact that they *are* males, to other males, over a distance, which saves them from having to land and investigate the sex of a conspecific individual at close range. There are also many other movements and behaviours – like proboscis waving, wing clapping and pretending to be dead – which occur in many different species, that I have grouped together in this chapter.

Refusal and Rejection

A female who has already mated and wants to put a stop to further attentions or harassment by males has several options. In some butterflies, like the pierids, there are stereotypical behaviours – the mate-refusal posture – which, as well as being a potent visual signal, also creates a physical barrier that effectively prevents the male from attempting genital contact. The wings are pressed downwards, the abdomen is elevated upwards – 45–60° above the body axis – and the genitalia are extruded (Fig. 6.1).

© Raymond J.C. Cannon 2020. *Courtship and Mating in Butterflies: Reproduction, Mating Behaviour and Sexual Conflicts* (R.J.C. Cannon)

Fig. 6.1. Green-veined white *(Pieris napi)* mate-refusal posture by female on leaf. Male hovering above, Galicia, Spain.

Nevertheless, males in some species persist in their attempts at courtship, in the face of such signalling of non-receptivity by females, as rejection can sometimes change to acceptance. However, unreceptive females may fly away if males remain insistent. Females also adopt mate-refusal postures when courted by males of another, closely related species. For example, female Asian green-veined whites, *Pieris melete*, exhibit mate-refusal postures to courting *Pieris rapae* males, but become receptive to conspecific males, perhaps seduced by a species-specific sex pheromone. If all else fails, females can simply disperse to other habitats to avoid persistent harassment, particularly in high-density populations.[1]

In the Julia butterfly, *Dryas iulia alcionea* (Fig. 6.2a, b), which is found in South America, females react to persistent males attempting to alight on their wings by kicking them! Or if this fails, they fly away. The males often persist however, grabbing the females by the wings and knocking them to the ground. The rejection posture by female *D. iulia* involves pressing her wings against the substrate on which she has alighted, and raising her abdomen – like pierids – which prevents the male from attempting copulation. Males can intensify their courtship in response to non-receptive signals by females. A sudden shutting of the wings and lowering of the abdomen signals receptiveness, and this posture allows the male to approach and begin copulation, bending his abdomen towards that of the female, starting from a position alongside her.[2]

In some species, such as the common eggfly, *Hypolimnas bolina* (Fig. 2.1), females may repeatedly resist male attempts to copulate – i.e. flying off

Fig. 6.2. (a) Julia butterfly (*Dryas iulia*), Iguazu Fall, Argentina. (b) Julia butterfly (*Dryas iulia alcionea*) dorsal and ventral view. (From Muséum de Toulouse [CC BY-SA 4.0 (https://creativecommons.org/licenses/by-sa/4.0)].)

when the male curls his abdomen towards them – before finally accepting the male's advances, after three or more tries, and allowing him to mate. Virgin female orange tip butterflies, *Anthocharis cardamines* (Fig. 3.5), can also exhibit an elevated abdomen, mate-refusal display, but this behaviour does not last long and is quickly followed by mating. In this example, the raised abdomen is more a way of *detaining*, rather than *rejecting*, courting males. Virgin females are courted much more persistently by males than mated females, which suggests that male butterflies can somehow tell whether a female has already mated. Male *A. cardamines* spend about 60 s courting receptive virgin females, but only a few seconds before giving up on unreceptive females that have already mated. Female graylings, *Hipparchia semele* (Fig. 6.11a, b), have also been observed by Enrique García-Barros in Spain raising their abdomens at an angle of c. 30° to the perpendicular, whilst keeping their wings half-folded. Wing flapping is the more common mate-refusal posture in this species.[3]

In the Japanese small copper, *Lycaena phlaeas daimio*, females can curtail courtship attempts by males simply by closing their wings and remaining still. This wing-closing behaviour helps to protect females from physical damage and deters unwanted attention by males; females usually only mate once in this species and the persistent attentions of courting males can be harmful. The closure of the wings effectively hides the highly attractive orange patterns on the dorsal surfaces of the females' wings (Fig. 6.3) and makes them considerably less visible to patrolling males. A high proportion of newly eclosed, virgin female *L. phlaeas* exhibit wing-closing behaviour on the day of their emergence, but the behaviour declines markedly over the following 2 days as they become more receptive. Turning face to face with the courting males, and wing fluttering, were also considered to be mate-refusal behaviours in a minority of unsuccessful courtship attempts involving *L. phlaeas*.[4]

In other species, however, a closed-wing posture can signal receptivity. In the pale grass blue butterfly, *Pseudozizeeria maha argia*, males are attracted to individuals of both sexes if they are resting motionless with closed wings. Odours, or pheromones, determine whether the courting male is repelled – by other males – or attracted, towards a female. Females can respond in two ways: remain motionless and allow the males to mate with them, or exhibit a flutter response, which in this species serves as the mate-refusal behaviour. Large skippers, *Ochlodes sylvanus* (Fig. 6.4), also exhibit similar behaviour: mated females flutter their wings and sometimes raise their abdomens to deter courting males. The artist David Measures, in his beautifully illustrated 1976 book on butterflies, described the 'no' signal for female skippers as being a 'violent whirring of the wings'. He also described the male skipper running across vegetation 'like a water hen' chasing the female 'who stops now and then to let him get near'! When the female is ready, she stops and stands still, writes Measures, 'the signal that she is ready to accept him'.[5]

Some species, like the Chinese windmill, *Atrophaneura alcinous* (Fig. 9.6) – and wood white butterflies (see below) – lack any overt mate-refusal postures, but, in the case of *A. alcinous*, can signal their receptivity by exposing their abdomens to approaching males. Similarly, receptive female eastern pale clouded yellows, *Colias erate* (Fig. 6.5), expose their

Fig. 6.3. Small copper (*Lycaena phlaeas*), Galicia, Spain.

Fig. 6.4. Large skipper (*Ochlodes sylvanus*) nectaring on a dog rose showing sex brand on underside.

Fig. 6.5. Eastern pale clouded yellow (*Colias erate poliographus*). (Photo by 池田正樹 (talk)masaki ikeda – 池田正樹, CC BY-SA 3.0, https://commons.wikimedia.org/w/index.php?curid=4731804)

abdomens towards courting males, but signal their unreceptivity by hiding their abdomens, or flying away in an ascending flight. It is physically impossible for males to mate with females adopting an unreceptive position, so copulation only occurs if the female chooses to expose her abdomen towards the male, perhaps after assessing his suitability, vis-à-vis his age or other factors. Ascending flights – lasting for 17 s on average – also occur between males and mated females of the *Colias eurytheme–Colias philodice* species complex, with the females rising to 20 m above the ground, to curtail courtship attempts by the male. The females initiate the ascending phase, although many males break off before this occurs. In this way, a female can avoid wasting time in unwanted courtship and spend time laying her already fertilized eggs instead.[6]

Another way that females can avoid harassment by males is to look like them. In some polymorphic species, such as the African swallowtail, *Papillio dardanus* (Fig. 2.20), male-like female morphs exist, which escape the unwanted attentions of males by mimicry. The price these so-called andromorphs pay is that they suffer increased predation because they lack the protection afforded by Batesian mimicry, which the most common female morphs benefit from. The fact that such male-like female forms have evolved, however, emphasizes the fact that harassment by males has a significant negative effect on mated females, to the extent that they, in effect, trade off potential predation against a hassle-free life![7]

The dotted checkerspot, *Poladryas minuta* (Nymphalidae, Melitaeini), has a rather unusual rejection behaviour, which females use to shake off the unwanted attentions of males. Although rarely performed, unreceptive females slowly fly vertically upwards, for about 3 m, and then drop rapidly downwards and fly away. This rejection dance can be repeated if the male persists in courting. The vertical manoeuvre is very similar to that which occurs between two perching males when they encounter each other. On the ground, *P. minuta* females can also flap their wings in a rejection dance, followed by spreading the wings and raising the abdomen to exclude the male. However, determined males can sometimes overcome moderately unreceptive females by persistent hovering, nudging and attempting copulation. Unreceptive female mountain checkered skippers, *Pyrgus xanthus* (Hesperiidae), may also fly up in the air for about 1 m, repeatedly, until the male departs. *Papilio glaucus* (Fig. 9.5) females also reject males by dropping out of the air, in a 'relaxed freefall' into the vegetation below. If the male follows them down, they close and depress their wings, also pressing their abdomen against the substrate, to avoid genital contact.[8]

Playing possum: thanatosis

Another, slightly more drastic mate-refusal tactic is adopted by some satyrine butterflies, like the speckled wood, *Pararge aegeria* (Figs 3.24 and 3.25), where the females reject mating by 'playing possum', i.e. feigning death. They just close their wings and pretend to be dead; if the male persists, the female lets go of the substrate on which she is standing and falls to the ground. In some cases, even this is not enough to deter the male, who will pursue the female onto the ground and persist in trying to force copulation. These attempts are, however, always unsuccessful and the male eventually gives up and departs. Possum behaviour usually only occurs as a last resort, when other mate-rejection behaviours, such as abdomen raising or wing fluttering/closing, have failed to stop the attentions of a male.[9]

Thanatosis is also utilized as a predator escape mechanism in some species, and can be induced by gently pinching the wings together at their bases near the thorax. For example, thanatosis was simply induced by grasping the base of the folded wings in the Illioneus giant owl butterfly, *Caligo illioneus*. The butterfly usually comes out of the trance – called 'tonic immobility' – after a couple of minutes. According to the late lepidopterist Torben Larsen, thanatosis is particularly prevalent amongst danaid butterflies, especially those in the genus *Euploea*, such as the common crow, *Euploea core* (Fig. 6.7). Larsen suggested that thanatosis has coevolved with unpalatability and the development of a tough integument, and is only therefore found in robust, distasteful species like these. There are, however, reports of thanatosis in a great many butterfly species, especially amongst fritillaries – including the marsh (*Euphydryas aurinia*) (Fig. 6.8), heath (*Melitaea athalia*) and meadow (*Melitaea parthenoides*) fritillaries – and in some Pieridae and Nymphalinae, including the common brimstone, *Gonepteryx rhamni* (Fig. 3.2), and the peacock, *Aglais io* (Fig. 3.27).[10]

Fig. 6.6. Small heath (*Coenonympha pamphilus*) resting on bell heather, Galicia, Spain.

Fig. 6.7. Common crow (*Euploea core*) possibly *E. core haworthi*, Bali.

Fig. 6.8. Marsh fritillary (*Euphydryas aurinia*) on fern, Galicia, Spain.

Female Solicitation

In highly receptive females, some of the later courtship phases may be shortened. Most virgin female butterflies are highly receptive and often mate soon after emerging from their pupae. In some species, such as the large grass yellow, *Eurema hecabe* (Fig. 2.9), this receptivity then decreases, before increasing once again a few days later. As unmated females mature, they often become much more active, in terms of soliciting males for courtship – flying up in front of them or engaging in elaborate posturing behaviour. This activity correlates with their production of eggs and their need to fertilize them; or in the case of already mated females, as the contents of the previous spermatophore become depleted. Females in the *C. eurytheme–C. philodice* species complex solicit male courtship, to remate, in order to receive another spermatophore. Virgin female barred sulphurs, *Eurema daira* (Fig. 5.6), can fly directly towards passing males and grab hold of their wings. Using the tarsi on the ends of their legs, they 'aggressively tug' the males downwards, and then land nearby with wings folded. The females then tilt sideways slightly, away from the males, and expose the tips of their abdomen. The males respond by conducting their wing-waving, courtship displays, and if all goes well this results in courtship. Researcher Jaret C. Daniels noted that females may repeat this courtship solicitation behaviour if the males show little initial interest![11]

There are numerous other examples where female butterflies do the chasing. For example, female tiger swallowtails, *Papilio glaucus* (Fig. 9.5), exhibit a high incidence of courtship solicitation, flying towards males and then turning upwards, and flying away as the males give chase. Female *P. glaucus* have been seen pursuing males and forcing them down into the vegetation! Wing contact can occur during these flights and solicitation is repeated if the males do not respond. These courtship flights last for up to a minute before the female lands, with the male briefly hovering nearby, then alighting beside her. There is a report of a female Rajah Brooke's birdwing, *Trogonoptera brookiana* (Fig. 10.4), attempting to solicit courtship with a male, nectaring on *Ixora* flowers in Sarawak, Malaysia. The female *T. brookiana* hovered above the male, moving up and down and buffeting him from above; both hovering with outstretched wings throughout.[12]

Virgin female small heath butterflies, *Coenonympha pamphilus* (Fig. 6.6), search out male territories and then carry out lengthy solicitation flights, but this behaviour ceases once they have mated. Virgin female pale grass blues, *Pseudozizeeria maha* (Lycaenidae, Polyommatinae), will also sometimes fly up from the ground towards a patrolling male, who then gives chase, resulting in copulation. Both virgin and mated female checkered white butterflies, *Pieris protodice*, approach and chase males, trying to solicit courtship, which is very short in this species. There is no prolonged aerial phase and once the female has landed the male does little more than alight on her thorax and couple with her. Despite the absence of prolonged courtship, they do engage in post-nuptial flights, moving off together from the area in which they first encountered each other. Interestingly, virgin female large heath butterflies, *Coenonympha tullia*, solicit passing males and are much more eager to mate than virgin females of the longer-lived, small heath, *C. pamphilus*. The increased longevity of the latter appears to be associated with a different, more 'laidback' strategy, whereby they remain quietly perched and do not approach any passing males. When *C. pamphilus* females arrive at the male lekking sites, however, they engage in lengthy, conspicuous circling flights, aimed at attracting the attention of the competing males, a much lengthier process which results in a small loss of fecundity. Females of species that occur in widely dispersed, low-density populations are often short-lived, and are therefore, more likely to actively seek out males to mate with.[13]

Clinging on and Standing Guard

Pair clinging

Males of the Chinese windmill, *Atrophaneura alcinous* (Fig. 9.6), often cling onto a pair of butterflies who are *in copula*. These competitors are opportunists, waiting for a chance to mate with the female before the mating plug – secreted by the first male in a vain attempt to prevent remating – hardens sufficiently to prevent such events. Up to five other males may be clinging on to the copulating couple at the same time, waiting for an opportunity to mate again. The second male can sometimes push aside the freshly formed mating plug of the first suitor, or tunnel through it using his aedeagus! This rather ungallant behaviour seems to have evolved because it is more profitable, in terms of mating success, for males to cling on to a pair *in copula* than it is to find another virgin female. Such pair-clinging behaviour is commonly seen in a variety of other species, such as apollos (*Parnassius* spp.), which also produce mating plugs or sphragides (see Chapter 8, this volume).[14]

The presence of a sexual competitor can influence the behaviour of butterflies of either sex in a triad. For example, the presence of an additional male on courting pairs of *Bicyclus anynana* was to reduce the likelihood of copulation within a given time period. Mating pairs also copulated less in the presence of a sexual competitor, of either sex. This inhibitory effect of another male (or female) on the courtship activities of a pair is subtle in this species. Professor Erica L. Westerman and her colleagues describe it thus:

> male-male competition in the presence of a female is a very quiet affair where males are aware of each others' presence and moves, and become distracted from courting females.[15]

Mate guarding

Guarding of potential mates *prior* to copulation has been observed and studied in *Heliconius* butterflies (see Chapter 7, this volume) and other pupal mating species. Some male butterflies also guard females *after* they have mated with them; this usually occurs in species in which a mating plug

Fig. 6.9. Cairns birdwing (*Ornithoptera euphorion*) male. Melbourne zoo. (Photo by Fir0002/Flagstaffotos – own work, GFDL 1.2, https://commons.wikimedia.org/w/index.php?curid=1388954)

is used by the male to seal the female genital opening (see Chapter 8, this volume). Since mating plugs can be penetrated or pushed aside before they have hardened, it makes sense for the male to guard the female and prevent her from remating. Females of some species, such as the Richmond butterfly, *Ornithoptera richmondia*, will sometimes accept mating attempts whilst the mating plug is still soft; for example, if the first male has only provided a small spermatophore, or if the next male is especially persistent. Females of this Australian butterfly can usually avoid mating simply by remaining immobile, but if population densities are high, they may be subjected to intense courtship attempts, lasting up to an hour, immediately after mating. In the closely related Cairns birdwing, *Ornithoptera euphorion* (Fig. 6.9), males have been observed guarding the female they have just mated with, following her around for about a day and driving off other males.[16]

An occurrence of both pre- and post-copulatory mate guarding of female pupae was inferred by studying the territorial behaviour of Taylor's checkerspot, *Euphydryas editha taylori*, a North American fritillary of great conservation concern. The males – which emerge up to 2 weeks before the females – aggressively defend small territories containing virgin females but vacate these soon after copulation. Interestingly, they spend a short time (11 min on average) guarding the females – by positioning themselves directly behind them and preventing other males from getting close – *after* they had mated with them. This post-copulatory mate guarding behaviour may have evolved as a way of ensuring that the sperm plug – which prevents the female from ever mating again – remains firmly in place in the copulatory duct. A form of sperm-guarding.[17]

Antennal Contacts: Courtship Bowing

Common graylings

In Chapter 1, this volume, the classic experiments carried out by Dutch ethologist Niko Tinbergen on the grayling butterfly, *Hipparchia semele*

(Fig. 6.10), were reviewed. These pioneering observations of courtship in this species have been widely disseminated, and are commonly referred to, so it is worth describing them here in detail. The interaction begins when a perching male flies up and intercepts a receptive female. This approach is straighter and faster than normal flight and is elicited by any passing insect, butterfly or even a falling leaf. The female lands, and the male alights near her, usually close behind (Fig. 6.11a, b). The male then begins to move around, in a series of short jerky movements and side steps, until he is facing her. This circling takes about 3 s, followed by *wing quivering*, where the male slowly raises and spreads his forewings, then quickly closes and lowers them, about twice a second. During this

Fig. 6.10. Grayling (*Hipparchia semele*) male on a tree, Galicia, Spain (14 September 2017).

Fig. 6.11. (a) Grayling (*Hipparchia semele*) courting pair, female above, Galicia, Spain (14 September 2017). (b) Grayling (*Hipparchia semele*) courting pair, female on right, Galicia, Spain (14 September 2017).

behaviour, the males' antennae are spread out at right angles to the body. The wing movements then become more pronounced and the forewings gradually protrude more from beneath the hindwings. Two other movements then occur simultaneously. *Fanning*: where the forewings are raised up, with their leading edges separated from each other, and then closed again 'rhythmically and at a fast tempo'. *Antenna spinning*: as the male starts fanning, the antennae are held stiff, and together they are rotated in a circular motion, such that they trace out a conical shape, which gradually becomes an ellipse, with its centre some 45° to the ground. These bouts of fanning and antenna spinning can last for several minutes.[18]

The final, and most iconic behavioural element in the courtship sequence is *bowing*. The male spreads both pairs of wings and inclines the forewings into a raised and forward pointing position. The wings are then slowly closed, so that they contact the antennae of the female, who is facing the male (see photographs for rock graylings: Fig. 6.12a–d). During this bowing phase, the female bends her antennal knobs outwards so that they lie parallel to the upper surface of the male's forewings. After the male has shut his wings and usually contacted the female's antennae, they are drawn backwards again; the whole bowing act taking about a second. After bowing – during which it is assumed that male sex pheromones are transferred via contact with the androconia – the male attempts to copulate,

Fig. 6.12. (a) Rock grayling (*Hipparchia hermione*) courtship, Galicia, Spain. (b) Rock grayling (*Hipparchia hermione*) courtship bowing, Galicia, Spain. (c) Rock grayling (*Hipparchia hermione*) male watching courtship bowing pair in background, Galicia, Spain. (d) Rock grayling (*Hipparchia hermione*), Galicia, Spain.

bending his abdomen forward and to the side in the usual manner. An attempt to copulate may be made after a single bowing, but in other cases, the courtship sequence is repeated, one or more times. During the male's display, the female usually remains motionless, but she may start to raise her wings slightly so that her genitalia are more exposed. Movement, and wing flapping, by the female has an inhibitory effect on the male. Copulation itself lasts for between 45 min and 2 h, according to Tinbergen.[19]

Rock and tree graylings

Manuela Pinzari investigated the courtship behaviour of the tree grayling, *Hipparchia statilinus* (Fig. 6.13), and compared it with that of common grayling, *H. semele* (Fig. 6.10), in flight cages and in the wild – using video recordings – in the Central Appennines, Italy. Building on the work of Tinbergen on *H. semele*, Pinzari showed that, in general terms, sexual behaviour in *H. semele* and in *H. statilinus* is for the most part very similar. What differences there are between these congeneric species relate mainly to the order and performance of steps. A total of 37 complete *H. statilinus* courtship sequences were observed, and six principal behavioural events were identified, which were common to both species. I have summarized these as follows:

- *Fanning.* The male lands on the ground behind the female with his wings closed but revealing the large post-discal eyespot. The forewings are rapidly opened and closed in a rhythmical, vibration-like motion.
- *Circling.* The male walks around the female – in a semi-circular track – so he ends up facing her head on, whilst fluttering his wings.
- *Bowing* (Fig. 6.12a–c). The male starts bowing even before he is in front of the female. Initially, his wings are closed; he then rotates his body forward on the sagittal axis, levering upwards on the posterior legs, whilst his head is brought close to the ground. Simultaneously, both wings open and close with great speed. Bowing ends when the male closes his forewings and moves them back to the initial position, along with the body. During very deep bowings, the male kneels on the ground. Whilst bowing, the male can take hold of the female forewings and antennae, letting them slide between his forewings.
- *Antenna orientation.* Immediately after bowing, the male antennae are waved back and forth very rapidly, in a circular motion, but this can only be seen clearly in slow motion recordings.
- *Copulation attempt.* This resembles an unfinished bowing movement with an 'embrace'. The male remains bowed forward with the wings open and, in this bowed position, walks towards the side of the female, then behind her, until they both face the same direction. The male then places his left forewing under the right forewing of the female, then, moving laterally, he places himself parallel to her.
- *Clasping.* The male attempts copulation by flexing his abdomen sideways and making contact with the female genitalia.

Fig. 6.13. Tree grayling (*Hipparchia statilinus*) female, Asturias, Spain.

Whether or not courtship in both *H. statilinus* and *H. semele* progresses as far as copulation largely depends on the female's reaction to the male: her refusal or rejection – by wing flapping and then flexing her abdomen downwards – can interrupt the courtship or can produce fragmented sequences. The whole courtship sequence is often interrupted and restarted again at an earlier point in the cycle. For example, the *bowing* and *antennal orientation* behaviours can be repeated up to 25 times in a row! The female communicates her availability – or otherwise – to the male, therefore, by her body language. As well as wing flapping, she may just become inactive and fail to respond, thereby rejecting or discouraging persistent males. Pheromones (scents) produced by the male probably play a major role in the courtship, and bowing behaviour is thought to be a way of exposing the female to these scents and thereby stimulating her into a state of receptivity. This process may also allow the female to evaluate the quality of the male and, if he is not to her liking, she can break off the affair![20]

Manuela Pinzari, together with co-worker, Valerio Sbordoni, from the University of Rome Tor Vergata, went on to investigate courtship behaviour in a subspecies of the rock grayling, *H. hermione genava*, with the aim of investigating whether there were any differences in behaviour between this species and the woodland grayling, *Hipparchia fagi*, which may lead to reproductive isolation between the two species. It turned out that, although the overall pattern of sexual behaviour was similar in the two species, there were some significant differences in the bowing behaviour. For example, bowing occurred less frequently in *H. fagi* than in *H. hermione*; whereas the duration of bowing was shorter in *H. hermione* than in *H. fagi*. In summary, these detailed and I think, equally pioneering studies, showed that, although the courtship sequence in the three grayling species (*H. fagi*, *H. hermione* and *H. semele*) is similar, there are important differences in terms of when, and in what order the steps are carried out, and also in how frequently the critical bowing behaviour occurs.[21]

A key finding of these studies, I think, is that when interrupted, a male can restart the courtship from any step of the sequence. Only about 10% of the time was a single courtship sequence followed by copulation.

In other words, the males must court the females repeatedly – repeating certain sequences dozens of times in a row – before reaching copulation. This means that females are repeatedly 'rejecting', or at least interrupting the courtship – and are also presumably swapping partners many times – before eventually accepting, i.e. choosing, a partner. Persistence appears to pay off for some males, and reluctant or initially rejecting females appear to gradually become more receptive via the stimulation of the courtship process.[22]

Other species

Other butterfly species, particularly satyrids, e.g. members of subtribe Satyrina, such as *Pseudochazara beroe* – a grayling called Freyer's tawny rockbrown, which occurs in Western Turkey – and probably many other grayling species, have similar bowing behaviours aimed at bringing the females' antennae into contact with the scent-producing glands of the male. For example in the meadow brown, *Maniola jurtina*, the male stands in front of the female and opens his wings suddenly (Fig. 6.14c). Wall brown, *Lasiommata megera* (Satyrinae, Elymniini), butterflies also have an elaborate courtship ritual in which the male engages in bowing movements, as described by Adrian Hoskins:

Fig. 6.14. (a) Meadow brown (*Maniola jurtina*) female, Galicia, Spain. (b) Meadow brown (*Maniola jurtina*) male showing tufted androconia on forewings. (c) Meadow brown (*Maniola jurtina*) pair, male on the left, engaging in courtship bowing, Galicia, Spain.

> When a male intercepts a female, a short but elaborate courtship ritual takes place, in which the male chases the female until she settles on the ground, and then flies around her, eventually settling in front of her, face to face. The female then responds by quivering her open wings, at which point the male half opens his own wings, and 'bows' several times to his prospective mate.[23]

Silver-washed fritillaries, *Argynnis paphia* (Fig. 1.17a, b), also engage in a ground-based courtship where the male 'bows' in front of, or beside the female, clasping her body between his wings and closing them suddenly. The male also vibrates his antennae and middle legs against the female's hindwings, which she holds still, and vertical. After bowing, the male places one antenna on the top of her head. This may then lead on to mating, but, as always in butterflies, the sequence may be interrupted and resumed, for example if the female flies off to do some nectaring.[24]

Proboscis Waving: Wood Whites

Courtship in wood white butterflies, *Leptidea* species, is particularly interesting on account of the peculiar, and somewhat endearing, head-swaying behaviour of the males. After alighting directly opposite a receptive female, a male wood white, *Leptidea sinapis* (Pieridae, Dismorphiinae), will uncoil his proboscis until it is fully extended and pointing forwards (Fig. 6.15a–c). Then, with his antennae pointing downwards, he repeatedly sways – or oscillates – his head back and forth about 0–60 times/min, according to research carried out by Professor Christer Wiklund of Stockholm University in the 1970s. The female may or may not respond to this display, depending on her predilection, but males are not thought to initiate copulation unless the female signals that she is receptive. If she accepts him, she makes her abdomen available for copulation, but unlike many other butterflies who have evolved an unambiguous mate-refusal posture (see above), it seems she has no way of overtly rejecting him. Consequently, if he fails to woo her, she can only sit tight, waiting for him to give up and fly away! This is one reason why these courtship displays go on for so long in wood whites, up to 30–35 min, for example. When it does occur, the act of mating lasts between 25 and 55 min, before the male releases himself and flies away.[25]

Newly emerged virgin female wood whites are usually highly receptive, and courtship is brief, but females that have already mated are unreceptive as they are generally monogamous. Already mated females can usually evade amorous males by carrying out a special male avoidance flight, where they fly upwards to a height of 10–20 m, then suddenly stop flying and descend to the ground with widespread wings; only starting to fly again when they are close (c. 0.5–1 m) to the ground. Continuing like this, females try to shake off the attentions of the males, but if they land somewhere, and the males alight next to them, then courtship commences again. The males do give up eventually – and mating never takes place – but

Fig. 6.15. (a) Wood whites (*Leptidea sinapis*) courting, Galicia, Spain (21 August 2017). Extended proboscis of the courting male (right) shown by white arrow. Second male arriving on the left. (b) Wood whites (*Leptidea sinapis*) courting, Galicia, Spain (21 August 2017). Extended proboscis of the courting male (right) shown by white arrow. (c) Wood whites (*Leptidea sinapis*) courting, Galicia, Spain. Male (left) uncoiling proboscis, female (right) with antennae angled backwards. Extended proboscis of the courting male (right). (d) Wood whites (*Leptidea sinapis*) in copula, male on left, Galicia, Spain.

they can persist in their courtship of already mated females for a long time. However, being detained by courting males in this way does not seem to harm the females in any way, with no effects on their fecundity, and only a very tiny (2%) impact on their fitness. Very few male courtships do result in mating: only two out of 238 attempts in a study of *Leptidea sinapis* in Sweden. In many (28%) of these courtships, it was the male who terminated the courtship, after 20 s or less.[26]

Although they are not readily distinguishable by eye and can only be separated by genital preparations or DNA sequencing, there are now understood to be three reproductively isolated wood white species in the Eurasian genus *Leptidea* (Fig. 6.16). In 1988, the common Eurasian wood white butterfly was split into two species (*L. sinapis* and Réal's wood white, *Leptidea reali*), and in 2011, a third cryptic species (*Leptidea juvernica*) was discovered. There are, however, some differences, regarding the habitats they prefer, e.g. meadows or forests, and elements of the courtship behaviour. For example, the cryptic wood white, *L. juvernica*, is said never to perform any wing flapping behaviours during courtship, like the wood white, *L. sinapis*. Crossbreeding experiments have shown that all three species are reproductively isolated, meaning that, although there may be

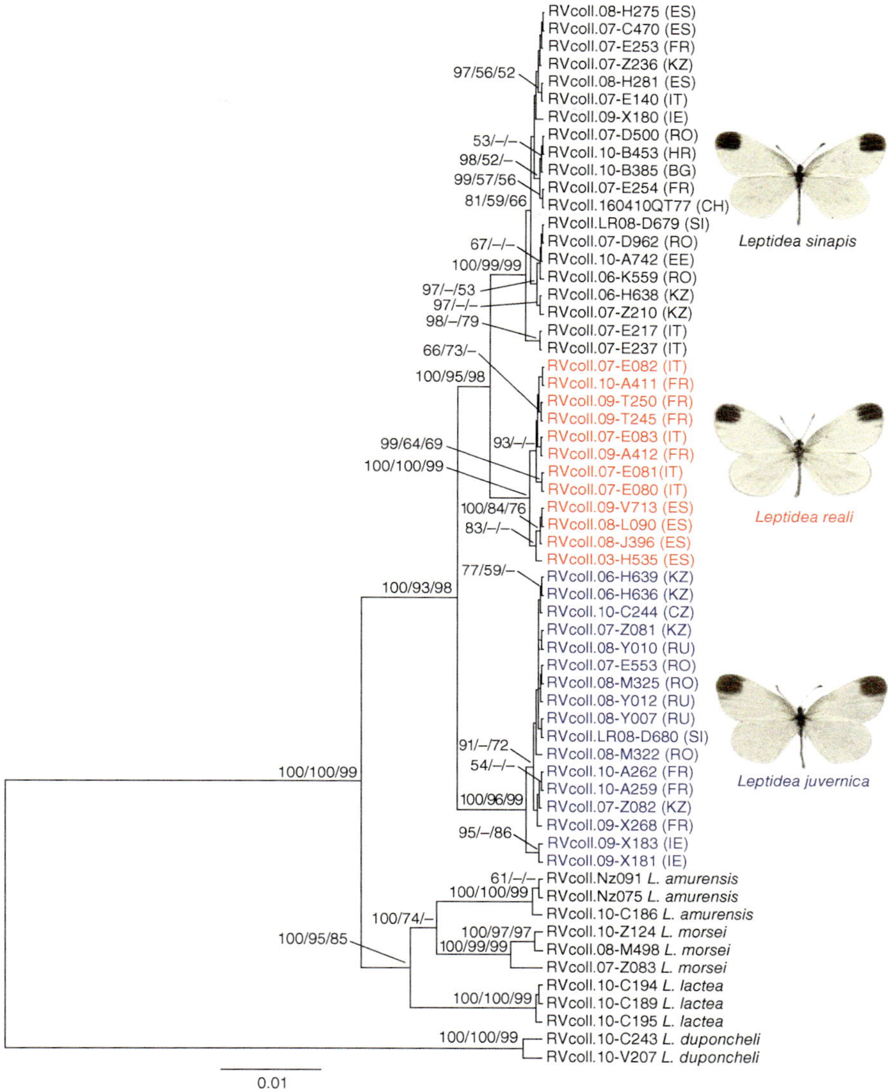

Fig. 6.16. *Leptidea* molecular phylogeny; Bayesian ultrametric tree based on the combined analysis of mitochondrial and nuclear loci. (From Dincă *et al.*, 2011, fig. 1.)

courtship between individuals of different species, it is always unsuccessful, and no mating results: the species remain separate. Pairs of these species do occur together – in sympatry, as it is called – in Europe. For example, *L. sinapis* and *L. reali* overlap in north-east Spain, and *L. sinapis* and *L. juvernica* are found in areas of Sweden and in the Vento region of northern Italy. In greenhouse flight cage experiments, males courted females of both species, but *L. reali* males courted *L. sinapis* females disproportionally rarely, whilst *L. sinapis* males courted females of a different species even more frequently than their own species![27]

There are differences in courtship behaviour between *L. reali* and *L. sinapis*, regarding the wing strokes performed by the males, but it is thought more likely that they use chemical signals, species-specific volatile pheromones, to tell each other apart. Males and females can distinguish each other via their antennae: the males have a large white patch on the underside of theirs (Fig. 6.15d). Unfortunately, the whole system is not foolproof, since males are unable to tell whether a given female is of his species or not, and court both con- and heterospecific females. Rather unfairly, it seems to me, the females *do* appear to know the specific identity of the courting male and will only mate with conspecific males. As we have seen, they have no way of letting the wrong sort of male know – via a rejection posture – when a cross-species courtship is taking place. In other words, a premating reproductive barrier exists, whereby females only accept males of the same species, and it is the female mate preference that enforces this restriction. The situation is not helped by the fact that some females – especially those of the spring generation – can also take a long time, over 11 min in one case, deciding whether or not to accept a courting male, even when he is of the right species, but other, less choosy females accept their suitors in a matter of seconds. The reason for the poor communication skills between these species is that they have only recently diverged and are still evolving behavioural differences. The common ancestor of all three wood whites probably originated in central or western Asia. Scientists are still debating when the speciation occurred, with one estimate placing it as recent as 270,000 years ago, and another suggesting that they diverged well before this, perhaps 1.5 or 1.6 million years ago (mya).[28]

It has been suggested that this rather unusual arrangement in wood whites, where the decision whether to mate or not lies entirely with the females – if they accept the male they stop hiding their abdomens between their wings and make their genitalia available – may have evolved to maintain a state of monogamy. The males are unable to force themselves upon unreceptive females who are therefore able to avoid costly and time-consuming remating, simply, by 'remaining quiescent', to quote Professor Wiklund. Male butterflies are generally able to mate much more frequently than females, which explains why, on average, most courtships involve eager males and unwilling females.[29]

Sonic Courtship: Do You Like My Click?!

Neotropical *Hamadryas* butterflies are unusual in that they produce a loud clicking noise during flight. Typically, on encountering a perched individual, patrolling males of the *Hamadryas feronia* group (Fig. 6.17) – species that can produce sounds – carry out a semi-circular flight below the perching individual whilst constantly making sounds for 5–10 s. If the other individual is a female, the male positions himself behind her, slowly opening and closing his wings, whilst inching forwards a few steps at a

Fig. 6.17. Epinome cracker (*Hamadryas epinone*), Iguazu Falls, Argentina.

time. In interactions between males, sounds are generated continuously, whereas they are only made by males during the initial phase of the courtship pursuit, probably as a way of communicating to the female that he is a male. It has been suggested that the intensity of the sonic signal might be used by the females as a way of assessing the male's fitness. Indeed, sounds may have replaced pheromones in these butterflies, although courtship remains a silent affair in some species in this genus, such as *Hamadryas februa*.[30]

Hamadryas butterflies have a multicoloured, patterned wing coloration, which gives them a cryptic appearance on bark (Fig. 6.17). They frequently alight and perch on tree trunks, which they use as a base for courtship territories. Although not overtly aggressive, they are territorial in the sense of ejecting intruders. A small number of other types of butterflies emit sounds, which are thought to play a defensive function, scaring off predators for example.[31]

Notes

[1] Ohguchi and Hidaka (1988) and Kan and Hidaka (1997). The mate-refusal posture is common to three pierid species: *Pieris rapae, Pieris protodice* and *Colias eurytheme* (Shapiro, 1970).
[2] Mega and Araújo (2010).
[3] García-Barros (1989), Rutowski (1992), and Wiklund and Forsberg (1985).
[4] Ide (2010, 2011, 2014) and Watanabe and Nishimura (2001).
[5] Wago *et al.* (1976), Wago (1978), and Dennis and Williams (1987). Measures (1976) contains some beautifully illustrated and graphically described examples of courtship and mating in British butterflies.

[6] Rutowski (1978), Suzuki and Matsumoto (1992), and Watanabe and Irie (2011). Taylor (1973b) showed that no contact was made with the female antennae by the male in *Colias* courtships.
[7] Cook *et al*. (1994).
[8] Based on Scott (1974c, 1975a, 1986a) and Krebs (1988).
[9] The occurrence of thanatosis as a mate-refusal behaviour is taxonomically quite restricted within the tribe Satyrini (Shreeve *et al*., 2000).
[10] Dudley (1990), Larsen (1991), and Humphreys and Ruxton (2018). Dennis (1984, 1998) was able to induce thanatosis in brimstone and peacock butterflies by squeezing them close to the wing bases.
[11] Rutowski *et al*. (1981), Hiroki and Kato (1996), Hiroki *et al*. (1998) and Daniels (2007).
[12] Panchen (1981).
[13] Rutowski (1979, 1980a), Wickman (1986, 1992), Krebs (1988), Krebs and West (1988), Wickman and Jansson (1997), Wickman and Rutowski (1999), and Imafuku and Kitamura (2018).
[14] Suzuki and Matsumoto (1990, 1992).
[15] Westerman *et al*. (2014).
[16] Orr (1999b) and Orr and Kitching (2010).
[17] Male Taylor's checkerspots, *Euphydryas editha taylori,* remain near female pupae, defending them and waiting for them to emerge (Bennett *et al*., 2012).
[18] Tinbergen (1972).
[19] Tinbergen (1972).
[20] Pinzari (2009).
[21] Pinzari and Sbordoni (2013).
[22] Pinzari and Sbordoni (2013).
[23] https://www.learnaboutbutterflies.com/Britain%20-%20Lasiommata%20megera.htm
[24] Magnus (1950) and Bastock (1967). Kemal and Koçak (2013) present photographs of courtship behaviour in *Pseudochazara beroe* (Satyridae) taken in the Artos Mountains (at 2900 m) in eastern Turkey.
[25] Wiklund (1977) and Friberg *et al*. (2008a). The photos of courting wood whites (Fig. 6.15a–c) were taken in Galicia, north-west Spain. There are records of *Leptidea reali* in northern Spain (Dincă *et al*., 2011, 2013) so I cannot be sure if they are *L. sinapis* or *L. reali*.
[26] Berger *et al*. (2012).
[27] Réal (1988), Lorković (1993), Friberg *et al*. (2008a), Gallo *et al*. (2015) and Lewington (2016). There is also the eastern wood white, *Leptidea duponcheli*, in southern Europe, and other *Leptidea* species further east in Asia.
[28] Freese and Fiedler (2002), Friberg *et al*. (2008b), Dincă *et al*. (2013), Friberg and Wiklund (2007) and Talla *et al*. (2017). *Leptidea reali* is behaviourally polyphenic: females of the summer generation are less discriminatory towards courting males than those of the spring generation (Friberg and Wiklund, 2007).
[29] Wiklund (1977, 2003).
[30] Courtship is silent in *Hamadryas februa* and noisy in *H. feronia* (Monge-Nájera *et al*., 1998).
[31] Marini-Filho and Benson (2010) conjectured that in *Hamadryas feronia* clicks are used as an early recognition of sexual partners and that females may use sound intensity to assess the male's fitness. Once sexual recognition occurs, however, there is a shift to courtship, which is accompanied by the cessation of sound production (Garzón-Orduña, 2012).

7 The Mating Game

Reproduction: A Battle of the Sexes, or a Tug of War?

Mating may be the most important activity in an animal's life, as it provides an opportunity to propagate genes into the next generation. For females, some of whom will only mate once, it is imperative to choose the best male in terms of his fitness – genetic potential – and quality in providing for her and her offspring. In some species, like the green-veined white, *Pieris napi* (Figs 6.1 and 9.1), females effectively control mating: their optimal mating rate is lower than that of males and they do not increase the rate even when males are more persistent, i.e. in terms of the intensity of their courtship. It seems they cannot be hurried into making their choices. In some rare, but well-known cases – such as the monarch butterfly – males can, in certain circumstances, force themselves upon the female (see Chapter 5), but for most butterfly species, females are in control of whom they mate with. Female choice is the most common driver of sexual selection, but in a minority of species and/or circumstances, males become the choosy sex (see Chapter 2). Ultimately, many of the characteristics of a species' mating system – the level of competition and who gets to choose their mate – depend on the ratio of the numbers of males and females attempting to mate at any given time.[1]

Females are usually choosier than males because they produce fewer gametes, i.e. eggs, and have more resources invested in each one, i.e. relative to the millions of sperm that males can potentially produce. Courtship and mating provide the means by which the female can assess and select a high-quality mate, based on the attributes he displays to her. The information he provides includes chemical (pheromones), physical (colour, shape and size) and behavioural (courtship displays) signals and features, which both communicate his health and vigour and arouse her

© Raymond J.C. Cannon 2020. *Courtship and Mating in Butterflies: Reproduction, Mating Behaviour and Sexual Conflicts* (R.J.C. Cannon)

interest. What stimulates her senses and determines her choice might be a flashy colour or an exaggerated feature, which acts as a proxy signal for some underlying quality. The process of selection – in terms of individual gametes – can take place before, during and after the act of copulation. Males can increase their chances of fertilization in several ways, including providing more sperm, or monopolizing the female so that she has less of a chance of remating with another male. Sexual reproduction is a cooperative venture, but that does not mean that there is always perfect harmony in terms of the needs and interests of the two different sexes. Whilst many sexual traits and features coevolve with each other – such as genitalia, for example – it is now recognized that there can be conflicting forces in operation between males and females, and that these result in many of the interactions we observe in the courtship and mating biology of different species. Males and females have evolved certain adaptations aimed at shifting the outcome of any sexual conflicts in their own favour.[2]

In butterflies, the main conflicts appear to arise as a result of the costs to the female of mating, particularly of repeated matings, compared with the male. From her disproportional investment in gametes, and more generally, from the fact that males and females contain a different set of genes. In other words, the genomes of males and females may be, to a certain extent at least, in conflict with each other. Another way of looking at the extent to which the interests of the two sexes are in harmony or conflict is to examine the costs and benefits (see below) associated with the different activities involved in courtship and mating. The balance of this type of accounting differs markedly from species to species, as sexual antagonism has resulted in the evolution of different solutions to the conflicts, in terms of biology and behaviour. The ways and means involved in resolving these on-going conflicts – because ultimately males and females need to exist in some form of dynamic harmony – operate before and after insemination. The mechanisms involved in enabling females to make choices *after* they have mated, i.e. cryptic female choice, have received a lot of attention and interest, although less is probably known about post-copulatory sexual selection in butterflies than in some other species.[3]

Butterflies are relatively successful in terms of mating success, compared with moths for example, with a very high proportion (95–100%) of females ending up mated in many cases. For example, the mating rate of overwintering female Asian comma butterflies, *Polygonia c-aureum*, was 100%, which suggests that females rarely experience problems in obtaining mates. Nevertheless, some individuals do inevitably die mateless, i.e. without reproducing, particularly those that emerge late in the season when the availability of the opposite sex is low. In populations with such asynchronous reproduction patterns, the proportion of females who will die without having mated can be relatively high, up to 18%, for example, in low-density populations of the snow apollo butterflies, *Parnassius clodius* and *Parnassius smintheus*. Mating success in males on the other hand, can be much lower, as a result of both male–male competition and female choice. For example, 73% of males failed to mate in a Japanese population of the Chinese windmill, *Byasa* (= *Atrophaneura*) *alcinous* (Papilionidae).[4]

Mating Strategies

The availability of females is usually the key component in determining the mating systems of butterflies, but either sex can be limiting, i.e. in short supply. Males do not as a rule, either monopolize females or defend resources or mating territories per se, although they do defend territories and sites where females may pass through (see Chapter 3). So, the duration of time when females are receptive and potentially fertilizable, in the habitat occupied by both sexes, will usually determine the breeding strategy employed, which may be short or prolonged. The term *explosive breeding* has been applied to species where males are under selective pressures to acquire a mate as rapidly as possible. The bog fritillary, *Boloria eunomia* (Heliconiinae, Argynnini) (Fig. 7.1), is an example of a butterfly with a short reproductive period and an explosive mating system, as are many *Euphydryas* species.[5]

Heliconius butterflies are particularly long-lived: adult *Heliconius charitonius*, for example, have potential life spans in excess of 100 days and both sexes can mate multiple times. The exception being amongst 'pupal mated' females (see below), who rarely mate again. However, for prolonged breeders, such as non-pupal mating *Heliconius* butterflies, males have access to receptive females for extended periods, and as such are less solicitous than males in explosive breeding species. The time constraints of explosive breeders, where access to fertilizable species is extremely limited, can lead to mistakes, e.g. with males trying to copulate with other males. There is probably a continuum between the two ends of the spectrum, in terms of the time available for breeding in a given species.[6]

The ratio of males to females in naturally occurring populations is often skewed significantly in favour of males, in many species. Indeed, females are often very difficult to find in some butterflies. Although, such

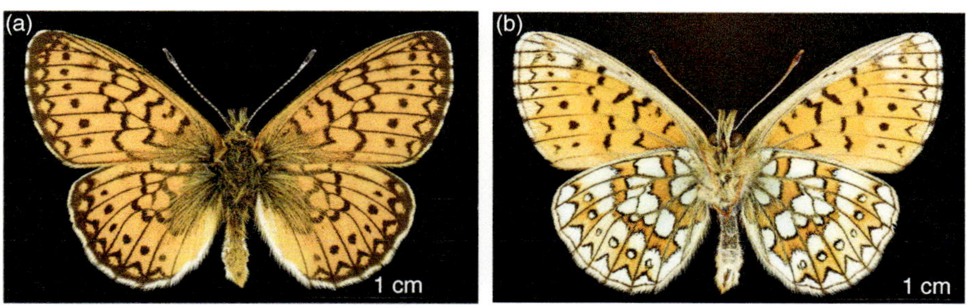

Fig. 7.1. (a) Ocellate bog fritillary (*Boloria eunomia*) dorsal side, Porté-Puymorens, Pyrénées-Orientales. (b) Ventral side. (Photos by Didier Descouens – own work, CC BY-SA 4.0, https://commons.wikimedia.org/w/index.php?curid=41733954)

male bias can occur as an artefact of sampling – for example, females may be under-represented because they spend less time flying – it is also considered to be a real phenomenon resulting from several possible causes. These include higher rates of pre-adult mortality in females; faster development of males; higher rates of emigration by females; and higher adult mortality in females. Nevertheless, the underrepresentation of females can simply occur because they are less mobile, often sitting hidden in the vegetation somewhere, and therefore harder to find or observe.[7]

Copulation

Pair carrying *in copula*

Mating pairs may be stationary and cryptic, or relatively mobile: flying about whilst locked together in a nuptial flight. One individual, usually – but not always – the largest sex in those species where there is a pronounced gender size difference, does the carrying. In some species, where the sexes are about the same size – like the plain tiger, *Danaus chryssipus* (Fig. 7.2) – the individual flying, and carrying the other butterfly, can be of either sex, although one sex usually predominates in this role. The individual being carried often appears to be in a state of catelepsis, remaining limp, motionless and frozen, thereby allowing themselves to be transported by the active partner. In some butterfly families however, such as the Satyridae, where the female appears to always do the carrying, the role may be fixed, probably having been determined in their evolutionary past. Similarly, in pierids, males usually do the carrying, even though they may be smaller than the females. In monarch butterflies, *D. plexippus*, the male always carries the female whilst *in copula*, and males also take on this role in the queen, *Danaus gilippus*, and tropical queen, *Danaus eresimus*, butterflies. There are even some reports of the role of carrier alternating between the male and female, during a single copulation – for example, in

Fig. 7.2. Plain tiger (*Danaus chrysippus bataviana*) male and female in copulation, Bali, Indonesia.

the silver-washed fritillary, *Argynnis paphia* (Figs 1.17 and 1.18) – something that can also occur if they are continuously disturbed.[8]

There are some conflicting reports and many gaps – i.e. species without any documented records – but there is a general consensus in the literature that males usually carry the female in Pierinae (whites) – but not in the black-veined white, *Aporia crataegi* (Fig. 7.3), where the females are considerably larger than the males – Coliadinae (sulphurs and yellows), Danainae (milkweeds and clearwings) and *Heliconius* species. On the other hand, the female mainly carries the male in Papilioninae (swallowtails), Parnassiinae (snow apollos), Satyridae (browns), Riodininae (metalmarks), Theclinae (hairstreaks) and Hesperiidae (skippers). Either the male or female may be the carrier in Nymphalidae (brush-footed butterflies) – except females are the carrier in checkerspots (*Euphydryas* species) – Argynnini (a tribe including many fritillaries) and Lycaenidae. There are, however, many variations; for example, males are the carriers in *Polyommatus* species, but in lycaeids as a whole, the role is probably more often taken by the female.[9]

Copulation duration

Butterflies remain *in copula* for very variable amounts of time, from the relatively brief (c. 15–20 min) to the very lengthy: nearly a week in one pair of brimstones, *Gonepteryx rhamni*, reportedly! Lengthy copulations are a way in which males can increase their fitness by preventing the female from remating for as long as possible. Butterflies are considerably more vulnerable to predation whilst *in copula*, and their behaviour is often very 'flighty': landing and taking off again several times if disturbed, before eventually settling down again, perhaps deep in the vegetation. Another reason for movement whilst *in copula* may be to allow the individual who

Fig. 7.3. Black-veined white (*Aporia crataegi*) mating pair, Gorzente, Genova, Italy. (Photo by Hectonichus – own work, CC BY-SA 4.0, https://commons.wikimedia.org/w/index.php?curid=56238615)

is doing the carrying to feed, or even to bask, as seen with a mating pair of gatekeepers, *Pyronia tithonus* (Satyrinae, Satyrini) (Fig. 7.4), where the female is on top, basking, in order to maintain a thoracic temperature that will enable her to take off quickly, in the event of a threat.[10]

Male butterfly genitalia have a relatively large structure called a valva (plural valvae) – commonly called claspers – which are paired, moveable clasping structures used to hold the female during mating. The valvae – together with other parts of the male genitalia – grip the terminal segment of the female's abdomen, but the exact position is variable. The phallus is then introduced into the bursa copulatrix. Claspers are often strongly sclerotized and their complex shapes are sometimes used in taxonomic descriptions of species. Male Japanese swallowtails, *Luehdorfia japonica*, use the scales from their claspers to form a sphragis on the female abdomen during copulation, and degree of scale loss from this organ can be used to assess how many times they have mated![11]

Copulation times appear to be relatively short in small butterflies, such as 'blues' and skippers. For example, copulations lasted between 24–34 min ($n = 3$) in the dusky large blue, *Phengaris nausithous*, and for at least 35 min ($n = 1$) in the scarce large blue, *Phengaris teleius*. Copulation lasted from 4–72 min in the fiery skipper, *Hylephila phyleus*, with the female carrying the male if they were disturbed into flight. However, the duration of copulations in the relatively large Japanese swallowtail, *L. japonica*, ranged from 32–154 min; and in the slightly smaller pearl crescent, *Phyciodes tharos*, copulation lasts from 28–158 min, with an average of 72 min.[12]

Mating in *Parnassius* species usually takes place with little or no courtship and the males usually only copulate once; a lengthy act that can last for several hours. Strangely, there is often a surplus of males relative to females in *Parnassius* species, although this may be an artefact caused by the more cryptic behaviour of the females, who sit or stay closer to the ground during egg laying. Males, on the other hand, spend most of the time in flight. Nevertheless, it appears that there are genuine surpluses of males in some *Parnassius* butterfly populations, such as in the clouded

Fig. 7.4. Gatekeeper butterflies (*Pyronia tithonus*) mating pair. Female on top, basking, Galicia, Spain.

apollo, *P. mnemosyne*, not caused by differential mortality or emigration of females (Fig. 7.20).[13]

Post-nuptial flights and mate-guarding

Remaining locked together *in copula* has been shown to severely limit the ability of mating butterflies to take off and fly, and one only need observe the flight of a mate-carrying butterfly to see how impeded they are (Fig. 7.5). Feeding, or nectaring as it is called – sitting on top of brightly coloured flowers, for example – potentially exposes a butterfly to predation, but carrying a mate around must put them at much greater risk. Nevertheless, it is not unusual to see pairs *in copula*, moving about together in their habitat, despite their impaired ability to escape from any would-be predator. Dr John Brackenbury has photographed copulating pairs in flight, including a male common blue, *Polyommatus icarus*, carrying a female (Fig. 7.6), and green-veined whites, *Pieris napi* (Fig. 7.7). A mating pair of common blues, *P. icarus*, with the larger female on the upper right are shown standing on a leaf (Fig. 7.8) for comparison with the pair in flight. In *P. napi*, single males usually take off at steep angles and fly upwards, whereas copulating couples have a very low take-off angle. Once mating commences, it is usually the male who decides when to terminate copulation, i.e. when he has finished forming and transferring the spermatophore and accessory gland secretion. If the male dies whilst *in copula*, the female is reportedly unable to rid herself of him![14]

Pairs often remain *in copula* for extended periods – beyond that necessary to transfer a spermatophore – as a result of the male's desire to monopolize the female and guard against additional matings. He acts as sort of living mating plug! Nevertheless, the huge variation in the lengths of copulation between different species suggests that there are other forces acting to constrain this sort of behaviour. Mate-guarding must become unproductive for males at some point, and there are increased risks of

Fig. 7.5. Plain tigers (*Danaus chrysippus bataviana*), male and female in copulation, Bali, Indonesia.

Fig. 7.6. Post-nuptial flight common blue (*Polyommatus icarus*) male carrying female. (Photo by Dr J.H. Brackenbury.)

predation whilst *in copula*, so terminating the union eventually becomes advantageous for both sexes. There are occasions when the copulation is either interrupted or terminated prematurely when no spermatophore is transferred. However, such premature 'detachments' from copulation are probably uncommon in all species. At the other extreme, there have been recorded instances of more than one spermatophore being transferred in a single copulation, in the lycaenid butterfly, *Callophrys xami*, although this is presumably rather rare. These could be categorized as Type I and Type II reproductive errors![15]

Sperm competition

Female butterflies are often ready to mate and engage in copulation within minutes of emerging from their chrysalises. In some cases, however, they wait for 30 min or so whilst their wings expand and harden, keeping males at bay via mate refusal posturing. Females in some species appear to have little or no choice, in some instances, getting mated

Fig. 7.7. Post-nuptial flight green-veined whites (*Pieris napi*). (Photo by Dr J.H. Brackenbury.)

Fig. 7.8. Common blue (*Polyommatus icarus*) mating pair, Asturias, Spain.

before they eclose (see below, pupal mating). In these circumstances, it is easy to imagine how postcopulatory, so-called 'cryptic' means and mechanisms may have evolved to enable females to make choices about which males' sperm fertilize their eggs. Gaining control *after* the event, as it were. Similarly, males may in effect continue to compete *within* the female reproductive tract – a phenomenon called 'sperm competition' – i.e. competing to fertilize the female ova. It may sound unusual to describe individual sperm as competing; in most cases it is a numbers game, where the chances of achieving a successful fertilization can be increased by simply ejaculating more sperm. The effect of this increased quantity of ejaculate is usually to boost the fecundity of the female as well, albeit sometimes having a negative effect on her longevity. Non-fertile sperm can also be used to fill up the female storage organ, rendering her unreceptive and unlikely to copulate with other males (see Chapter 8, 'Sperm and Spermathecae').[16]

In contrast to females, males often must wait for 2–3 days before they are able to copulate again successfully, which is one reason why emerging before the females is an advantage (see below, 'Protandry and female receptivity'). Male monarchs are somewhat unusual in being able to copulate again soon after they have mated, even though it takes them about 5 days to replenish the accessory gland material used to make their large spermatophores.[17]

Protandry and female receptivity

Male butterflies often develop faster and emerge earlier than females: a phenomenon known as protandry. For example, male *Pieris protodice*, *Pieris rapae* and *Colias eurytheme* all develop faster as pupae than females, and are ready to mate within 24 h of eclosion. In practice, this usually means that males emerge a couple of days before the females do, as was shown for example, for the wood white, *Leptidea sinapis*, in central Sweden. Protandry may have evolved as a result of intense competition amongst males for mates, i.e. sexual selection, especially in species where the females only mate once. Indeed, protandry is negatively correlated with polyandry, meaning that in species where females mate repeatedly, males are less likely to emerge early. This is because there is a trade-off between selection pressures for large males – which produce large ejaculates – and selection for early emergence. In monandrous species however, males that emerge late – i.e. after most or all the females have mated – are clearly going to be at a severe disadvantage! The orange tip butterfly, *Anthocharis cardamines* (Fig. 9.3), is a good example of a protandrous, and monandrous, species.[18]

The initial, high receptivity of virgin female *Eurema hecabe* just after adult eclosion declines in subsequent days – if they do not mate – before increasing once again, after 3 days. The window of opportunity for mating is, therefore, often quite short for female butterflies; e.g. only 3 days in the Rocky Mountain Apollo, *Parnassius smintheus* (Papillionidae). In many species, females can mate immediately after eclosion, or after expanding their wings, but usually males need at least a day or so before they are ready to copulate. Mating pairs are, however, frequently harassed by other males, which they either ignore or attempt to avoid by moving off somewhere else. Copulating pairs of the sulphur butterfly, *Colias erate* (Fig. 6.5), were harassed by single males, which resulted in longer times *in copula*, but no reduction in spermatophore size. In other words, the harassment *in flagrante* slowed down the copulation process but did not affect the outcome. Older males came in for more harassment than younger males in copulating pairs. So, although the females appeared to show no age-related preferences in their acceptance of partners, lone males strongly discriminated against older competitors – N.B. older males generally deposit smaller spermatophore packages – and this occasionally led to a take-over of the mating female by the younger usurper. It is worth noting that sperm is only ejaculated in the latter phase of copulation.[19]

Sexual Conflicts and Confluences

Sexual conflict arises because what is best for one sex is not always best for the other. Another way of putting it is that some genes expressed in females will conflict with other genes expressed in males. This modern view of the battle of the sexes is far removed from the traditional idea that mating involves a harmonious cooperation between males and females, but in polygamous species there will always be a divergence between the two sexes, especially in terms of how they cope with competitors. For example, in polyandrous species there is a conflict of interest between males who first mate with a female – and would prefer her not to mate again – and the females who strongly benefit from multiple matings. This conflict arises because of sperm precedence: some of the nutrients provided by the first male can be recycled and used in the eggs fertilized by the second male. As we shall see, this has led to the evolution of ways and means of preventing further matings, in the form of chastity plugs, for example.[20]

One theory proposes that sexual conflicts over mating or reproduction have resulted in the evolution of female choice. In this scenario, females have evolved a certain resistance to males, e.g. becoming much choosier, in order to reduce the burden – economists talk about 'costs' – which falls on them as a result of the asymmetrical investment in reproduction between the sexes. Thus, female choosiness over mates evolves in tandem with elaborate traits, such as ornamentation and other secondary sexual characteristics, which have been evolved by males to entice or attract the female into mating. In the absence of natural selection – i.e. where the male may fall prey to predators – such a system will coevolve costly and elaborate trails: what biologists call 'a runaway feedback' system.[21]

Polyandry – the Risks and Benefits of Multiple Matings

It is easy to understand why males would want to mate more than once: by doing so they leave more offspring and increase their fitness. Females on the other hand, might be better off mating just once, and then concentrating on egg laying; choosing the best sites for the survival and development of her offspring. So, there appears to be a conflict of interest between the sexes in terms of how many times they need to mate in order to maximize their reproductive success. As we shall see, in the section on 'Sperm and Spermathecae' in Chapter 8, this conflict is resolved to some extent via nuptial gifts from the male to the female. The frequency with which a sexually reproductive species mates is also of interest from an evolutionary perspective, as it determines the rate at which the genetic material is shuffled, and thus how well it can respond to natural selection.[22]

The number of times a female butterfly has mated can be determined in wild-caught butterflies by counting the number of spermatophores, or residues, inside her reproductive tract. Such counts are thought to be a reasonably good estimate of mating frequency, but in some species, males occasionally transfer more than one spermatophore per copulation, and some old females lose – or absorb – all traces of spermatophores. Contrary to what was once thought, females in most butterfly species mate more than once, but there is probably a continuum between, on the one hand, species where females almost all mate just once, and on the other, species where females almost all mate multiple times. In between, there are species that exhibit a mix of mon- and polyandrous behaviour. In other words, there are some species – like the peacock, *Aglais io* (Fig. 3.27), the orange tip, *Anthocharis cardamines* (Fig. 9.3) and ringlet butterflies, *Aphantopus hyperanthus* (Fig. 7.9), which are largely monandrous, but where a small proportion of females mate more than once; thus, producing an average mating frequency slightly greater than one. Females of these monandrous species spend the remainder of their lives laying eggs and resting in the vegetation, avoiding courting males. The reason why such monandrous species, including the wood white (*Leptidea sinapis*;

Fig. 7.9. Ringlet (*Aphantopus hyperanthus*), Scarborough, UK.

Fig. 6.15) as well, for example, do occasionally remate may be because they received an unusually small spermatophore in the first copulation.[23]

The comma butterfly, *Polygonia c-album* (Fig. 7.10), is a good example of a polyandrous species, where females mate between two and three times on average per lifetime. Females who mate more than once have a refractory period of several days – usually dependent on the size of the spermatophore – after mating, during which they are unreceptive to males and engage in oviposition of fertilized eggs. Large spermatophores may have evolved in some species to maximize this period and thereby delay mating with a subsequent male. Females can maximize their reproductive success – i.e. increase the survival of their offspring – by spending time searching for the best oviposition sites. Researchers in Sweden found that the average number of matings, based on the number of spermatophores per female, in 11 different pierid species, was approximately 1.5: i.e. ranging from 1.1 in the orange tip, *A. cardamines* to 2.1 in *P. rapae.* The number of lifetime-matings per female was even higher in the Japanese cabbage white, *Pieris rapae crucivora* (Fig. 7.11): up to three times on average. The mating season is usually longer in polyandrous species like the comma, *P. c-album*, than in monandrous butterflies, like *A. io.*[24]

Fig. 7.10. Comma (*Polygonia c-album*), Galicia, Spain.

Fig. 7.11. Japanese small whites (*Pieris rapae crucivora*) mating pair. (Photo by 池田正樹 (talk)masaki ikeda [CC BY-SA 3.0 (https://creativecommons.org/licenses/by-sa/3.0)].)

Satyrine butterflies are essentially monandrous, or weakly polygamous, and multiple matings are usually much less common. For example, three bushbrown species studied in Queensland, Australia – common bushbrown, *Mycalesis perseus* (Fig. 7.12), *Mycalesis terminus* and *Mycalesis sirius* – had average mating frequencies of 1.2, 1.2 and 1.4, respectively. Despite the relatively high proportions of individual females that mated only once – between 57 and 69% were monandrous – all three species had a small number who mated up to three, and sometimes four times. The worn – presumably older – females, with chipped wings and missing scales, were likelier to contain more spermatophores than the more pristine ones.[25]

Females gain directly from multiple matings in several ways, including the receipt of nutrients, increasing the genetic diversity of their offspring and replenishing their supply of sperm. The overall effect of multiple matings is generally to increase the production of offspring during a female's lifetime, as well as her own longevity, but some experiments have suggested that there can be a physiological cost associated with successive rematings. For example, the longevity of twice-mated female windmill butterflies, *Atrophaneura alcinous* (Fig. 9.6), was significantly shorter than that of once-mated ones. Yet, despite the overall advantages from multiple matings – e.g. from male ejaculates – the average number of female matings remains relatively low in the field, indicating the existence of an optimum mating rate, or a balance between risk and reward. Mating is a risk: copulating pairs are relatively inactive and therefore vulnerable to attack by predators. Nevertheless, in butterflies – as in most insects – most species are polyandrous (mate multiple times) and there are advantages to both sexes in being so. Female reproductive output is generally significantly increased in polyandrous species.[26]

Fig. 7.12. Common bushbrown (*Mycalesis perseus*), wet season form, Kadavoor, Kerala, India. (Photo by © Jee & Rani Nature Photography (CC BY-SA 3.0) derivative work Mario Link, CC BY-SA 3.0.)

Despite the advantages of polyandry, there are some situations where monandry is advantageous. For example, in *P. napi*, individual females with a low mating rate – i.e. those that were slow to mate again – produced more eggs during the early period of reproduction than those with a high mating rate. The benefits of multiple mating only accrue about 1 week after the first mating. So, in unfavourable conditions where the female life span is short – less than 5 days on average in *P. napi* – monandry is the more favourable option. In certain circumstances, particularly in butterflies at the northern edge of their ranges, conditions favourable for egg laying may be limited. Heavy rain and/or low temperatures may curtail opportunities for multiple matings – by limiting egg laying periods – such that the benefits of polyandry cannot be realized. Females generally allocate proportionally less of their resources to reproduction in polyandrous species, presumably in anticipation of receiving nuptial gifts from the males, who contrastingly, allocate proportionally more resources in multiple mating species.[27]

In polyandrous species, the males usually deliver heavier and more nutrient-rich ejaculates – spermatophores – than in species that only mate once. In these situations, the sperm from different males are, in effect, in competition with each other, and, like buying lottery tickets, a male's chance of winning – in this case fertilizing the female's eggs – are increased if he transfers more sperm during mating (i.e. more tickets!). So-called sperm competition results in adaptations – both behavioural and physical – which maximize an individual male's ability to achieve fertilizations, i.e. increase his evolutionary fitness. These adaptations include an increase in the relative size of the testes and the length of the sperm in species where sperm competition can occur. Longer sperm are presumably faster swimmers and can, therefore, compete in terms of speed and/or endurance![28]

The females that will be selected by evolution are those that maximize their genetic reward in the next generation, regardless of the consequences for individual males. This illustrates the fact that the reproductive interests of the sexes are not necessarily identical. For example, the optimal rate of mating is usually higher for males than for females. This can lead to conflict between the sexes, a situation that has been called an 'evolutionary arms race', or sexually antagonistic coevolution. There must, however, be a continual balancing of conflict and cooperation between the sexes over time as they struggle to reconcile their somewhat different needs.[29]

The adaptations seen in polyandrous species include the overall larger size of males and their increased ability to produce sperm and accessory substances. This comes in the form of larger spermatophores, which take longer to be utilized by the females, thus increasing the period of unreceptivity. Remarkable as it may seem, males can determine the mating history of the female during the process of copulation. Somehow, during genital contact – whilst they are forming their own spermatophore – male small whites, *P. rapae* (Fig. 7.11), can sense the size of residual spermatophores from previous matings in the bursa copulatrix of the female, and tailor the amount of sperm they deliver accordingly. They provide more sperm to females when they are competing with other males, i.e. when many sperm are present. Male *P. rapae* produce larger spermatophores – but with fewer sperm – when they first mate, then smaller, sperm-rich spermatophores when mating a second time. They do not use up all their sperm in a given mating, but hold some back in a special storage organ called the duplex for subsequent matings. Male monarchs, *Danaus plexippus*, are also able to adjust the quantity of eupyrene (large nucleate) sperm they transfer during mating, in response to competition. They transfer more of these fertile sperm if the bursa copulatrix of the female contains large amounts of ejaculates from other males. Female monarchs are thought to sense the presence of spermatophores by means of bristles within the bursa copulatrix and this organ is generally able to expand sufficiently to receive multiple ejaculates. The ejaculate packets are opened by means of the contraction of muscles within the bursa copulatrix, which bring large 'teeth' into contact with the spermatophore wall (see Chapter 8).[30]

In monandrous species, males usually produce only a relatively small investment in terms of ejaculate, with small spermatophores representing as little as 2%, or less, of the body weight in species like the imperial hairstreak, *Jalmenus evagoras* (Lycaenidae) (Fig. 7.13). This species has an *explosive mating system*, in which there is intense competition amongst clumps of males for access to newly emerging females. Female *J. evagoras* pupae give off a chemical signal before they are about to emerge (eclose), which attracts dense 'mating balls' of males, which 'scramble and butt each other in their attempts to ward others off and copulate with the newly emerging butterfly' according to researchers who studied this species in Australia. The competition is fierce, and only a relatively small proportion of males – the larger and older ones – achieve a mating. The females live for only a short time and 'vigorously reject further attempts at copulation'.[31]

Virgin males usually transfer a much larger spermatophore than males that have mated previously, but these may contain many less sperm, and they do not adjust the sperm number in relation to the females' mating

Fig. 7.13. Imperial hairstreak (*Jalmenus evagoras*), ventral side, Melbourne, Australia. (Photo by Benjamint444 – own work, GFDL 1.2, https://commons.wikimedia.org/w/index.php?curid=13277599)

history. The mass of the spermatophore is up to a half smaller in males of the Asian swallowtail butterfly, *Papilio xuthus*, at the second mating. Older virgin males also produce larger spermatophores than younger virgin males in some butterflies, although larger, rather than older, males generally deliver the heaviest ejaculates. Thus, the effect of remating depends on the type of male the female chooses to mate with in subsequent copulations. Since virgin males deliver larger spermatophores than those who have just mated, a female would benefit more from mating with these types of males, in order to fully realize her full potential fecundity. When female *P. napi* could mate at liberty only with recently mated males (i.e. males that delivered a small ejaculate), they increased the number of matings they made – to average 5.1 times over their lifetime – as if to compensate. Female *P. napi* that only mated with virgin males (i.e. males that delivered large ejaculates), on the other hand, only mated on average 2.8 times. They presumably had enough ejaculate material and had no need to mate again. However, under normal circumstances in the field, females do not always discriminate between virgin males and sexually experienced males (see section on 'Costs and benefits'), so do not necessarily benefit in this way. In summary, the mating history of the male has a profound influence on the outcome – in terms of the survival and fitness of their offspring – of mating.[32]

Experience counts? Old male mating advantage

Despite their declining powers, worn wings and lighter body mass, older males are sometimes more successful at mating than younger ones. This is not because females are deliberately choosing older males – as we have seen, such matings are not to their advantage – but instead, it is put down to the fact that older males are usually more active, aggressive or persistent during courtship. Females suffer a fitness cost associated with mating with older males whose ejaculate quality has diminished, but these males are often more eager to mate. For example, in *Bicyclus anynana*, older males are more active than younger males, spending considerably more time flying around and courting females, despite being in poorer condition. It is thought that because butterflies lack any way of effectively fighting each other – they are weaponless (see Chapter 3) – traditional advantages such as size and weight are lately nullified, and determination – or persistence – may be more important in terms of settling disputes or winning contests. The high intrinsic motivation of older males to reproduce could explain why they are more aggressive and persistent than younger males during courtship, even though they are in poorer condition, often with worn wings (Fig. 7.14).[33]

Some long-lived male butterflies have a high lifetime mating success as a result of their increased experience in finding females. For example, the higher mating success of older males – i.e. those that had already mated more than once – of the large Asian papilionid, *Byasa (= Atrophaneura)*, the

Fig. 7.14. Clipper (*Parthenos sylvia apicalis*) old male with worn wings, Chiang Dao, Thailand.

Chinese windmill (Fig. 9.6), was because they were better at finding females and more skilful at copulation. In other words, the most successful males had, over the course of mating with up to five females, become more skilled at the mating game. Experience correlates with age in this case: the success of these individuals was not simply because they were old, but because they were experienced. Indeed, the majority of older butterflies failed to find any mates at all. There is the possibility, though, that females were choosing to mate with persistent and aggressive males because of the attractiveness of this behaviour, or some other feature, which outweighed the fact that they were old and presumably worn.[34]

In the squinting bushbrown, *Bi. anynana* (Figs 2.12–2.14) – a so-called model species because it has been reared in several laboratories around the world and used in many experiments on different aspects of butterfly biology – females prefer to mate with what were classed as 'mid' aged, rather than younger males. It is thought that the females can differentiate between males of different ages by assessing changes in one specific pheromone component – hexadecimal – and can be tricked into mating with younger males if they have the smell, i.e. the pheromone titre, of a more mature male! Sensory cues other than pheromones may also be used by females to assess the quality of male butterflies. Older males of *B. anynana* courted females more often and for longer periods than younger males; surprisingly, in this species, the older males also have larger spermatophores and more fertile sperm, although curiously this does not translate into more offspring, presumably because the quality of the ejaculate has diminished. The reproductive success of females was highest when mating with young males, which had only mated once, even though in this case, they produced small spermatophores with low sperm numbers.[35]

Cryptic female choice: rewarding generous suitors

Cryptic female choice, which occurs within the reproductive tract of the female, utilizes devices that are not fully understood, but appear to involve the female making assessments of male quality *during* and *after* copulation. A variety of mechanisms may then be employed to delay, remove or select sperm, according to the current predilection of the female. In some species, females appear to be able to adjust their reproductive investment in relation to the quantity of nutrients received from the male. Female small white butterflies, *P. rapae*, may be able to adjust the rate at which sperm is transferred from the spermatophore to the spermatheca, potentially increasing or decreasing the probability with which sperm is used for fertilization. Other ways in which female insects can potentially exert some degree of cryptic choice include: the premature interruption of copulation; not storing the sperm at all; failing to produce any eggs for the sperm to fertilize; and even the selective abortion of fertilized eggs. Cryptic female choice probably has an important role to play in butterflies, particularly in highly polyandrous species and in situations where females have little or no pre-mating choices. The latter includes pupal mating, mating immediately or soon after female eclosion, and situations where males coerce or force females into mating without any obvious courtship.[36]

The extent to which butterflies utilize all these covert mechanisms, and others, is largely still to be elucidated, but some species are undoubtedly able to make postcopulatory choices regarding a mate, and to allocate their resources accordingly. For example, female commas, *Polygonia c-album*, vary the amount of resources they put into eggs – lifetime egg production – based on the amount of nutrients they receive from a male. The more protein he gives to her in his spermatophore, the more – i.e. a larger proportion – she puts into egg production. Another benefit of rewarding high-quality males in this way is that the female also lives longer, at least in *P. c-album*. It has recently been demonstrated that female Asian swallowtails, *Papilio xuthus* (Fig. 2.2a, b), can eject the sperm obtained from a prior copulation, if the spermatophore delivered by the male in the second mating is larger than the first one. Sperm displacement also depends on the time that has elapsed between matings. Virgin females can store sperm from a first mating for more than 10 days, but flush out sperm from the first male if remating occurs within 7 days; unless, that is, the female receives a small spermatophore from the second male. The females appear to be using spermatophore size as a measure of male quality and shift the paternity of their eggs towards the most generous suitor.[37]

Costs and benefits

It is not too surprising perhaps, to learn that there are costs associated with mating, for both males and females. As we have seen, the detrimental effect for females of mating per se comes in the form of reduced longevity, compared with females who remain unmated throughout their lifetimes. For example, in *B. anynana*, even females that mate only once have shorter lifespans than unmated, virgin ones. Mated female orange sulphurs, *Colias eurytheme*, also suffer a reduction in longevity relative to virgins. It is not altogether clear why these female butterflies suffer these costs, but it might be associated with the toxic effects from some component of the males' ejaculate.[38]

Male butterflies can also incur costs as a result of reproduction, particularly when an individual has been very actively engaged in courting and mating activities over an extended period. For example, in *B. anynana*, the mating success of males with unrestricted access to females for up to 15 days was greatly reduced: two to three times less than males who had not had prior access to females. Almost as though they were worn out! Clearly, a male's reproductive success – his ability to compete with other males and attract receptive females – can suffer if he has already expended large amounts of energy and donated considerable quantities of protein in the form of sperm. Indeed, the costs of mating, in terms of longevity for example, are probably greater for males in butterflies, such as *P. napi*, for example, which make very large contributions – i.e. in the form of a big spermatophore – compared with those that produce very small packets. N.B. Spermatophores are up to 18.5 times larger, proportionately, in *P. napi* compared with *B. anynana*. However, the costs of courtship and the demands of producing pheromones may be more important than nutrient provisioning in males that have substantial reserves as a result of larval feeding. The negative effects of multiple matings, i.e. reduced longevity, might be proportionately greater in smaller males for this reason. For example, the negative effects of multiple matings in the common green-eyed white, *Leptophobia aripa*, were most apparent in small males, even after only one copulation![39]

The resources required to make a spermatophore are finite and are largely derived from nitrogenous material obtained when the butterfly was a larva, i.e. feeding on plant material. Since these stores of nitrogen cannot be replenished simply by nectaring – nectar contains mostly sugars – the male butterfly must use its sperm strategically, donating large spermatophores only in circumstances where they will be utilized for reproduction, and not be flushed out by the female in favour of another male. The costs of producing spermatophores become apparent when resources are limited, and can result in reduced longevity, but there may also be genetic differences between individual males in terms of their ability to tolerate the demands of reproduction. Male checkered white butterflies, *Pieris protodice*, were depleted by copulation – they pass about 7–8% of their body weight in the form of ejaculate material to the female – but recovered their potency within 24 h.[40]

The mating costs for the female can be greater than those for the male, particularly when mating with recently mated males, who provide small spermatophores and take a much longer time to copulate. For example, nutrient provisioning decreases with increased copulations in male green-veined whites, *P. napi*, but females do not make preferential choices over virgin or recently mated males. So, they may incur the costs of increased time spent *in copula* – 7 h on average, with recently mated *P. napi*, for example, compared with 2 h for virgin males – as well as receiving smaller ejaculates. Female orange sulphur butterflies, *C. eurytheme* – die sooner and have a lower output of eggs over their lifetimes, as a result of receiving ejaculates from recently mated males, which are substantially smaller than those donated by males that have not mated.[41]

Pupal Mating

In some species of *Heliconius* butterflies – and one other, non-heliconiid species, the Australian lycaenid butterfly, *Jalmenus evagoras* – males carry out a behaviour known as pupal mating. Males use a combination of olfactory and visual cues to find pupation sites, where they perch on the backs of female pupae and mate with them as soon as, and sometimes even before, they emerge from their pupal case. In *J. evagoras*, males are said to engage in a 'frenzied scramble' as the females emerge. Mating often occurs before the females have even had a chance to spread their wings. Unlike most other situations, where female butterflies can reject males and choose whom they mate with, in this case they are unable to refuse, or prevent the male's advances. This is a clear example of sexual coercion, with potentially severe consequences for the females, in that they may be injured or knocked to the ground. At the very least, pupal mating deprives the female of the ability to choose a mate, which could be costly in terms of reproduction if 'poor-quality' males are successful at this type of forced mating. In monandrous species, mating usually take place immediately after the female emerges, but in polyandrous species, it tends to take place after a period of maturation.[42]

Heliconius males are attracted by pupal pheromones produced by the females and jostle with each other for access: the largest one usually being the winner in terms of preventing the others from gaining a foothold, or 'seat', on the developing female pupa. This size advantage is reversed during the actual mating event – the males puncture the cuticle and insert the ends of their abdomens – when smaller, more agile males are more likely to be successful. In pupal-mating *Heliconius* butterflies, the spermatophores are small (c. 2% of dry weight) and, in contrast to almost all other species, they break down and degrade relatively rapidly, within 2–3 weeks. It is notable that *Heliconius* species can also derive amino acids and proteins via pollen feeding, which allows females to produce a constant daily production of eggs throughout their relatively long lives. Nevertheless, nutrients derived from the male spermatophore are still important in terms of egg production, especially in the early life of adult females, in species like the zebra longwing, *Heliconius charitonius*. In one so-called pre-copulatory mate-guarding species – the Sara longwing, *Heliconius sara* (Fig. 7.15), it is the larger males who exhibit this behaviour; smaller males employ a different strategy and compete for more mature females on small, dispersed territories, which are thought to function as mating stations. In these territorial contests, it is the smaller males – whose wings are 3% shorter, on average – who consistently defeat the larger individuals. Why the smaller males have this territorial advantage

Fig. 7.15. Sara longwing (*Heliconius sara*), butterfly zoo, Sayn, Germany. (Photo by Pro2 [CC BY-SA 3.0 (https://creativecommons.org/licenses/by-sa/3.0)].)

yet are at a disadvantage when they compete with larger males for access to the pupae is something of a mystery. It is tempting to think that the agility and manoeuvrability associated with being small is an advantage in one type of contest, but not another, but this is unproven.[43]

Some mate-guarding males start searching for potential mates when the females are still in the larval stage. They scour their host plants for developing larvae of their own species, using scents produced by the caterpillar-damaged plants, to home in on, and then track the progress of the fifth instar larvae, until they start to pupate. In the zebra longwing, *Heliconius charithonia*, males recognize which pupae are female by means of sex-specific compounds – volatile monoterpenoids – produced towards the end of pupal development. In Hewitson's longwing, *Heliconius hewitsoni*, females exist as pupae for about 9 days, and males compete for access, and start to sit on them, about 1–2 days before they emerge as adults. The pupae change in appearance – becoming more transparent, with visible wing patterns – 1 day prior to eclosion, and mating can take place as soon as the female begins to emerge. Once again, only the largest males sat on and competed successfully for these female pupae.[44]

Male zebra longwings, *H. charithonia*, have been seen attending female pupae at least a week prior to them becoming sexually mature, or at least receptive, and fighting with other males over their position on the pupa. Males mated with the pharate – that is the uneclosed or unemerged – females by inserting their abdomens into the chrysalis or pupa. The female emerged after 2–3 h, still attached to the male, *in copula*, and they remained like this for a further hour or so. The mating male appeared to repel the attentions of other males by opening his wings. Males of another heliconiid butterfly – *Heliconius melpomene* – transfer antiaphrodisiac pheromones to females during mating that prevent other matings, at least for a while (see Chapter 9).[45]

About half of *Heliconius* butterflies engage in pupal mating, although how frequently it occurs in the wild is a matter of conjecture. DNA analysis suggests that this unusual mating strategy only evolved once, in a 'clade' of about 19 such species within this group. Strangely enough, in at least one of these pupal-mating species, *H. hewitsoni*, the male butterflies prefer to sit and compete with another male who is already guarding a pupa, rather than find a pupa that is unoccupied by other males. The females of the pupal-mating species are all thought to be monandrous, whereas the so-called adult-mating species – which are courted by males in the usual way of butterflies – are polyandrous. The incidence of pupal mating appears to be variable. Adult mating was the most prevalent form of reproduction in *Heliconius erato*, in the field, in a study carried out in Panama. The incidence of pupal mating may be related to factors such as the availability of food, the density of the population and the synchronous emergence of females.[46]

Heliconius species are particularly long-lived butterflies, with adults surviving for up to 4 months or more, so opportunities for guarding and copulating with new cohorts presumably arise frequently. In most seasonal species, with a well-defined flight season, males typically emerge from their cocoons earlier than females. Protandry is thought to arise as a result of competition for access to females: pressure to emerge earlier selects for rapid development and this produces smaller males as a result.[47]

Strange or Enigmatic Liaisons

Incest and inbreeding

Inbreeding occurs when reproduction takes place between closely related individuals. The consequences of such unions, especially between full siblings – individuals sharing the same parents – can be severe, including the risk of extinction, but may be unavoidable in species that are unable to recognize close relatives. Inbreeding caused a marked decrease in mating success in male *B. anynana* and a decrease in egg hatching. True kin recognition, as it is called, is the ability to distinguish relatives from non-relatives, even if the individuals have never met before. The ability to avoid inbreeding by recognizing your own kin in some way is probably relatively common in the animal world, and has been demonstrated in bumblebees, bees, ants, termites and cockroaches. Bumblebees (*Bombus terrestris*), for example, have been shown to exhibit kin recognition, at least to the extent that they are less willing to mate with their siblings than with unrelated individuals.[48]

Relatively few studies have directly addressed the question of whether such inbreeding avoidance occurs in butterflies however, and the results are varied. Researchers studying the green-veined white butterfly, *P. napi*, in Finland concluded that neither males nor females discriminated against close relatives during mating, despite having the opportunity to do so during courtship. There was, however, evidence of 'postcopulatory inbreeding avoidance' – in other words, actions that occur after mating – because females who had mated with a full sibling mated again sooner and laid fewer eggs after the incestuous mating than the ones who had mated with a non-relative. Somehow, the females detected that they had mated with a sibling and took appropriate action to avoid depositing too many of the inbred eggs, which were less likely to hatch successfully. Another study, on the Glanville fritillary, *Melitaea cinxia* (Fig. 7.16a, b), failed to find any preferences for mating with non-siblings, although postcopulatory avoidance of the type described above was not ruled out. One generation of brother–sister matings in this species can reduce the egg hatching rate by up to 46%, and such inbreeding increases the risk of extinction in small, isolated populations. The best way for butterflies originating from such colonies to avoid inbreeding is to emigrate. Metapopulations – consisting of numerous small, relatively isolated, local populations – can persist despite the extinction of some component subpopulations, in any one year, and may be more resilient to the effects of inbreeding, but nevertheless remain vulnerable to its effects.[49]

Some good evidence for kin recognition in butterflies comes from studies on *B. anynana*, which suggest that they can recognize kin, at least to some extent, and appear to discriminate against siblings for mating, under some circumstances. It is possible that they can do this by means of cuticular hydrocarbons: diverse and variable long-chain molecules that form a lipid layer on the outside of the cuticle, protecting the insect from drying out and being abraded. There is also some indirect evidence that *Heliconius* caterpillars can recognize kin, in that the frequency of cannibalism was considerably less for sibling eggs than for non-sibling ones. In other words, the caterpillars were less inclined to eat their unhatched brothers and sisters than unrelated ones![50]

Same-sex interactions

Sexual interactions between members of the same sex are relatively common in butterflies. These same-sex interactions include males guarding pupae of the same sex, same-sex courtship attempts and – less

Fig. 7.16. Glanville fritillary (*Melitaea cinxia*) dorsal, Avilly-Saint-Léonard, Oise, France. (b) ventral. (Photos by Muséum de Toulouse [CC BY-SA 4.0 (https://creativecommons.org/licenses/by-sa/4.0)].)

commonly – homosexual copulations. Various hypotheses have been put forward to try and explain these observations, including the *practice hypothesis*, where sexually immature adults were thought to be just practising on members of the same sex! For example, there are some reports of older males directing their courtship and copulation attempts towards teneral (newly emerged) males. The *intrasexual conflict hypothesis* puts forward the rather drastic proposal that same-sex interactions are sometimes used to inflict damage on sexual competitors. Male monarch butterflies may fall into this category (see Chapter 5). The *mistaken identity hypothesis*, as it is rather beguilingly called – and a version of this, called the 'sexual selection by-product' hypothesis – is thought to provide a good explanation for many same-sex interactions in butterflies. This favoured explanation makes the case that males with a 'strong male drive' frequently make category mistakes and fail to discriminate between appropriate and inappropriate partners – the latter including other males, flowers and inanimate objects! An unusual pairing between two males of closely related species – a hackberry emperor, *Asterocampa celtis*, and a tawny emperor, *Asterocampa clyton* – in Texas, USA, was presumably an example of mistaken identity. In species where there is intense competition for mates, males who are highly motivated to mate – what we would call a strong sex drive – who are continuously and actively searching for mating opportunities, will be at a selective advantage. Mistakes in the choice of potential mates could be a consequence of this sexual selection pressure.[51]

In a short note to the *Entomologist's Record and Journal of Variation*, entitled 'A note on the apparent lowering of moral standards in the Lepidoptera', W.J. Tennent amusingly described 'aberrant sexual behaviour' in the mazarine blue, *Polyommatus semiargus* (Polyommatini, Lycaenidae), in Morocco. Males were observed clustered around other newly emerging males, attempting to mate with them and ignoring females displaying in the same arena. The short piece finishes by stating, 'to set the reader's mind at rest', that several 'normal', i.e. male–female pairs, were seen *in copula*.[52]

Interspecific mating and hybridization

Examples of mating between males and females of a different species are not uncommon in butterflies – particularly so-called congeneric pairing, between species in the same genus – and have been reported for a wide range of species from different families/tribes, including nymphalid, lycaenid, pierid and papilionid butterflies. Interspecific or inter-generic pairings (i.e. between species in different genera, but usually the same family) are occasionally reported in the literature. For example, interspecific courtship and mating was observed by Peter Hardy (UK), between a male *P. rapae* and a female *P. napi*. Findings of closely related, conspecific species *in copula* can be very useful to taxonomists in confirming theories regarding the relatedness of species. Landscape changes wrought by man have brought together closely related species that previously occupied separate geographical ranges, including, for example, the sulphurs, *C. eurytheme* (Fig. 2.15) and *Colias philodice*, which now hybridize widely. These closely related North American pierids, with a high degree of genetic compatibility, can hybridize where both species occur together. Interspecific matings in the *C. philodice–eurytheme* complex commonly occur in the first hour after eclosion, when the females are less able to discriminate or make any selection regarding mates. The females are courted soon after they emerge and offer little resistance at this stage; however, their ability to select conspecific mates increases as they age. Such newly emerged females are more likely to be found by males when numbers are high, but hybrid offspring may be at some sort of disadvantage because hybrid frequencies rarely rise above 10% and the two species maintain their separate identities. There are numerous other examples of hybridization, including the swallowtails, *Papilio zelicaon* and *Papilio machaon* (Fig. 3.9). The genus *Limenitis* (Fig. 7.17) also readily undergoes interspecific hybridization; e.g. *Limenitis archippus* occasionally hybridizes with other *Limenitis* species in N. America.[53]

Interspecific mating between two closely related species – the common plain tiger, *Danaus chrysippus*, and the common tiger, *Danaus genutia* (Fig. 7.18) – was reported on two occasions from India (Kolkata). Both cases involved a male *D. genutia* and female *D. chrysippus* (Fig. 7.5) and a spermatheca was transferred. There are several other reports of intergeneric mating between lycaenid species from India, including: between *Rathinda amor* and *Anthene emolus*, in Kolkata; between *Castalius rosimon* and *Talicada nyseus*; and between *Chilades parrhasius* and *Zizina otis*, both in Maharashtra. Courtship between different, but closely related species can sometimes occur, without progressing to copulation, presumably because they still have some traits associated with attraction (colours, patterns, odours?), or mate choice, in common. Courtship attempts with individuals of a very different type of species are probably uncommon but can occur when they look superficially similar. For example, a male skipper – the northern cloudywing, *Thorybes pylades* (Hesperiidae, Eudaminae) – repeatedly approached a male moth, *Euclidina cuspidea* (Erebidae) of vaguely similar size and colour (brown!), which reportedly

Fig. 7.17. White admiral (*Limenitis camilla*) basking, Galicia, Spain.

flies in 'a skipper-like manner'. Clearly, another example of mistaken identity.[54]

Finally, I will mention a strange courtship attempt I came across in the forest on Doi Sutep, a mountain near Chiang Mai, Thailand. A male common earl (*Tanaecia julii*) was chasing after a female archduke (*Lexias pardalis dirteana*); the pair settled on a leaf and the male started bending his abdomen towards the much larger female butterfly (Fig. 7.19), who eventually flew off.

Fig. 7.18. Common tiger (*Danaus genutia genutia*) male underside, Kumarakom, Kerala, India. (Photo by Charles J. Sharp [CC BY-SA 4.0 (https://creativecommons.org/licenses/by-sa/4.0)].)

Fig. 7.19. (left) Common earl (*Tanaecia julii*) male courting an Archduke (*Lexias pardalis dirteana*) female; (right) Common Earl (*Tanaecia julii*) male, abdomen bending. Thailand.

Fig. 7.20. Clouded apollo (*Papilio mnemosyne*) newly eclosed, Tahitótfalu, Pest, Hungary. (Photo by Ádám Gór.)

Notes

[1] Bergström and Wiklund (2005).
[2] Chapman *et al.* (2003) and Hosken *et al.* (2008). The appearance and aroma of a butterfly may be a form of 'shorthand notation' indicating their health and vigour (Rutowski, 1998).
[3] Andersson and Simmons (2006). Competition between the sperm of different males can occur within the female reproductive tract, 'aided and abetted by the female herself' (Birkhead and Pizzari, 2002). Wedell (2001) suggests that there is often a conflict between the sexes over female mating rate, because females remate in order to obtain male ejaculatory donations, whilst males try to prevent females from remating, in order to monopolize fertilization.
[4] Suzuki and Matsumoto (1992), Calabrese *et al.* (2008), Rhainds (2010), and Hiroyoshi and Reddy (2018).
[5] Emlen and Oring (1977) and Odendaal *et al.* (1985).
[6] Boggs (1981) and Odendaal *et al.* (1985).
[7] Suzuki (1980), Tabashnik (1980) and Ehrlich *et al.* (1984).
[8] Scott (1973a), Shields and Emmel (1973), Calhoun (1996), Miller and Clench (1968), Shull (1978), Wiklund and Forsberg (1991) and Wells *et al.* (2011). Steve Cheshire reported an observation of mating *Argynnis paphia*, where the role of carrier switched whilst *in copula*, see https://www.britishbutterflies.co.uk
[9] Miller and Clench (1968), Shields and Emmel (1973), Scott (1986a), Shull (1978). John Brackenbury (2019) photographed different species in flight, whilst in copula, including the common blue, *Polyommatus icarus*, with the male carrying the female (fig. 7.6). The female carries the male in *Hylephilia* skippers (Benyamini, 1992). The male carried the female in the Turkish fiery copper, *Lycaena ochimus* (Schurian and Fiedler, 1996).
[10] Hughes *et al.* (2000). Mating pains of the thecline butterfly, the western brown elfin, *Incisalia augustinus iroides*, (Lycaenidae), remained in copula for 4–6 h, until after dark (Powell, 1968). Copulation lasted for between 2 and 6 h in *Hipparchia semele* (García-Barros, 1989). Copulation lasted about 45 min between virgin male and female sulphur butterflies, *Colias erate* (Watanabe *et al.*, 1997). Copulation lasted between 30 min and 1 h in the banded peacock, *Anartia fatima* (Nymphalidae), but can continue for as long as 12 hours (Silberglied *et al.*, 1979). Copulation lasted about 16 min in the small copper, *Lycaena phlaeas* (Watanabe and Nishimura, 2001), and an average of 27 min in the buckeye butterfly, *Precis coenia* (Scott, 1975b). A copula duration of nearly a week was recorded for a brimstone, *Gonepteryx rhamni*, pair (Almbro and Kullberg, 2009, quoting Lindfors, 1998). Copulations in *Pieris rapae crucivora* lasted for 85 min on average (n = 10) (Ohtani, 1985).

[11] The male clasps the end of the female's abdomen with his valvae, uncus, gnathos and falces, before inserting his aedeagus (Shields and Emmel, 1973). Tsubaki and Matsumoto (1998).
[12] Data on copulation duration for *Phengaris nausithous* and *Phengaris teleius* from Pfeifer *et al.* (2000) and Gao *et al.* (2016), respectively. *Hylephila phyleus* copulation data and pair carrying observations, from Shapiro (1975). Copulation durations, based on hand pairings of *Luehdorfia japonica* from Tsubaki and Matsumoto (1998). In *Phyciodes tharos*, the female often (the male seldom) basks during mating, but if disturbed, the female flies, carrying the male (Scott, 1986b). Scott (1975b) reported that females were always the carrying sex in common buckeye butterflies, *Junonia coenia*, and frequently spread their wings to bask, during copulation.
[13] Adamski (2004). Surpluses of adult males in clouded apollo, *P. mnemosyne,* populations are real, according to Vlasanek *et al.* (2009). Harassment by males causes females to increase their rate of movement (Odendaal *et al.*, 1989).
[14] Wickman (1985c), Almbro and Kullberg (2009) and Välimäki *et al.* (2011). For more photographs of mate carrying, see Brackenbury (2019).
[15] Cordero (2002), Almbro and Kullberg (2009) and Ng *et al.* (2017).
[16] Edward *et al.* (2015), South and Lewis (2011) and Wedell (2001).
[17] Oberhauser (1988).
[18] Wiklund and Fagerström (1977), Wiklund and Solbreck (1982), Nylin *et al.* (1993) and Zonneveld (1996).
[19] Shapiro (1970), Hiroki and Kato (1996) and Watanabe *et al.* (1997). Female *Parnassius smintheus* can mate immediately after expanding their wings, and their mating success decreases with age, possibly due to the loss of a pheromone produced in the pupal stage (Matter *et al.*, 2012).
[20] Sexual conflict between males and females is the rule rather than the exception, according to Andersson and Simmons (2006). Conflict between the sexes occurs whenever there is not strict genetic monogamy (Hosken and Stockley, 2004). See Alexander *et al.* (1997) for a survey of male–female conflicts and confluences in insect mating.
[21] Gavrilets *et al.* (2001) and Arnqvist and Rowe (1995).
[22] Wiklund (1982) and Drummond (1984).
[23] Ehrlich and Ehrlich (1978), Wiklund (1982), Wiklund and Forsberg (1991), Cordero (1999), Wiklund *et al.* (2001, 2003) and Watanabe (2016b).
[24] Suzuki (1979), Svärd and Wiklund (1989, 1991), Watanabe and Ando (1993) and Wiklund *et al.* (2003).
[25] Braby (1996).
[26] Kawagoe *et al.* (2001), Wedell *et al.* (2002), Välimäki *et al.* (2006), Kivelä and Välimäki (2008), Watanabe (1988). Wiklund *et al.* (2001) demonstrated the positive effects of polyandry on female reproductive fitness; however, Svärd and Wiklund (1988b) found that the multiple matings had no effect on longevity, lifetime fecundity or egg weight in female monarchs, *Danaus plexippus*, in a laboratory study.
[27] Karlsson (1995) and Välimäki *et al.* (2006).
[28] Karlsson (1995), Kemp and Rutowski (2004) and Wiklund *et al.* (2001). Gage (1994) showed that the lengths of both eupyrene (fertilizing) and apyrene (non-fertile) sperms are positively correlated with body size.
[29] Kemp and Rutowski (2004).
[30] *Pieris rapae* studies: Wedell and Cook (1998, 1999a, b). *Danaus plexippus* studies: Drummond (1984), Rogers and Wells (1984), and Solensky and Oberhauser (2009).
[31] Male *J. evagoras* transferred 2.2% of their eclosion weight during their first mating and this percentage decreased with subsequent matings (Hughes *et al.*, 2000).
[32] Svärd and Wiklund (1989), Lindfors (1998), Forsberg and Wiklund (1989), Watanabe and Hirota (1999), Andersson *et al.* (2000), Arnqvist and Nilsson (2000), Torres-Vila and Jennions (2005), Cardoso *et al.* (2009) and Duplouy *et al.* (2018). Recently mated *Colias eurytheme* males produced only c. 40% of the ejaculate material produced by those that had not mated recently (Rutowski and Gilchrist, 1986). Female *C. eurytheme* who mated with recently mated males died sooner and produced fewer eggs (Rutowski *et al.*, 1987).

[33] Older *B. anynana* males achieved higher mating rates than younger ones, by being 'more active, aggressive and persistent during courtship' (Karl and Fisher, 2013). Older males were more active and eager to mate (Karl *et al.*, 2013). Intrinsic motivation was more important in terms of mating success than body size and condition (Fischer *et al.*, 2008). Female *B. anynana* suffer fitness costs as a result of mating with older males (Fischer *et al.*, 2018). Male *Hypolimnas bolina* become more willing to persist in territorial contests as they age, and are more accepting of injury risks (Kemp, 2002a).
[34] Suzuki and Matsumoto (1992).
[35] Old male mating advantage has been repeatedly demonstrated in *B. anynana* (Fischer *et al.*, 2018). Nieberding *et al.* (2012), Kehl *et al.* (2013, 2014, 2015a).
[36] Firman *et al.* (2017). Sexual selection during mating is called *syncopulatory* mate choice; afterwards is called *postcopulatory*.
[37] Eberhard (1996), Wedell (1996), Wedell and Cook (1998), Watanabe and Sasaki (2010) and Sasaki *et al.* (2015b).
[38] Kemp and Rutowski (2004) and Fischer (2007).
[39] Ferkau and Fischer (2006), Janowitz and Fischer (2010), Wedell (2010), and Caballero-Mendieta and Cordero (2013).
[40] Rutowski (1984d), Cordero (2000), Wedell (2010) and Takeuchi (2012).
[41] Rutowski and Gilchrist (1986), Rutowski *et al.* (1987), Kaitala and Wiklund (1994a, b), and Wedell and Karlsson (2003). Kaitala and Wiklund (1994b) consider that *P. napi* males are more successful in the sexual conflict over mating decisions because females do not minimize their mating costs by choosing to mate preferentially with virgin males. On the other hand, Watanabe *et al.* (1997) suggested that female Japanese sulphurs, *Colias erate*, can identify the age of males, and may choose mates based on their pheromones and wing scales.
[42] Elgar and Pierce (1988), Ridley (1989b), Estrada (2009), Estrada *et al.* (2010) and Thurman *et al.* (2018).
[43] Gilbert (1972), Boggs and Gilbert (1979), Boggs (1981, 1990), Deinert (2003), Walters *et al.* (2012) and Jiggins (2016). The territorial advantage shown by small *H. sara* males is a conclusion that must be considered provisional according to Hernandez and Benson (1998).
[44] Mendoza-Cuenca and Macías-Ordóñez (2009). The main compounds used in sex recognition by *H. charithonia* males are the monoterpenoids, linalool and linalool oxide, found in male and female pupae, respectively. These are common semiochemicals released by flowers or leaves in response to insect feeding (Estrada *et al.*, 2010).
[45] Gilbert (1976). See Sourakov (2008) for photographs and descriptions of pupal mating in zebra longwings (*Heliconius charithonia*). So-called *true* pupal mating is where the males break open the pupal cases and insert their abdomens into the chrysalis, to start mating before the females have fully emerged (Thurman *et al.*, 2018).
[46] Deinert *et al.* (1994), Walters *et al.* (2012), Thurman *et al.* (2018) and Beltrán *et al.* (2007, 2010).
[47] Rutowski (1984c) and Nève and Singer (2008).
[48] Saccheri *et al.* (1996), Joron and Brakefield (2003) and Whitehorn *et al.* (2009).
[49] Saccheri *et al.* (1998) and Haikola *et al.* (2001, 2004) and Välimäki *et al.* (2011).
[50] Kather and Martin (2012) and Fischer *et al.* (2015). Kin recognition (caterpillar–egg) occurs in *Heliconius erato phyllis* (De Nardin and De Araújo, 2011).
[51] Clark (2011) and Caballero-Mendieta and Cordero (2012).
[52] Tennent (1987).
[53] Taylor (1972), Downey (1962), Silberglied and Taylor (1973), Platt *et al.* (1978), Sperling (1990), Baidya *et al.* (2018) and Hardy (2018).
[54] Heitzman (1964), Bhakare and Smetacek (2010), Ramos and Francini (2014), Ray and Dutta (2016) and Baidya *et al.* (2018). Ramos and Francini (2014) provide a summary table of examples of inter-generic mating, and mating attempts in butterflies and moths.

8 The Inside Story

The Mechanics of Coupling

The genitalia of butterflies are complex organs that come in all shapes and sizes (Figs 8.1c and d). The traditional view was that each species has its own unique combination of genital structures that fit together like a lock and key. In theory, this might explain why species remain genetically isolated: males of other species simply have the wrong 'key' and cannot open the female lock! Evolution is thus the driving force behind the diversification of genital structures in order to prevent hybridization of related species. In practice, however, it is difficult to prove this theory one way or the other, i.e. to falsify. Unless a pair are taken in the act of copulation – *in flagrante delicto* – it is impossible to distinguish whether a male and female are physically unable to form a union, or are just unwilling, or uninclined, to do so. Although not completely discarded, the lock-and-key hypothesis is now seen as too simplistic, and there are several other competing theories to explain genital evolution. For example, according to the so-called sperm competition theory, evolutionary selection may have favoured males with genitalia that are able to remove sperm deposited by a rival in a prior mating. It is easy to imagine how such competition might have driven the evolution of variation and diversity in male genitalia. It has also been argued that the female *locks* do not exist, or at least not to the same extent as that for male *keys*. In other words, there is not the equivalent diversity of female genital structures to match that seen in males; although, this may be because they have not been studied to the same extent as male genitalia. Regarding insects as a whole, there is evidence supporting the coevolution of male and female genitalia, with equivalent degrees of complexity, shaped by sexual selection.[1]

The genitalia of male butterflies are extremely diverse in shape and size and have long been used by taxonomists to separate species (Fig. 8.1).

© Raymond J.C. Cannon 2020. *Courtship and Mating in Butterflies: Reproduction, Mating Behaviour and Sexual Conflicts* (R.J.C. Cannon)

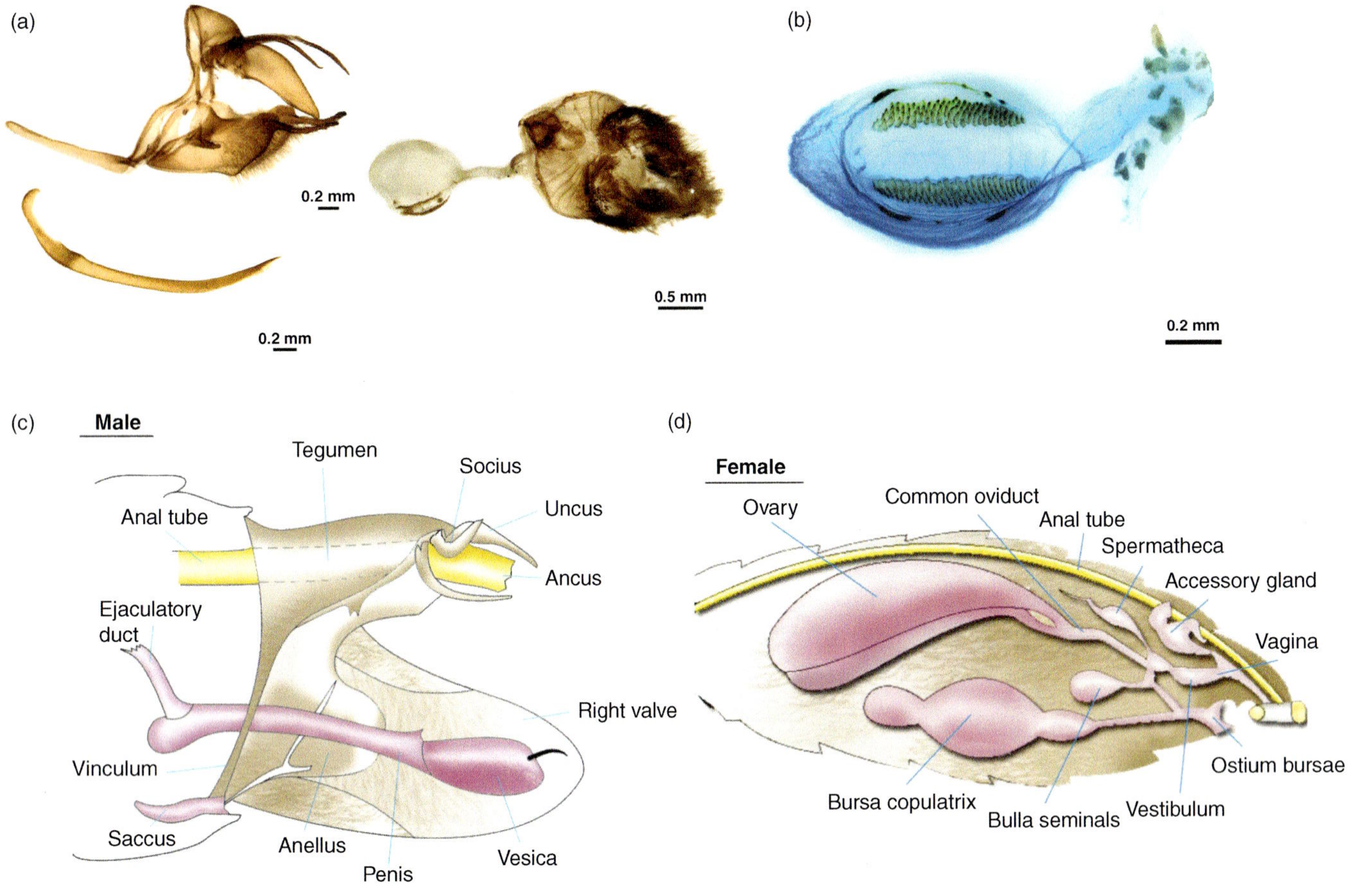

Fig. 8.1. (a) Male and female genitalia of *Sepona punctata* (Nymphalidae: Satyrinae); top left, male genitalia in lateral view; bottom left, male aedeagus (lateral view); right side, female genitalia in ventral view (corpus bursae on left) (from: Freitas *et al.*, 2016, fig. 3.); (b) female genitalia of *Sepona punctata* (Nymphalidae: Satyrinae): detail of the signum in corpus bursae. (from: Freitas *et al.*, 2016, fig. 3.); (c) stylized diagram of male genitalia in Lepidoptera (produced by Indian graphic artist Savita Murmure); (d) stylized diagram of female genitalia in Lepidoptera (produced by Indian graphic artist Savita Murmure).

Traditionally, male genitalia have been viewed as more informative in terms of taxonomic studies, but this may be because they are generally more heavily sclerotized and therefore easier to study (Figs 8.1a and 8.2) Female genitalia are also complex and diverse, and one reason why they have been less well studied than those of male butterflies could be that they are generally 'soft and membranous', hence trickier to extract and examine. Lepidopterists still do not fully understand how all the pieces of the complex genitalic structures function during mating, but the anatomical details of coupling are beyond the scope of this book, other than to mention that the site of insemination in the female is the corpus bursae (Figs 8.1a and d). This is the bag-shaped organ where the male deposits his ejaculates and produces the spermatophore. Note that the spermatophore is formed by the male, inside the female. The inner surface of the corpus bursae is covered with complex folds and sharp processes for holding the spermatophore and ensuring the release of sperm (see below).[2]

Tickle sticks!

The phallus – also called the aedeagus in insects – is an intromittent organ, which consists in part of a membranous tube called the endophallus, which may bear sclerotized structures called cornuti: thorny or spiny projections (Fig. 8.2e), which can become detached during copulation, as the endophallus is everted into the corpus bursae of the female. Male genitalia are considered by some researchers to be courtship devices, formed via the evolutionary process of sexual selection to enable the male to stimulate the female inside her reproductive tract. According to this hypothesis, originally proposed by William Eberhard, females choose males based on which individuals provide the most stimulating genitalia, and/or how well they fit her anatomy. In order to make such cryptic, i.e. internal, choices of mates or their gametes, females must be able to sense and discriminate between males based on their morphology – shape and size! – as well as their behaviour during mating. In monogamous species, where mating only occurs once, such choices appear limited, unless the female can reject the male at some point, early on during the copulation – before the male has produced a spermatophore – and start again with another individual. The function of the sclerotized structures – cornuti – on the surface of the everted endophallus is however, largely unknown, and many hypotheses have been put forward to explain their presence, depending on whether they break off during mating. In practice, they may have more than one function, but the fact that they vary so much in terms of number, position and structure suggests that they may have evolved as a result of sexual selection.[3]

Mating Plugs and Sphragides

Female butterflies are often ready to mate sooner than males, and for most species this is on the first day of emergence from the pupal cocoon. In species

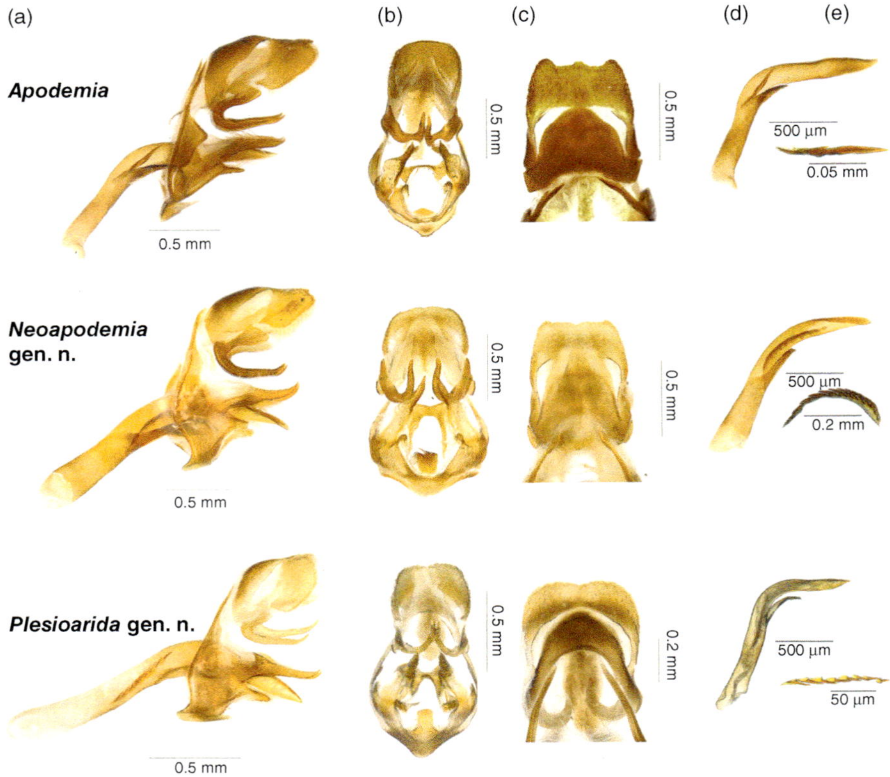

Fig. 8.2. Male genitalia of three metalmark butterflies of North and Central America (Lepidoptera, Riodinidae): *Apodemia*, *Plesioarida* and *Neoapodemia*. (a) lateral view, (b) ventral view, (c) dorsal view, (d) aedeagus and (e) cornuti. (From Trujano-Ortega *et al.*, 2018.)

where females mate with more than one male (see Chapter 7, this volume), the sperm from the most recent mating will, in theory, be enough to fertilize all of the available eggs produced by the female over the following few days – the so-called refractory period – until she is ready to mate again. In order to try and prevent another male from mating with her, the males of some species produce a small plug that is placed in the female copulatory orifice, which is called the gonopore or ostium bursae. This mechanical barrier does not prevent the female from depositing her fertilized eggs, because there are two genital openings in all higher Lepidoptera, but it may prevent her from remating. Unfortunately for the males, these butterfly chastity belts are not always very effective at preventing multiple matings, although one must assume that they are at least, partially effective – perhaps in delaying remating – to have evolved in the first place. Female graylings, *Hipparchia semele* (Fig. 1.13), for example, receive a postcopulatory plug during mating – a long, flat structure – which is not always obvious as it can get lost or abraded over time. Small mating plugs like this are generally less effective than larger ones.[4]

In some species, males secrete a much larger and more complex, external structure called a sphragis (plural: sphragides) – from the Greek word for a seal – to prevent further matings. Sphragides occur in widely disparate genera within the families Papilionidae, e.g. *Parnassius* (Fig. 8.3), *Luehdorfia*, *Cressida* (Fig. 8.4), *Euryades* (Fig. 8.5) and Nymphalidae (e.g. *Acraea* and *Pteronymia*). For example, male clouded apollos, *Parnassius mnemosyne*, attach waxy mating plugs during copulation, and female North American snow apollos, *Parnassius clodius*, can carry a very large sphragis. These large, external sphragides have a range of species-specific shapes, with the most complex forms occurring in some papilionid genera, such as *Cressida* and *Luehdorfia*. In the Japanese swallowtail, *Luehdorfia japonica* (Fig. 8.6), females mate early on in their adult life – probably on their first day of emergence – and never mate again, due to a very firmly attached sphragis. Males, on the other hand, may mate up to three times or more, although the physiological costs of producing such large sphragides, which contain proteins and lipids from accessory secretions, must be considerable. In some species, the males also utilize specialized abdominal scales to form the sphragis – giving it bulk and rigidity – during copulation; in other species, males incorporate scales from their valves into the sphragis. It can take a couple of hours to produce, time lost for continued mating in the male. In *L. japonica*, the sphragis is a broad, flat, slightly

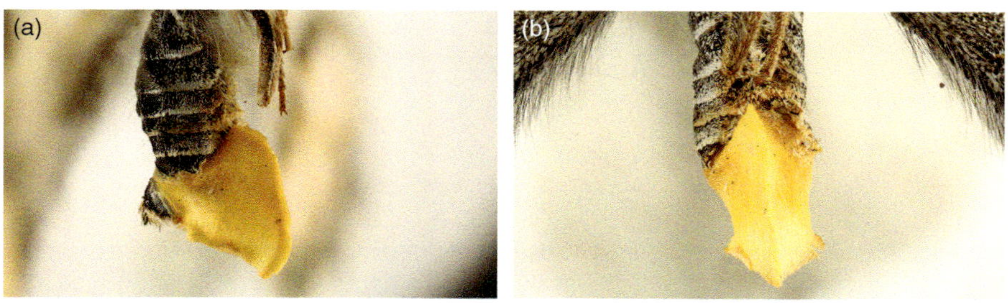

Fig. 8.3. (a) Sphragis on *Parnassius autocrator* (lateral view). (b) Ventral view. (From Carvalho *et al.* (2017), with kind permission of Ana Paula S. Carvalho.)

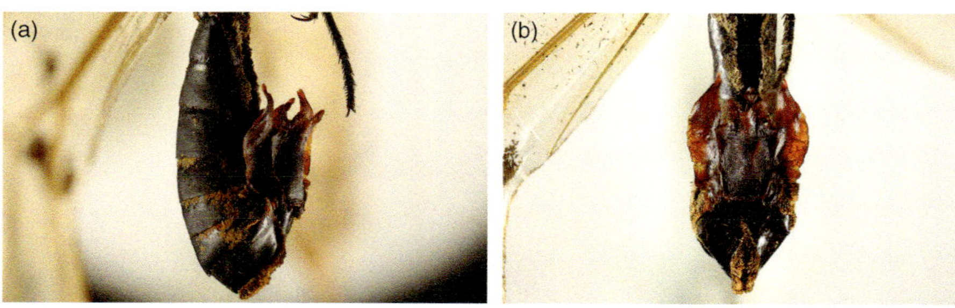

Fig. 8.4. (a) Sphragis on the clearwing swallowtail (*Cressida cressida*) lateral view. (b) Ventral view. (From Carvalho *et al.* (2017), with kind permission of Ana Paula S. Carvalho.)

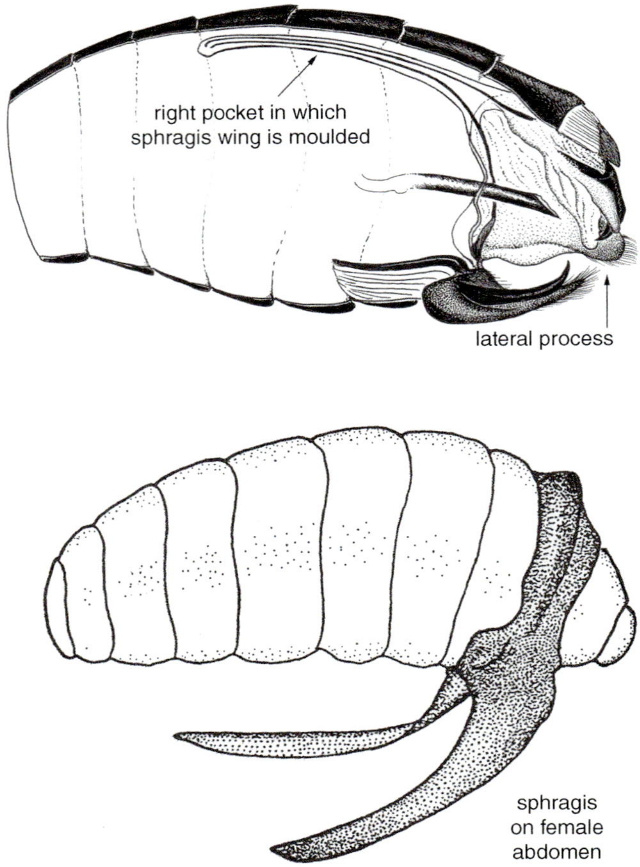

Fig. 8.5. Sphragis of *Euryades corethrus* (Papilionidae). (From Carvalho *et al.* (2017). *ZooKeys* 694, 41–70, fig. 34, with kind permission of Ana Paula S. Carvalho.)

Fig. 8.6. Japanese swallowtail (*Luehdorfia japonica*), Kasugai, Aichi prefecture, Japan. (Photo by Alpsdake [CC BY-SA 4.0 (https://creativecommons.org/licenses/by-sa/4.0)].)

keeled, shield-like structure (5–6 mm in length). Its large size probably prevents other males from even attempting to mate with the female and it remains in place presumably until she dies.[5]

Insemination plugs prevent subsequent inseminations by other males but are not totally reliable in all species. Some females avoid being plugged in the first place and a small proportion lose theirs altogether; furthermore, the males of some species have evolved specially modified genitalia with which they can remove the sphragis, for example, by levering it off or dissolving it. Freshly formed plugs can also be tunnelled though by the aedeagus of some male Chinese windmills, *Atrophaneura alcinous* (Fig. 9.6), which results in more multiple matings than would otherwise occur. Interestingly, the females of sphragis-bearing butterflies usually have a small, thin-walled bursa copulatrix and receive a small spermatophore. Competition between males seems to have selected for a large investment in terms of ways and means to prevent other males from mating with an already mated female, rather than in a large investment in sperm. Presumably, if a male can prevent other males from inseminating the female, there is no need to produce a lot of costly sperm. On the other hand, females may have evolved broader genitalia to prevent the plugs remaining in place.[6]

In the Chalcedon checkerspot butterfly, *Euphydryas chalcedona* – which is found in western North America – mating plugs prevent the intromission of new sperm and reduce the copulatory success of other courting males. The effect, however, is not just down to the mating plug; the females also exhibit mate-refusal behaviours – such as prolonged wing fluttering and turning away from courting males – which reduce the likelihood of remating. The evolution of mating plugs may have occurred as a result of intense sexual competition between males – often resulting in extreme forms of sphragis – together with conflicts between males and females. In some species, such as the Australian, shouldered brown, *Heteronympha penelope* (Nymphalidae, Satyrinae), males have specialized genitalia capable of removing sphragides. They make genital contact with a female and grasp the sphragis between the valves of their aedeagus, then with a twisting movement of their abdomen, wrench it off! As females evolved more 'externalized' genitalia, which were harder to plug, males countered by evolving larger and more effective structures that were difficult to remove, i.e. sphragides. However, the conflict between males and females may be relatively weak, as mating plugs ward off other males during the refractory period, when females are unreceptive. The benefits cease, however, when the females return to receptivity and would gain by remating.[7]

In Edith's checkerspot, *Euphydryas editha* – another butterfly of western North America – the spermatophore inserted in the bursa copulatrix by the male has a long neck, which forms a mating plug by sealing the female genital tract. Most mating plugs are connected to the spermatophore stalk. Once it has hardened, it prevents remating, although some sneaky males can apparently circumvent it before it has fully hardened, achieving a second mating. Females of this species usually only mate once, however, whereas the males mate multiple times with different females. In this polygamous arrangement, there are fewer females available to mate with because females usually actively reject male advances after mating, with or without mating plugs![8]

Sphragides are quite diverse in shape and size and have been found to be useful for separating closely related species – e.g. in the genus *Actinote* (Nymphalidae, Acraeinae), from south Brazil – which are difficult to tell apart using wing patterns. These are species in which the males carry out forced copulations, knocking the females down to the ground. Males of other sphragis-forming species – e.g. in the genera *Luehdorfia*, *Parnassius*, *Acraea* and *Cressida* – also capture females, either in mid-air or on the ground, and copulate with them by force, forcibly removing the sphragis if necessary (see Chapter 5, this volume). There are about 280 sphragis-bearing butterfly species known to science, and the trait probably evolved independently, a number of times during the evolution of different groups. Another feature of sphragis-bearing, female butterflies is that the genital opening (the *ostium bursae)* is often externalized, meaning that it is not located inside a pouch-like structure, as in most other butterflies. The evolution of such externalized genitalia may have occurred as a result of sexually antagonistic coevolution (see Chapter 2, this volume), driving the development of larger mating plugs by males, in order to seal the opening. Sphragides have also been 'lost' over time, for example in some clear-wing butterflies, in the genus *Pteronymia*. In other butterflies, including some *Bhutanitis* species, the sphragis is amorphous – e.g. a whitish coagulation of male secretions – and in most cases only partially covers the female genital plate. These crude and irregularly formed sphragides are much smaller and less structured than those of *Parnassius* and *Luehdorfia* species and are probably vestigial. Females with these poorly developed sphragides may have evolved ways and means – including sclerotized genital plates – of making it more difficult to plug their genital openings.[9]

Sperm and Spermatophores

One way in which males can increase their offspring's chance of survival – and subsequent reproductive success – is a parental investment in the form of a spermatophore. This is the only way in which male butterflies can contribute; it's no use providing edible prey items like some other insects, because adult butterflies can only feed on nectar. Females appear to treat these male nuptial gifts – accessory gland secretions that are deposited into the female during the act of copulation – as a sort of meal, in some species. In mating sulphur butterflies, *Colias erate* (Fig. 6.5), the spermatophore is only produced – 'structured' by the male in the female reproductive tract – some 40 min after the initiation of copulation. In monarchs, the transfer of sperm only occurs after nightfall, and consequently their lengthy copulations only end after darkness has fallen. These prolonged copulations in *Danaus plexippus* are thought to be an adaptation to ensure male paternity. In other words, the male is monopolizing the female during copulation and preventing her from mating again – at least until such time as his sperm has reached her seminal duct and cannot be usurped by another male's sperm. He is also trying to prevent her from inactivating or discarding his sperm by some form of cryptic female choice, after copulation has taken place. It is to the females' advantage – in terms of a higher reproductive output and extended longevity – to receive more than one spermatophore, but it is to the male's advantage to ensure that she utilizes all his sperm. This conflict of interests between the sexes has driven the evolution of larger contributions from the male. Larger spermatophores remain in the bursa copulatrix after the female monarch has mated again, whereas smaller ones (<15 mg) are degraded and removed within 2–3 days. About one half or so of larger spermatophores are still present in the bursa copulatrix after the female monarch has remated, some 3–4 days after the first mating. This suggests that the sperm of the first male could still be used to fertilize the female's eggs, even after remating, but there is some degree of last male precedence in monarch butterflies, perhaps because the sperm from the first male gets pushed to the back of the spermatheca where it is unlikely to be used to fertilize eggs after the female has remated. In the words of monarch butterfly researcher, Karen S. Oberhauser: 'Males control spermatophore size and content, but females exert control over spermatophore use and fate'.[10]

Spermatophores are balloon-like, or teardrop-shaped structures, with a swollen 'body' and a tube-like 'tail' with an opening at its tip (Fig. 8.9). The ejaculate transferred to the female during mating contains sperm, nutrients, antiaphrodisiacs (in some species) and gonadotrophic hormones, which regulate reproduction and gonad development. Protective cyanogenic glycosides – molecules that liberate cyanide – are also present in the spermatophores of *Heliconius* butterflies. Similarly, pyrrolizidine alkaloids (PAs) – obtained by male queen butterflies, *Danaus gilippus* (Fig. 5.17) from plants – are transferred to the female at mating. These defensive compounds increase the value of the nuptial gift as they are costly to obtain as adults – e.g. by pollen feeding – and are used by the female to protect her eggs from predators. Spermatophores tend to be larger, on average, in polyandrous species, probably because of the advantages of having more sperm to compete with the contributions of rivals, inside the female reproductive tract, after copulation. The size of spermatophores delivered by males in second and subsequent matings is often significantly smaller than that of the first one, and there may be an inverse relationship between the size and number of ejaculates that a male butterfly can produce in his lifetime.[11]

Most of the sperm in a spermatophore are so-called, apyrene sperm – without a nucleus – and play no role in fertilization: they measure about 300 μm in length, e.g. in *Euptoieta hegesia* (Nymphalidae), and can swim freely. A smaller proportion, 10–15%, are genuine (eupyrene) sperm that are capable of fertilizing eggs. Both types of sperm migrate to the spermatheca –

where sperm is stored in the female – but only eupyrene spermatozoa fertilize the eggs. In *Pieris brassicae* (Fig. 8.7), some apyrene sperm leave the spermatophore contained within the bursa copulatrix after about 5–8 h and appear to play a role in 'freeing' the eupyrene sperm, which then move out and into the female receptacle, the spermatheca, about 3 h later. The smaller anucleate apyrene sperm is a protein contribution from the male; i.e. his parental investment in the reproduction process. Another theory – the so-called *cheap filler* hypothesis – proposes that in polyandrous species, the apyrene sperm fill the spermatheca of the female and reduce her receptivity to mating again. Apyrene sperm may also play a role in displacing or inactivating any competing sperm from previous matings, particularly as the apyrene sperm move first – from the spermatophore to the spermatheca – before the eupyrene sperm migrate across to the female. However, such 'second male precedence' is not always the case in butterflies, and both 'first male precedence' and sperm mixing can also occur.[12]

The last, or most recent male that has mated with a female generally has what is called, *sperm precedence*, i.e. he is the father of all subsequent offspring. Although there are some exceptions – e.g. where the *largest*, rather than the *latest*, male has sperm precedence – the male who mates first is usually at a disadvantage if his precious nutrients are used by the female to boost the reproductive output of eggs fertilized by the next male she mates with. This is one of the selection pressures for the evolution of mating plugs. The proportion of offspring produced by a second male, i.e. in remating – across the order Lepidoptera – is of the order of 65%, but there is considerable interspecific variation amongst butterflies in the degree to which females use the sperm of different males, although *last male* sperm preference, appears to be the general rule. The longer the first male can prevent the female from remating, the better it is from his perspective. He will then be the father of all the eggs she lays until such time as she mates again. There are, however, some examples of *first* male precedence in butterflies, so the situation may be more complex, with females able to exert choices after copulation with more than one partner.[13]

Fig. 8.7. Mating Nepalese large cabbage whites (*Pieris brassicae nepalensis*), AP, India.

Females benefit from nuptial gifts in terms of increased longevity, fecundity and egg size. In other words, the male ejaculate boosts their reproductive output. For example, the amount of nitrogen transferred by virgin *Pieris napi* males – within a spermatophore – is equivalent to that in about 70 eggs. It is therefore, to the females' advantage, and to that of her offspring, to mate more than once. The fecundity of *P. napi* females who mate more than once can be twice that of those who have only mated once. For this reason, females of some species have been described as *actively foraging* for additional matings, in order to realize their potential fecundity. A female usually has enough sperm from just one mating to fertilize all her eggs; it is the additional nutrients that she needs from multiple matings.[14]

Spermatophores are physiologically costly to produce and most of the donated protein is obtained by the males when they were larvae, i.e. as caterpillars feeding on plants. Male and female butterflies both must build up their protein *capital* during their early herbivorous life stages. Additional amino acids can, however, be obtained as adults by mud-puddling (Fig. 8.8) on protein-rich sources – and in some species by feeding on pollen – and the importance of these nutrients in mating explains the eagerness of male butterflies to top up their protein levels in these ways. The protein content of ejaculates tends to increase with the degree of polyandry, with males of polyandrous species producing more than one high-quality spermatophore. This was demonstrated by Brazilian researchers Marcio Cardoso and Emmanuelle Silva from Universidade Federal do Rio Grande do Norte, Brazil; they showed that in *Heliconius* butterflies like *Heliconius melpomene* – where females mate several times in their lifetimes – the males invest more in terms of spermatophore quality than in monandrous species, like *Heliconius erato*. The males are, in effect, trying to prevent a female they have just mated with from mating again with another by endowing her with a large nuptial gift.[15]

Fig. 8.8. Stately nawab (*Polyura dolon*) mud-puddling on excrement (a rich nitrogen source!), Thailand.

Spermatophores can, in extreme cases, represent over 20% of a male butterfly's body weight, but 10% is probably more typical, and the mass is significantly reduced when a male comes to mate a second time, particularly if the next mating is soon after the last one. This does vary, however, depending on the life history and degree of polyandry in a species: ejaculate size is usually greater in species where females mate more than once. There is some suggestion that males can take account of their own individual mating history, and tailor the size and content of their ejaculates accordingly. Surprisingly perhaps, the number of both types of sperm generally increases in smaller, second spermatophores, as does the proportion of viable (eupyrene) sperm. This is a way of prioritizing mating success in second matings. Nevertheless, although males are keen to maximize their fitness by mating as often as possible, a delay of a few days between matings usually results in larger spermatophores with increased protein contents.[16]

The speckled wood butterfly, *Pararge aegeria* (Figs 3.24–3.25), is somewhat unusual in that males do not contribute very much to the female in terms of protein: spermatophores are less than 1% of body mass in this largely monandrous species. Males that have mated once already provide even smaller spermatophores than virgin males, and take much longer (up to five times!) over copulation to transfer their smaller packages. The spermatophore and accessory substances transferred by swallowtail butterflies (*Papilio xuthus*), are on average, only about 2% – 2.4% and 2.1%, respectively – of their body weight. A contribution of only 0.31% of body weight – the smallest relative investment in a spermatophore known to date – is made by males of the mountain white, *Leptophobia aripa*, a pierid native to Mexico, Central America and South America.[17]

Spermatophores are composed of a tough, indigestible outer envelope, an inner matrix, and a bolus of sperm. The spermatophore itself is formed inside the female in a receptacle called the bursa copulatrix during mating. The physical effect of the spermatophore in stretching the bursa of the female changes her behaviour, such that she is no longer receptive to other males, at least until the spermatophore becomes significantly deflated. For example, female *Pieris rapae crucivora* (Fig. 7.11) display a mate-refusal posture when they have, at least, a half-full spermatophore in their bursa copulatrix. In the green-veined white, *P. napi*, it is the non-fertile apyrene sperm that switch off female receptivity by filling the bursa, thus delaying remating. In monarch butterflies, the larger the spermatophore, the longer the female remains in an unreceptive state. So, for the female butterfly there is a trade-off between the size of the nuptial gift and the length of time she remains unreceptive. There is also a conflict between the sexes over the length of the refractory period, played out via the production and storage of infertile sperm.[18]

The interaction between the spermatophore and the bursa copulatrix encapsulates the conflict between the sexes. Although females benefit from the nutrients they receive from males, there is a cost in terms of being unable to remate until the spermatophore is almost fully digested. Researcher Camille Meslin describes how 'both conflict and cooperation are reflected in the morphological and physiological traits of the spermatophore and the bursa'. On the one hand, males have evolved thicker spermatophore envelopes to resist digestion inside the bursa, whilst, on the other, females have evolved mechanical devices – such as the signum – and digestive enzymes, to speed up the process of digesting and absorbing the spermatophore and its contents (Fig. 8.9). Meslin and others consider that there has been an overemphasis – a 'strong male bias' according to Professor Malin Ah-King, an evolutionary biologist and gender researcher at Stockholm University, Sweden – on the male side of this story, and the role of the female in genital evolution has been understudied. More research, focused on female reproductive adaptations is needed to fully understand the 'coevolutionary dynamics' between the sexes, regarding reproduction, writes Professor Ah-King.[19]

After copulation, the active sperm migrate to another storage organ in the reproductive tract, and the bursa processes the spermatophore in order to obtain the nutrients present in this nuptial gift. The process of rupturing the tough outer envelope is aided by the signum – a toothed structure on the inside of the bursa – but this can still take up to 36 h. The digestion of the inner matrix of the spermatophore takes an additional 24–48 h. So, on average, it takes butterflies like *P. rapae* – on which most of these studies have been carried out – about 72 h to process the spermatophore and absorb the nutrients. This means that the average mating cycle, from copulation to regained receptivity, is about 3 days in this species. Virgin female *P. rapae* avoid so-called precocious mating – just after they have emerged but before their wings have fully expanded and hardened – by adopting the mate-refusal posture. Male *P. napi* adjust the size of the ejaculates they transfer to a female based on the intensity of the competition: they increase the mass of their spermatophores by more than 20% when other males are around. The size of an ejaculate usually correlates with the size of the males delivering the package, so sexual selection may be favouring large males in polygamous species, i.e. to induce a longer period of unreceptivity in females. This relationship between mating success and male size is, however, probably counterbalanced by the advantages of emerging earlier than females (protandry) in order to mate with virgin females.[20]

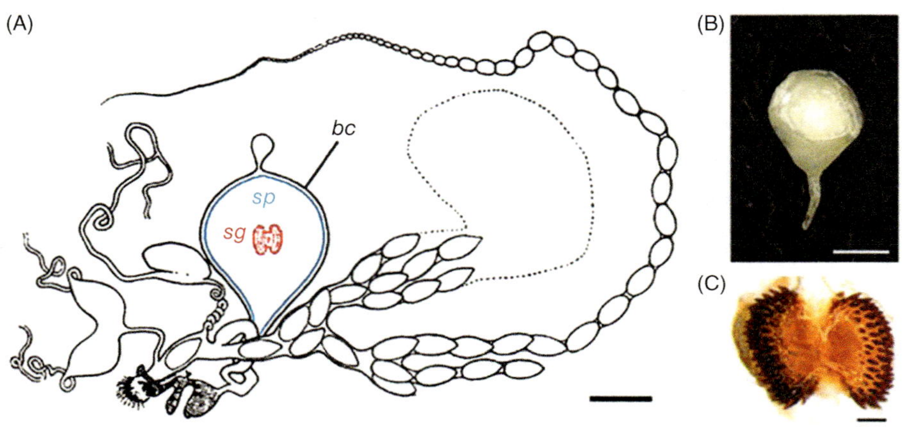

Fig. 8.9. Cardinal (*Argynnis pandora*) male. Sex brands circled, Galicia, Spain.

The Female Side of the Story

The bursa copulatrix is a stomach-like organ in the female reproductive tract that ruptures and then digests the male ejaculate, the spermatophore. The sperms leave the bursa copulatrix some time after copulation, via the ductus seminalis, and are then transported, to the spermatheca. As we have seen, it is to the male's advantage – in terms of fathering more offspring – to delay the female from remating for as long as possible. It is, however, to her advantage to receive another spermatophore, or more, in subsequent rematings, earlier than would be optimal for the male. Consequently, an evolutionary 'arms race' of sorts has evolved in polyandrous species, with heavily sclerotized, indigestible spermatophore envelopes on the male side, and highly effective thorny or spiny devices on the inner wall of the bursa (called a signum) to tear it open, on the female side (Fig. 8.1b). These structures have evolved as a result of what is described as 'sexually antagonistic coevolution', or more simply, as products produced in an ongoing 'battle of the sexes'. Or, as Dr Camille Meslin and her colleagues put it, 'a quarrel over ejaculate processing'. I prefer to see it as an evolutionary 'tug of war' between the somewhat different needs of the two sexes. Surely, such a dynamic tussle is at the heart of evolution (see Chapter 2, 'Sexual Conflicts', this volume)?[21]

In virgin *P. rapae* females, the bursa exhibits extremely high levels of proteolytic activity, equivalent to that found in the midgut of the larval stages. The protein-digesting enzymes, together with the mechanical grinding and rupturing by the signa, process the spermatophore and digest the contents. The sperm escape destruction and migrate to a special storage organ called the spermatheca, where they are available to the female for fertilization. The envelopes of spermatophores in polyandrous *Heliconius* species are thicker than in species that only mate once, providing additional evidence for the theory that these adaptations have evolved as a result of selection pressures arising from so-called, 'sexually antagonistic' interactions. Clearly, females benefit most by digesting the contents of spermatophores as quickly as possible, so they can remate and receive another ejaculate. Males on the other hand, benefit most from delaying their breakdown, so they can maintain paternity – 'father' the eggs produced by the female – for as long as possible.[22]

Large spermatophores are probably maintained by selection pressures on males to maximize the quantity of ejaculate they transfer in order to outcompete other males. The effect of the spermatophore is to inflate the bursa copulatrix in the female, which induces a change in behaviour – from mate acceptance to mate refusal. So, spermatophores delay remating, and in some species – such as the female monarch butterfly, *D. plexippus* (Fig. 5.19) – large spermatophores delay remating for longer than small ones. Hard spermatophore envelopes, which are difficult to break down, also increase the length of the female refractory period. Females who mate several times usually lay more eggs and live longer than those who only mate once, and in such polyandrous females, there is an inverse relationship between sexual receptivity and the quantity of spermatophore in their corpus bursae. There is, therefore, a balance to be had, between receiving the benefits from spermatophore-derived nutrients, and the incurred costs involved in remating, which might include less time for egg laying, for example.[23]

Notes

[1] Shapiro and Porter (1989), Eberhard (1985, 2010) and Simmons (2014). 'The value of genitalia for distinguishing species can hardly be overstated' (Burns, 1997). The results of recent studies show that sexual selection is a potent force in the evolution of genital diversity (Hosken and Stockley, 2004).
[2] Woo *et al.* (1990), Burns (1997) and Ah-King *et al.* (2014). According to Cordero and Baixeras (2015), it is impossible to 'even suggest an approximate function' for every anatomical structure of female genitalia in Lepidoptera. Female butterflies in the subtribe Polyommatiti (Lycaenidae) have a unique genitalic feature called the henia – an eversible, membranous, flexible tube, the end of which has an opening to the ductus bursae – the shape and morphology of which can be used to separate superficially very similar species in this group (Coutsis, 2012).
[3] Eberhard (1985, 2010) and Cordero and Baixeras (2015). For hypotheses on the function of both 'deciduous' and 'non-deciduous' sclerotized structures (=cornuti) of the endophallus, see Cordero (2010) and Cordero and Miller (2012).
[4] Scott (1973a), Ehrlich and Ehrlich (1978), García-Barros (1989) and Matsumoto *et al.* (2018).
[5] McCorkle and Hammond (1985), Matsumoto (1987), Tsubaki and Matsumoto (1998), Matter *et al.* (2012) and Matsumoto *et al.* (2018). A comprehensive review of the occurrence and diversity of sphragis in butterflies was carried out by Carvalho *et al.* (2017), who found that they occurred in about 13% of Papilionidae and 3% of Nymphalidae; accounting for c. 1% of all known butterfly species.
[6] Matsumoto and Suzuki (1992). Vlasanek and Konvicka (2009) found that 74 out of 579 marked female clouded apollos, *Parnassius mnemosyne*, did not bear sphragis and a small proportion (3.3%) lost this structure during their lifetimes. Matsumoto (1987) showed that, although males of *Luehdorfia japonica* attempted to mate forcibly with previously mated females, the presence of sphragis, and/or their escape reactions prevented copulation.
[7] Orr (1995, 2002), Matsumoto (1987), and Dickinson and Rutowski (1989).
[8] The reproductive biology of intensively studied bay checkerspot (*Euphydryas editha bayensis*) is described in Boggs and Nieminen (2004), and Ehrlich and Hanski (2004).
[9] Paluch *et al.* (2003), Francini *et al.* (2005), Carvalho *et al.* (2016, 2019), Freitas *et al.* (2018) and Matsumoto *et al.* (2018).
[10] Marshall (1982b), Watanabe *et al.* (1997), Svärd and Wiklund (1988), Bergström and Wiklund (1992) and Oberhauser (1992).
[11] Wiklund (1993), Wiklund and Forsberg (1991), Svärd and Wiklund (1986), Mellström and Wiklund (2009), Zagrobelny *et al.* (2008) and Watanabe (2016b). Cardoso and Gilbert (2007) found cyanides in the spermatophores of nine *Heliconius* species (*charithonia, cydno, erato, ethilla, hecale, hewitsoni, ismenius, melpomene* and *numata*) at concentrations ranging from 0.08 to 4.41 µg/mg cyanide. Male *Danaus gilippus* transfer 63% of their acquired alkaloids (pyrrolizidine alkaloids, PAs) to the female, who in turn transfers about 90% of this gift to the eggs (Eisner and Meinwald, 2003). Male Ithomiinae spermatophores contain up to 23% dry weight of PAs, and large amounts (up to 1.7 mg) are transferred to the female (Brown, 1984). Clearwing butterflies such as the ithomiine, *Aeria olena*, 'intensively seek and sequester' PAs from plant sources, as freshly emerged adults (Trigo *et al.*, 1996).
[12] Silberglied *et al.* (1984), Tschudi-Rein and Benz (1990), Cook and Wedell (1996), Watanabe *et al.* (2000), and Mancini and Dolder (2001).

[13] Labine (1966), Drummond (1984), Dickinson and Rutowski (1989), Ridley (1989a), and Swallow and Wilkinson (2002). The *largest* male fathers the most offspring in *Pieris rapae,* and *P. napi*, regardless of the order in which mating takes place (Wiklund and Kaitala, 1995), Bissoondath and Wiklund, 1996a; Boggs and Nieminen, 2004). Platt and Allen (2001) found that *first* male precedence occurred in only 29% of cases involving multiple matings, in two subspecies of the North American white admiral, *Limenitis arthemis.* Instances of both complete *first* and *last* male sperm precedence, as well as cases of sperm mixing, also occurred in cage experiments with *Bicyclus anynana* (Brakefield *et al.*, 2001).
[14] Wiklund *et al.* (1993), Kaitala and Wiklund (1994b), Karlsson (1998), Wiklund *et al.* (2001), and Stjernholm and Karlsson (2000).
[15] Cardoso and Silva (2015).
[16] Spermatophores are on average c. 15% of male body mass in *Pieris napi* (maximum: 23%) (Kaitala and Wiklund, 1994a). Ejaculates vary from 5–15% of body mass in pierids (Wiklund *et al.*, 2001). *Pieris rapae* spermatophores are up to 10% of male body mass (Tigreros *et al.*, 2014), or c. 7% of the male's body mass (Watanabe and Sato, 1993). Male monarch spermatophores are c. 10% of their body mass (Oberhauser, 1988). Male contributions (including the mating plug) represent 7% body mass in *Euphydryas chalcedona* (Rutowski and Gilchrist, 1987), and 10–12% in *Euphydryas editha* (Boggs and Nieminen, 2004), but only <2% in *Euphydryas anicia*. Spermatophores weigh c. 6% of male body mass in *Battus philenor*. The size of the ejaculate is greater in polyandrous species (Svärd and Wiklund, 1989); the protein content of ejaculates increases with the degree of polyandry (Bissoondath and Wiklund, 1996b); spermatophore weight correlated with mating frequency in nine *Heliconius* species (Cardoso *et al.*, 2009); polyandrous *Heliconius* species invest more in spermatophore quality than the monandrous species, *H. erato* (Cardoso and Silva, 2015). The length of the time between first and second matings strongly affects the size and protein content of the ejaculates (Bissoondath and Wiklund, 1996a). Spermatophores were significantly smaller on second matings in *Pieris rapae*, but eupyrene and apyrene sperm numbers increased dramatically (Wedell and Cook, 1998). First ejaculates in *P. rapae* contained c. 46,000 apyrene sperm; successive ejaculates contained higher numbers (Watanabe *et al.*, 1997). Male Asian swallowtails, *Papilio xuthus*, may use their own mating history to tailor their ejaculates, probably assessing the probability of additional matings (Sasaki and Watanabe, 2015). *Bicyclus anynana* males make a relatively small investment in spermatophores, compared with other butterflies (Ferkau and Fischer, 2006). In many species, the size, or mass, of the spermatophore is positively correlated with the size of male butterflies (e.g. Svärd and Wiklund, 1986), and because large ejaculates induce a longer period of unreceptivity in females, sexual selection will favour larger males in polyandrous species (Wiklund and Fosberg, 1991; Kaitala and Wiklund, 1994a; Rajyaguru *et al.*, 2013).
[17] Svärd (1985), Bissoondath and Wiklund (1995), Lauwers and Van Dyck (2006), Watanabe and Kobayashi (2006), and Caballero-Mendieta and Cordero (2013).
[18] Sugawara (1979), Cook and Wedell (1996). Sexual conflict over female receptivity in *P. napi* could explain why the male ejaculate consists mainly of non-fertile sperm (Wedell *et al.*, 2009).
[19] Meslin *et al.* (2015) and Ah-King *et al.* (2014). Signa may have evolved to make cryptic choices of males, in order to reduce the post-copulatory refractive period (Cordero, 2005).
[20] Obara (1987), Wiklund and Forsberg (1991), Kaitala and Wiklund (1994b), Rajyaguru *et al.* (2013) and Meslin *et al.* (2015).
[21] Wedell (2005), Sánchez *et al.* (2011) and Meslin *et al.* (2017). The signum might also be used by females for inducing ejaculation (Galicia *et al.*, 2008).
[22] Oberhauser (1989), Sánchez and Cordero (2014b), Knight (2015) and Plakke *et al.* (2015).
[23] Sugawara (1979), Oberhauser (1989), Bissoondath and Wiklund (1995) and Torres-Vila *et al.* (2004). Large spermatophores are maintained by selection pressures to maximize ejaculate transfer (Vahed, 1998). Sánchez *et al.* (2011) found an inverse relationship between sexual receptivity and spermatophore quantity in the corpus bursae, in Lepidoptera, in general.

9 Scents: Chemical Communication

Due to our own olfactory limitations, we are largely unaware of the sophisticated chemical conversations taking place between organisms all around us. Chemical ecology is a relatively new science: it was only in 1959 that the first sex pheromone – bombykol – was discovered, by the Nobel Prize-winning German biochemist, Adolf Butenandt. The word itself, pheromone, was only coined in 1959, although the concept of chemical communication goes back to the ancient Greeks. Since 1959, pheromones have been discovered throughout the animal kingdom, including many amongst the Lepidoptera. A huge diversity of chemical compounds – the composition of which varies enormously from species to species – is gradually being elucidated, but much more remains to be discovered. Sex pheromones have evolved to enable organisms to exchange information between individuals in a direct and targeted manner – more specifically, to accurately identify members of the same species, of the opposite sex, who are ready and willing to mate. Pheromones have the advantage of producing an immediate, innate response in the receiver, which usually consists of stereotypical courtship behaviour. In this respect, they act as aphrodisiacs. They may also convey information on the quality, age and genetic relatedness of an individual, although how they do this is only just beginning to be understood.[1]

Chemical signalling occurs over a relatively close range in butterflies, unlike the long-distance pheromone communication used by moths, and most butterflies probably lack the necessary olfactory structures needed for long-range detection of chemical signals. Pheromones are detected by receptors and associated sensory neurons found on the opposite sex; typically, on female antennae, in the case of male sex pheromones (MSPs). A wide range of sex pheromones, and putative MSPs, have been identified to date – including alkaloid derivatives, terpenoids, aromatics

© Raymond J.C. Cannon 2020. *Courtship and Mating in Butterflies: Reproduction, Mating Behaviour and Sexual Conflicts* (R.J.C. Cannon)

and carboxylic acids – which are used by male butterflies during courtship. These compounds and pheromone systems have probably evolved multiple times during butterfly evolution.[2]

Aphrodisiac Pheromones

When a male finds a potential mate, the outcome is not a foregone conclusion and depends on the predilection – receptivity and choosiness – of the female butterfly. The male cannot force himself on a female, she must be reproductively receptive, and in some cases, may demonstrate this by specific 'rejection' behaviour; for example, spreading her wings and lifting her abdomen (see Chapter 6, this volume). For a successful courtship, however, the female must respond positively to the bouquet of chemical signals released by the male when he comes close to her. His courtship and scent serve to increase her receptivity. The pheromones produced by the male provide signals containing a wealth of information by which females can determine his health and status – and hence his desirability. In this way she can make her choice from the pool of available suitors. Most of the pheromone scents emitted by butterflies are unique to each species, but unlike moths, where thousands of species are known to have sex attractants, the chemical compositions of MSPs have only been determined for a relatively small number of species, although the list is growing steadily. Detailed studies on butterfly pheromones have been carried out on some milkweed butterflies (Danainae), some *Heliconius* species and several pierids, including: *Colias eurytheme*, *Colias philodice*, *Pieris napi*, *Pieris rapae* and *Pieris brassicae*. Although a great many compounds have been identified from the wings of butterflies, only a tiny proportion are specific to the male, and hence, are possible MSPs. For example, more than 120 chemical compounds were identified from the wings of *P. rapae* and *P. brassicae*, but only a small portion were male-specific, and therefore, potential pheromones. The concentrations and combinations of such compounds also differ widely, even between closely related species. For example, two subspecies of *P. rapae* were found to have diverged in their volatile chemical profiles over a short evolutionary period (~5,000 years, or roughly 10,000 to 20,000 generations), such that the levels of ferrulactone were significantly higher in *P. r. rapae* males than *P. r. crucivora* males.[3]

Examples of MSPs include: citral, a lemon or citrus-smelling aromatic released by males of the green-veined butterfly, *P. napi* (Fig. 9.1); ferrulactone in the small white, *P. rapae*; geranial and neral in *Pieris melete*; and brassicalactone in the large white, *P. brassicae*. Citral – which is a 1:1 mixture of the two isomers, geranial and neral – has the effect of making *P. napi* females more prone to adopt mate-acceptance behaviour. The relative concentrations of these compounds vary with the age and biological condition of individual butterflies, but they may be disseminated very rapidly. For example, ferrulactone is released by male *P. rapae* less than 1 s after seeing a potential partner, and reaches a maximum concentration

Fig. 9.1. Green-veined butterfly (*Pieris napi*), Galicia, Spain.

within 30 s. Female butterflies exposed to the aphrodisiacal effects of these sex pheromones are more willing to acquiesce and accept a courting male. Furthermore, because males emit chemicals such as citral – in *P. napi* – whilst flying around, the concentration of the sex pheromone in the air provides the male with a good indication of the number of other suitors in the vicinity. So, males can adjust the size of the ejaculates they transfer to a female, based on the intensity of the competition. *Pieris napi* males increased the mass of their spermatophores by more than 20% when lots of other males were flying about in mating cages.[4]

Detectable odours

The scents of many male butterflies are perceptible to humans. I think this phenomenon was probably better known in the past, when people collected butterflies using nets and handled them (prior to killing them!). Nowadays, most people observe butterflies from a distance, or follow them using binoculars or a camera, so may not be so familiar with their scents. Although individual butterflies differ in their production of pheromones, these volatile essences are said to be 'almost uniformly pleasant' to us, as well as presumably, to female butterflies! Some of the more strongly smelling species are said to include: the Cleopatra butterfly, *Gonepteryx cleopatra* (freesia); the brown argus, *Aricia agestis* (chocolate); common blue butterfly, *Polyommatus icarus* (chocolate); and the green-veined white, *P. napi* (lemon verbena). Males of a subspecies of the holly blue butterfly, *Celastrina argiolus ladonides* (Lycaenidae, Polyommatinae, Polyommatini) (Fig. 9.2), emit a faint odour, composed of lavender lactone and δ-decalactone. Fan-shaped, scent-disseminating androconial scales – called 'battledore' scales because of their shape – occur on the upper surfaces of both fore- and hindwings in this species and are smaller than the normal wing scales. The mountainous duskywing, *Erynnis montanus* (Hesperiidae), also emits a pungent odour, stronger in the male, which is thought to be due to a compound called *p*-cresol. Males of the eastern Asian yellow tip, *Anthocharis scolymus* (Pierinae, Anthocharidini) – related to

Fig. 9.2. Holly blue butterfly (*Celastrina argiolus ladonides*).

Fig. 9.3. European orange tip (*Anthocharis cardamines*), Biatorbágy, Hungary. (Photo by Ádám Gór.)

the European orange tip, *Anthocharis cardamines* (Fig. 9.3) – emit a floral smelling scent containing volatile terpenoid compounds: (*S*)-(+)-linalool and (*E*)-*β*-farnesene. Unlike other pierine species, however, the males do not have specialized alar scent scales; instead ordinary scales – with round saccular structures connected to their sockets – found across the whole wing, release these MSPs.[5]

Males of many papilionid butterflies, including *Papilio machaon* (Fig. 9.4), *Papilio protenor*, *Papilio polytes* and *Papilio glaucus* (Fig. 9.5), also have MSPs distributed across both fore- and hindwings. The major components isolated from the wings of male *P. machaon* were n-dodecane, linalool and geranylacetone, which are also present on the wings of the females, but in lower quantities. These volatile scents have been described as sweet and flowery. In *P. polytes* (Fig. 5.21), the relatively faint scent is thought to be mainly due to linalool, which is also one of the compounds present on the wings of the Chinese windmill, *Atrophaneura alcinous alcinous* (Fig. 9.6), which have a strong musk-like odour. However, new

Fig. 9.4. Swallowtail (*Papilio machaon hispanicus*), Galicia, Spain (14 September 2017).

Fig. 9.5. *Papilio glaucus* (reared specimen). (Photo by Ádám Gór.)

species-specific volatiles have recently been discovered in this species. Some hairstreaks, including males in the genus *Atlides* (Eumaeini, Lycaenidae) have conspicuous odours, ranging from floral (*Atlides polybe*) to grape-like (*Atlides bacis*) and chocolate (*Atlides imperialis*) scents. Females lack these perceptible odours, which fade soon after death in males.[6]

Studies using *Bicyclus anynana*

MSPs may be isomeric mixtures or blends of more than one compound, and the contribution of the various fractions to the overall effect can be difficult to determine. The relative proportions of the different chemical constituents can also be highly variable, as can the overall concentrations between individuals. The MSPs are produced by the male secondary sexual structures – called androconia (see below) – and are transferred via specific behaviours carried out by the male during courtship, but can be released as soon as a male sees a prospective mate. Three chemical components constitute the putative MSPs of *Bicyclus anynana* and are probably transferred during the flickering and thrust phases of courtship in this species (see Chapter 5). Female antennae of *B. anynana* have been

Fig. 9.6. Chinese windmill (*Atrophaneura alcinous*) female at World Glasshouse of Gunma Insect World, Japan. (Photo by岡部碩道 [Public domain] (Okabe Shinmichi).)

shown to respond to these male-produced pheromones, which are structurally like some pheromones produced by moths. However, these three compounds produced in *B. anynana* exhibit large differences in relative concentrations between individual males – a consistent indicator of individuality – and females may select partners based on the concentrations of these aphrodisiacs, although much more remains to be discovered about their role in butterfly courtship and female choice.[7]

Sex pheromones can be thought of as private messages, in the sense that olfactory receptors work in a 'lock-and-key' manner, where a specific chemical key fits and activates a unique type of receptor lock. Only butterflies with the right sort of receptor(s) will receive the message, and the unique message can induce a specific sort of behaviour. There is scope for these messages to evolve rapidly as a result of sexual selection, i.e. involving female choice. Differences in the quantity and quality of mixtures of volatile MSP compounds allow closely related species to exist in reproductive isolation from each other, even though they share the same habitat. For example, up to 20 different, but morphologically very similar, *Bicyclus* (Satyrinae) species may coexist in the same patch of forest in Africa, encountering one another but remaining separate: a phenomenon called sympatry. Their MSPs are discrete signals that require specific one-to-one corresponding chemical receptors to allow communication to take place. MSPs are used over short ranges when males erect their androconial hairs.[8]

Differences in the pheromone 'dialects' – i.e. subtle differences in MSP components – of *Bicyclus* species are thought to first evolve within individual populations of a given species, before they become fully separate species. In other words, incipient species start to 'talk' their own unique chemical language, which eventually keeps them apart and avoids inbreeding. As well as helping butterflies to identify members of their own species, subtle differences in the quantities and ratios of MSPs also enable females to judge the age of males in these species; for example, *B. anynana* females prefer their mates to be in the mid-range – not too old, not too young. They use age-related changes in the concentration of one specific pheromone component – hexadecanal (MSP2) – to differentiate the males, suggesting that this compound may be an honest signal of male quality. There is some evidence to suggest that the pheromone bouquet of *young* males is not a reliable signal of their quality, beyond indicating species and age. However, MSPs are in general thought to tell the female something about the quality and desirability of the male as a potential mate, and males actively control the emission of MSPs when courting females, especially when faced with competition from other males. Females can learn preferences for novel pheromone blends after a brief, early exposure, and these learnt preferences are passed down to the next generation.[9]

Heliconius butterflies

In *Heliconius* butterflies, MSPs are complex blends of long chain hydrocarbons, fatty acid esters, aldehydes and alcohols, which are present in species-specific mixtures. Many different compounds have been identified from androconia on the wings of different Neotropical *Heliconius* species, although it is not yet clear which are the most biologically important ones. However, bouquets of chemicals of variable volatility are only found on the hindwing androconia of mature male *Heliconius* butterflies and are not present in females, or on other regions of the male wings. A small number of putative MSPs have been identified in *Heliconius melpomene* and other species, including octadecanal ($C_{18}H_{36}O$), which has been found in the pheromone bouquet of other lepidopteran species. Nevertheless, the most abundant compounds are not necessarily the most biologically important ones within a blend of putative MSPs. Pheromone mixtures are thought to evolve rapidly as a result of small changes in the biosynthetic pathways of the putative MSPs.[10]

Androconia

Androconia are modified scales found on the wings of male butterflies and are thought to be where MSPs are produced and disseminated. In some cases, the pheromonal function has not been established for certain, but is inferred by their structure – i.e. a very high surface area – and the fact that they are commonly brought into contact with female antennae during courtship. There is a wide variety of species-specific scent glands involved in the release of pheromones, including specialized scales in the shape of brushes, fans, clubs and so on. In some species, like those in the genus *Pieris* (Pieridae), scent scales are widely distributed across the wings of the males, but in other species they may be located on the wings, legs or abdomens. Androconia are often narrow, filamentous or seta-like in structure, ending in a meshwork of fine 'hairs', which helps to evaporate the volatile pheromones. There are, however, myriad different shapes and forms – often used by taxonomists to distinguish species – with different types of androconia sometimes being present on the same individual.[11]

Varieties of androconial structures

Male secondary sexual organs, androconia, are widely distributed and highly variable on the wings of some butterflies, such as male lycaenids, and have been lost, and perhaps regained – i.e. evolved again – in the evolutionary history of this family. Consequently, they are present or absent in closely related taxa. For example, different species in the Neotropical genus *Thereus* (Lycaenidae, Eumaeini) have forewing scent patches, hindwing scent patches, forewing scent pads and abdominal brush organs. Putative scent-disseminating hair-tufts have been identified on the *ventral* hindwings of some species in the subfamily Aphnaeinae (Lycaenidae), such as *Pseudaletis antimachus*. Similarly, circular androconial bulges appear on the ventral hindwings of some male lycaenids, such as the common red flash, *Rapala iarbus iarbus* (Fig. 9.7), indicating the presence of a circular brand on the upper side. Scent patches occur on the dorsal forewings of some New World hairstreaks, including some species in the genera *Arcas* and *Atlides* (Theclinae, Eumaeini); all male *Arcas* reportedly possess a scent pad on the dorsal forewing surface, except for *Arcas alleluia*. In other species, like the Hemon blue hairstreak, *Theritas hemon* (Lycaenidae, Eumaeini) (Fig. 9.8), the androconia are clustered together in patches, i.e. a ventral hindwing scent patch. In some species, these small patches of hair-tufts may have lost their original function, whilst in others their pheromonal function is suggested by the presence of androconia at the base of the hair scales.[12]

There are several long, distinctive patches of pheromone-disseminating structures located along the veins of the male forewing in most fritillary butterflies – such as the Gulf fritillary, *Agraulis vanilla* (Heliconiinae) – where patches of several scale types alternate with scaleless areas along the veins, the cardinal, *Argynnis pandora* (Heliconiinae, Argynnini),

Fig. 9.7. Common red flash (*Rapala iarbus*) with circular androconial bulge, Maharashtra, India. (Photo by Dr Raju Kasambe [CC BY-SA 4.0 (https://creativecommons.org/licenses/by-sa/4.0)].)

(Fig. 9.9) and the silver-washed fritillary, *Argynnis paphia* (Fig. 1.17), for example. Brush-bearing scales are located directly above porous areas in these species, from where the pheromones may be secreted.[13]

The androconial organs of male Ithomiine, clearwing butterflies (Nymphalidae, Ithomiinae) consist of fringes of long hairs at the costal edge of the hindwings (Fig. 9.10). These are erected – as if to disseminate scent – in different circumstances, including male–male interactions, aggregations and presumably to attract females. Alar patches, or pockets, of the type found on most danaid males (see below, and Figs 9.14 and 9.15) are lacking in species such as *Ideopsis similis* (Danainae), but there are specialized scent organs on the upper surfaces of the hindwings, including glandular, microscopic 'cushions' on the inside edge of the

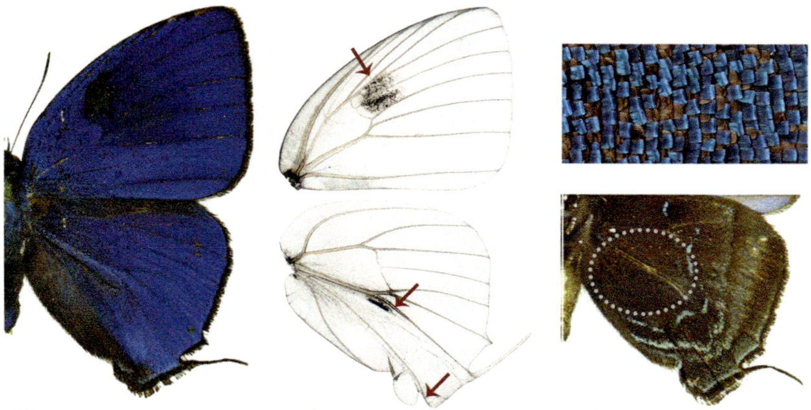

8 9 10

Fig. 9.8. Male secondary sexual organs on the wings of the Hemon blue hairstreak (*Theritas hemon*). 8. Dorsal wings showing the forewing scent patch and a hindwing pouch. 9. Cleared wings showing scent patch location distal of the discal cell (top arrow), location of androconia on the ventral hindwing (middle arrow) and the anal lobe cleft (bottom arrow), a synapomorphy for the Atlides Section. 10. Close-up of tan-coloured scent patch androconia interspersed with iridescent blue wing scales (top), and pouch location on the ventral hindwing (bottom). (From Martins *et al.* (2018), with kind permission of Dr Ananda Martins.)

Fig. 9.9. Cardinal (*Argynnis pandora*) male. Sex brands circled. Galicia, Spain.

Fig. 9.10. Courtship display of a male clearwing butterfly, *Napeogenes inachia*. All male ithomiine butterflies bear androconial 'hairs' on the hindwing, which are a cluster of highly modified wing scales, erected during courtship displays. (Photo by Keith Willmott.)

hindwing, which may produce the pheromone, danaidone. This chemical (a ketone) is a major component of the hairpencil bouquet in many milkweed species (see below), but a host of other compounds – hydrocarbons, terpenoids, aldehydes, alcohols, aliphatic acids, esters, and so on – also make up the species-specific odours of these butterflies.[14]

Sex brands in skippers (Hesperiidae)

In the subfamily Hesperiinae – grass skippers or banded skippers – there are patches, called stigmata, of pheromone-dispersing androconia on the forewings. The androconia of species in genera *Taractrocera* and *Telicota* – such as the Linna palm dart, *Telicota linna* (Hesperiinae, Taractrocerini) (Fig. 9.11) – comprise a linear series of discrete dull black patches of sex scales between veins. In the subfamily Pyrginae – spread-winged skippers – androconia look like long hairs and are in a 'costal fold' of the forewing and in 'tibial tufts' – specialized setae on the hind leg tibia of males. When the butterfly is resting, the tibial tufts are kept in a pocket under the thorax, where they are thought to absorb volatile compounds. These scent-producing organs have been found on the thorax and the abdomen of grizzled skippers in the genus *Pyrgus* (Hesperiidae) including *Pyrgus malvoides* (Pyrginae, Pyrgini) (Fig. 9.12). It is presumed that during courtship the males take the tufts out the pocket to release pheromones, although this has never been observed. Scent scales (androconia) are in patches – called stigmata or sex brands – on the forewings of skippers, such as the Essex skipper, *Thymelicus lineola*, and the large skipper, *Ochlodes sylvanus* (Hesperiinae, Hesperiini) (Fig. 9.13). These

Fig. 9.11. Linna palm dart, *Telicota linna*, male with sex brand, Chiang Mai, Thailand.

Fig. 9.12. Grizzled skipper, *Pyrgus malvoides*, Galicia, Spain.

Fig. 9.13. Sex brands on the forewings of the large skipper (*Ochlodes sylvanus*).

raised patches contain a high density of long, thin, hollow tubes, the tips of which fracture and break off. The tiny broken tips – called osmophores – have been found on the antennae of some female skippers, strongly suggesting that pheromones are disseminated in this way.[15]

Pierids

The main difference between androconia and ordinary wing scales is the presence of a swelling in the integument of the wing, behind the scale socket. Some of these specialized, pheromone-releasing scales are concealed beneath segments of the abdomen and are only exposed when the abdomen is extended. Others are present on overlapping areas of the wings – for example, in *Colias philodice* and *Eurema lisa* (Coliadini) – and are only exposed during courtship attempts. Their location in these overlapping, friction areas of the wings may be to minimize the evaporation of MSPs during non-mating periods. The scales in the sex brands of these sulphur butterflies are not very different from those found in other areas of the wing but are in discrete patches on the male hindwings, near the wing bases. In the small white butterfly, *P. rapae*, scent scales are evenly distributed within rows, between those of the ordinary wing scales.[16]

The squinting bushbrown

Androconia have long been used by taxonomists to classify different species as they are more reliable taxonomic features than, say, wing colour. For example, the number, position and shape of groups of specialized scales – androconial brushes – were used to identify the 80 or more very similar species in the genus, *Bicyclus* (Nymphalidae, Satyrinae). Somehow, via a combination of secretion and behavioural posturing, males use these specialized pheromone-producing patches – in combination with hairs and hairpencils – to disseminate the MSP during courtship. There are two androconial patches of shiny scales on the dorsal wing surfaces of *B. anynana* males: (i) a black spot of differentiated scales and a patch of yellow hairs on the hindwing; and (ii) a silver spot of scales on the forewing and a patch of dark hairs on the hindwing (Fig. 9.14). The groups of hairs – hairpencils – overlay the androconia when the wings are in a resting position and function to disseminate the pheromones. The transfer of pheromones from the androconia onto the antennae of females is essential for successful mating in this species. After landing behind a female, *B. anynana* males flick their wings and erect their androconial hairs. The two sets of pheromone-producing androconial hairs fan out above the male's wing surface during the thrusting phase of courtship (see Chapter 5, this volume), when the male touches the edge of the female's wings with his head.[17]

Androconia consist of specialized brush-like scales – often occurring in alternating rows with normal scales – in an elongated patch on the hindwings of male, *Heliconius melpomene* (Fig. 9.15). Not all butterflies rely on direct physical contact between the androconial-bearing wings of the male and the antennae of the female however, although bringing these body parts together is incorporated into the courtship behaviour of many species (see Chapter 5, this volume). In some species, transfer of MSPs

Fig. 9.14. Second androconial patch on the dorsal hindwing of *Bicyclus anynana* (after removal of the hair pencils). (From: Nieberding *et al.*, 2018, Gilles San Martin [CC BY-SA 2.0 (https://creativecommons.org/licenses/by-sa/2.0)].)

Fig. 9.15. *Heliconius melpomene* wings showing androconial dimorphism. (a) *H. melpomene malleti* (left) and *H. melpomene plesseni* (right). (b) Dissected wings from specimens of *H. melpomene plesseni* showing sexual dimorphism in the androconial region; male (left), female (right). For each sex, the left set of wings shows the ventral surface and the right set the dorsal surface. (c) Expanded view of the male forewing overlapping region (ventral side). (d) Expanded view of the male hindwing androconial region, with arrow highlighting the vein Sc + R1 (dorsal side). (From Darragh *et al.*, 2017, with kind permission of Kathy Darragh.)

probably occurs in the form of aerosol or vapours, disseminated by males when they hover over a female or vibrate their wings close to her during courtship.[18]

Abdominal Organs ('Hairpencils')

Male danaine (milkweed) butterflies have two types of scent-producing organs: (i) alar patches or pouch-like glands – sex brands – on the hindwings (Figs 9.16 and 9.17); and (ii) a pair of abdominal hairpencils (also called hairbrushes). During the final phases of courtship, the male expands his hairpencil organs (Fig. 9.18) and disseminates pheromones that act as aphrodisiacs on the females, inducing them to mate. Most butterflies in the tribe Danaini – including many iconic species colloquially known as tigers, queens and crows, for example – have androconia on the upper surfaces of their hindwings that produce the cuticular pheromone transfer particles

Fig. 9.16. *Danaus chrysippus bataviana* male dorsal view showing alar pockets (circled).

Fig. 9.17. *Danaus chrysippus bataviana* male nectaring; white patch of androconial (scent-producing) scales shown with white arrow.

Fig. 9.18. *Euploea* spp. found dead on the road showing androconial tufts. (Bernard DUPONT from FRANCE [CC BY-SA 2.0 (https://creativecommons.org/licenses/by-sa/2.0)].)

(PTP), sometimes called 'lovedust'. The form of the alar scent-producing glands varies: from that of a patch (in *Parantica* and *Amauris* species), a pocket (in *Danaus*, *Anosia* and *Limnas*) or a pouch (in *Tirumala*). No alar organs are found in species within the genera *Idea*, or in the rare species, the magpie, *Protoploea apatela*, endemic to the island of New Guinea. Male, *Idea leuconoe* butterflies release a complex mixture of volatiles from their hairpencils during courtship; they can also display these organs when threatened by a predator. The hairpencils are dipped in, or applied to, the alar organs – located on the hindwings of danaid butterflies (Fig. 9.16) – to collect the PTPs. The hairpencils are normally kept retracted within the abdomen but are extruded and opened by the male during courtship.[19]

During courtship, or sometimes during pre-copulatory pursuit, hairpencils are everted to form two round bundles of hairs, which are used to brush the female's antennae and transfer a fine dust of pheromone particles. Hairpencils vary in size, being relatively large in the queen, *Danaus gilippus* (about 4 mm long), but much smaller in the monarch, *Danaus plexippus*. Hairpencils are made up of many fine hairs – c. 1200 in *Danaus chryssipus* but only 400 in *D. plexippus* – with marginal hairs surrounding the particle-budding hairs. The abdominal hairpencils of male *I. leuconoe* consist of a pair of large brushes (c. 15 mm, fully elongated) and a pair of small brushes (c. 2 mm), each independently housed in its own sheath. If the female is responsive to a male's advances, she flies down to the ground, where the male hovers above her, continuing to brush her with his hairpencils. Differences in the courtship behaviour of danaine butterflies are not fully understood, however, but it is known that female queen butterflies, *D. gilippus*, will not mate unless they have been brushed on the head and antennae in this way. Males without hairpencils are capable of courting females but incapable of copulation.[20]

The hairpencils in danaines are everted and protruded by haemolymph (blood) pressure – rather like evaginating the finger of a glove – and can contain a relatively large amount of pheromone (up to 500 µg of danaidone in *D. chrysippus*), which represents a considerable investment on the part of the male, i.e. in terms of procuring toxic precursors for these molecules. In most milkweed butterflies, including the genera, *Amauris*, *Danaus*, *Ideopsis* and *Tirumala* (Fig. 9.19), the hairpencils are only expanded during courtship, but in some species in the genus *Euploea* – such as the common crow, *Euploea core* – they can be everted during patrolling flights. Five different types of hairs were identified in the binate hairpencils of the novice, *Amauris ochlea* (Danainae), each seemingly with a different function.[21]

Danaine hairpencil secretions are said to be 'strong but pleasant' odours to the human nose – although danaidone itself is odourless for humans – composed of species-specific mixtures of volatile compounds sequestered from plant chemicals such as pyrrolizidine alkaloids (PAs), often obtained by adult feeding (see below). The key components in the hairpencil secretion of the male queen butterfly, *Danaus gilippus*, were the first aphrodisiac pheromones to be identified in butterflies: danaidone, and a viscous terpenoid alcohol, which acts as a glue for the PTPs. An analysis of male hairpencil secretions from 10 African milkweed species in the genera *Danaus*, *Tirumala* and *Amauris* yielded a huge range of compounds belonging to 14 different chemical classes, with up to 59 compounds present in the secretions of a given species. The precise roles of all these compounds, however, remain a matter of speculation, but their diversity points to the possibility of subtle variations in the concentrations of individual components within a blend of putative pheromones.[22]

PTPs are c. 1–2 μm or more in size and vary in shape and form between species; each type with their own characteristic appearance (Fig. 9.20). PTPs are produced in several ways; for example, in the novice, *Amauris ochlea*, they are split off from the terminal portions of long white hairs, in the abdominal hairpencils. In *Danaus chrysippus* (Fig. 9.17) they are produced by distinct 'particle-budding hairs', also in the hairpencils, but in other species, they are generated within the alar organs by the fragmentation of extremely long 'cushion' scales.[23]

Fig. 9.19. Dark blue tiger (*Tirumala septentrionis*), Thailand.

Fig. 9.20. SEM of pheromone transfer particles.

The monarch butterfly, *D. plexippus*, is rather an exception compared with other danaine butterflies, in that it has largely abandoned complex courtship rituals and does not appear to rely on pheromones (see Chapter 5). Although the males do have extrudable, abdominal hairpencils and alar wing pockets, contact between these organs does not seem to occur, and hairpencilling behaviour is not required for successful courtships. It does occur in some instances, but the experimental removal of the hairpencils does not affect the mating success or vigour of the males! Monarchs do gather PAs (see below) but their hairpencils lack the PA derivatives found in other milkweed species and they probably use these plant-derived toxins more for defence than mating.[24]

Contact behaviour and perfuming

Interestingly, there have been relatively few observations – at least recorded in the literature – of the so-called, 'contact behaviour', when the hairpencils are rubbed against the alar wing patches in order to load them up with PTPs. The extent to which the hairpencils are expanded during contact with the alar organs varies. In *Danaus* and *Tirumala* species, the brush-like hairpencils are unexpanded when dipped into the open cavities of the hindwing alar pockets or pouches. In *D. chrysippus*, males are said to turn their heads, and twitch the ends of their abdomen, before inserting the hairpencils into the wing pockets! The abdomen is bent upwards so that the tip is near the pockets, and the two unexpanded hairpencils are together, pushed – synchronously – into the two cavities of the hindwing pockets, for a few seconds. Males of the chestnut tiger, *Parantica sita*, have

Fig. 9.21. Striped blue crow (*Euploea mulciber mulciber*) male, Thailand.

a large black patch, a sex brand, near the anal margin of the hindwing, which they rub their hairpencils against to collect danaidone. This contact behaviour was called 'perfuming' by researchers studying this butterfly in Japan. Males of the striped blue crow, *Euploea mulciber* (Danaini) (Fig. 9.21), however, unlike *Danaus* and *Parantica* males, never exhibit contact behaviour between the two scent organs prior to courtship.[25]

Pyrrolizidine alkaloids (PAs)

Males of the queen butterfly, *D. gilippus* (Fig. 5.17), feed on certain toxic plants such as *Heliotropium senecio* (Boraginaceae), which contain PAs such as lycopsamine. PAs, or closely related compounds, occur in danaid species in all three subfamilies (Danainae, Lycoreinae, Euploeinae) and probably play a similar role in most of these butterflies, i.e. making them unpalatable to potential predators. Males transfer PAs to the females as nuptial gifts contained within the spermatophore (see Chapter 8), and females may be able to assess the gift-giving or alkaloid-donating capacity of the male via his pheromone, danaidone (see below). Some species – such as *I. leuconoe* (Danainae), *Tithorea harmonia* (Ithomiine) and some *Euploea* spp. – derive precursors for pheromones via feeding in the larval stage on their host plants, rather than sucking up liquid excrescences as most adult Ithomiine butterflies do. The total quantity of these toxic alkaloids in some butterflies, such as *I. leuconoe*, may be relatively high (>3 mg/insect), which is why they are highly unpalatable to potential predators.[26]

Males of many nymphalid species in the subfamilies Ithomiinae and Danainae visit and congregate on plants containing toxic alkaloids, which are used both for defence – making themselves unpalatable – and as precursors to produce aphrodisiac pheromones. These butterflies have, therefore, cleverly evolved ways and means of using toxic compounds – produced by plants for their own defence – in both avoiding predation and in reproduction – a strategy that has been described as arming themselves with prefabricated, low-cost defences! They are also called 'pharmacophagous', because they literally eat their own chemicals. Remarkably, danaine butterflies – they are always male – obtain their PAs by means of tiny scratches to the surface of PA-containing plants, using their mid-legs (the forelegs are vestigial) to damage the leaf surface. They then imbibe the oozing sap using their proboscis. On dry, withered plants they apply droplets with their proboscises, which dissolve the PAs, and then imbibe the fluids. They can also take advantage of previous damage, for example scratching at the damaged sides of holes made by flea beetles. Some danaids even obtain dihydropyrrolizines – the putative sex pheromones found in hairpencils – from 'feeding' on the carcasses of dead butterflies![27]

Antiaphrodisiacs

Male butterflies usually make a significant contribution to the reproductive process in the form of a large spermatophore, which has been called, somewhat prosaically, a 'nuptial gift' (see Chapter 8, this volume). The size of the male's contribution is commensurate with that of the female's, in some species, and because of this large investment – and to deter rival males – some butterflies produce chemicals that make the female appear unreceptive and unattractive to other suitors; i.e. an 'antiaphrodisiac'. Such antiaphrodisiac pheromones, produced by males and transferred to females during mating, were first found in the genus *Heliconius* (Nymphalidae), and were subsequently identified in other butterflies, including at least three pierids: *P. napi*, *P. brassicae* and *P. rapae*. The semiochemicals involved however, differ from species to species. In the green-veined butterfly, *P. napi*, the antiaphrodisiac pheromone is a volatile ester called methyl salicylate (MeS), which is transferred by the male to the female, via the spermatophore. Remarkably, once transferred, this chemical is then emitted by female *P. napi* butterflies if they are courted by other males and released whilst she is performing the 'mate refusal posture'. The female's store of antiaphrodisiac is gradually depleted by such activity and mated females gradually become more attractive to would-be suitors. Her receptivity, therefore, eventually returns and she is ready to mate again after a period of between 3 and 7 days in *P. napi*. Male *P. napi* are highly sensitive to MeS and use the chemical to assess the mating history of the female as well as the degree of sperm competition.[28]

These antiaphrodisiac chemicals have been called 'honest signals' by some biologists, because they accurately communicate the reproductive status of the female and allow other males to curtail their courtship of unreceptive females and focus on receptive ones. Others, however, have queried whether antiaphrodisiacs are indeed honest signals, since the female lacks any direct control over their release and males do not modify the amount that is transferred according to the potential for competition. In this respect, they may be more of 'an instrument of male control' over the subsequent reproductive activity of the female. Nevertheless, it seems likely that these chemicals are honest signals of female receptivity that benefit both sexes by (i) freeing the females from male harassment and (ii) saving the males from unproductive courtships.[29]

In *Heliconius* species, on which many studies have been carried out, antiaphrodisiac compounds are produced in glands located in the last abdominal segment of the males, and then transferred to a specialized abdominal gland in females during mating, where they are stored. The chemicals that the *Heliconius* males transfer to the females vary considerably between species but consist mainly of a few dominant volatile compounds – esters, lactones and terpenes – which are thought to be the antiaphrodisiacs, together with a matrix of less volatile esters, which regulate the evaporation rate of the repellent volatiles during female rejection displays. Female *Heliconius* butterflies have tiny 'stink clubs' at the end of their abdomens, close to scent glands, which may help disseminate the antiaphrodisiacs from special 'scent-wafting' scales.[30]

In *H. melpomene*, the volatile chemical that is synthesized by the male and transferred during copulation to the female – and which makes her unattractive to other courting males – is β-ocimene. It is a common plant semiochemical that is released by plants in response to insect feeding. The butterflies are using this molecule – in combination with other volatile chemicals – in a different way from plants, but they are also utilizing it as a messenger molecule with a potent odour, which in *H. melpomene* renders females less attractive to courting males. The mixture of odours produced by these butterflies probably has a defensive function as well, discouraging predators from attacking them. The scent secretions of the great orange tip butterfly, *Hebomoia glaucippe* (Pierinae, Colotini) (Fig. 9.22), are also E-β-ocimene. Two compounds – α-farnesene and 1-dodecyl-acetate – are transferred from the male to the female in silver-washed fritillary, *Argynnis paphia*, and are thought to act as repellents for other males.[31]

It is not difficult to imagine why such systems to deter subsequent suitors evolved, because it is in the interests of both sexes to defer rematings, at least for a while. The male benefits by preventing the female from remating, and thus utilizing only his sperm. The female benefits by

Fig. 9.22. Great orange tip (*Hebomoia glaucippe*), Thailand.

Fig. 9.23. Large cabbage white (*Pieris brassicae nepalensis*), Thailand.

avoiding having to fend off the attentions of other males whilst she deposits her already fertilized eggs. Egg laying and mating make conflicting demands on the female; both require a substantial amount of time – pairs can for instance remain locked *in copula* for substantial periods of time in some species (see Chapter 7) – and both activities need warm sunny weather. The mutual arrangement breaks down after a while, however, as both males and females remate in polyandrous species.[32]

The chemical compounds used as antiaphrodisiacs differ between species. In the large cabbage white butterfly, *P. brassicae* (Fig. 9.23), it is benzyl cyanide, whilst in the small white, *P. rapae*, it is a blend of MeS and the aromatic compound, indole. Remarkably, some tiny parasitoids – *Trichogramma brassicae* and *Trichogramma evanescens* – exploit the pheromone benzyl cyanide in *P. brassicae*. On detecting the antiaphrodisiac, the wasps ride on the body of the mated female butterfly as she flies to a host plant, on which she will oviposit. Once they are laid, however, the wasps parasitize the freshly deposited eggs. This innate behaviour is charmingly called: 'espionage-and-ride' behaviour![33]

Notes

[1] Gomez-Diaz and Benton (2013) and Nieberding *et al.* (2012). 'Pheromones': a new term for a class of biologically active substances (Karlson and Lüscher, 1959). A total of 500,000 female silkworm moths were needed to produce 12 mg of the pheromone, bombykol (Butenandt *et al.*, 1961; Wyatt, 2009).
[2] Myers (1972), Gomez-Diaz and Benton (2013) and Liénard *et al.* (2014).
[3] Vane-Wright and Boppré (1993), Eisner and Meinwald (2003), Wang *et al.* (2014) and Yildizhan *et al.* (2009). Males of the two subspecies, *Pieris rapae rapae* and *P. rapae crucivora*, differed in the concentration of ferrulactone (McQueen and Morehouse, 2018). Viridifloric β-lactone and danaidone are the putative sex pheromones in *Idea luconone* (Nishida *et al.*, 1996).
[4] Details on pheromones from: Hayaski *et al.* (1978), Kuwahara (1979), Yata *et al.* (1986), Ohguchi and Hidaka (1988), Sappington and Taylor (1990b), Nakai *et al.* (2001), Andersson *et al.* (2007), Mellström and Wiklund (2009), Yildizhan *et al.* (2009), and Li and Mathews (2016). Citral is a mixture of the two isomers: geranial (E-isomer, (trans-3,7-dimethyl-2,6-octadien-1-al)

and neral (Z-isomer, trans-3,7-dimethyl-2,6-octadien-1-al); both of which are necessary to produce an effect (Larsdotter-Mellström *et al.*, 2016). Three chemical components n-heptacosane (C27), 13-methylheptacosane (13MH) and n-nonacosane (C29) of the male courtship pheromone of male *Colias eurytheme* butterflies must be present for mate acceptance by females (Grula *et al.*, 1980; Rutowski, 1980b).

[5] Ford (1945) summarized the scents of many UK butterflies, based on a survey by Dr F.A. Dixey and Dr G.B. Longstaff. Males of the yellow tip, *Anthocharis scolymus*, lack alar scent scales (Okumura *et al.*, 2016). Ômura *et al.* (2013) reported two lactones in males, but not females, of *Celastrina argiolus ladonides*. Ômura *et al.* (2016) investigated the morphology of androconial scales in 13 *Celastrina* species. *p*-Cresol (4-methylphenol), is a volatile low-molecular-weight phenole compound, present in *Erynnis montanus* (Ômura and Honda, 2011).

[6] Honda (1980), Ômura *et al.* (2001), and Ômura and Honda (2005). Males of *Papilio polyxenes* and *Papilio machaon* were said to have faint sweet odours, resembling that of carrot flowers (Clark, 1926). Benzaldehyde, phenylacetaldehyde (major component), 2-phenylpropenal, n-heptanal, 6-methylhept-5-en-2-one and linalool were present in the scent of male *Byasa* (=*Atrophaneura*) *alcinous alcinous* (Honda, 1980). Recent research has however, discovered male-specific volatiles (mainly γ-Himachalen-4-yl acetate) in *A. alcinous alcinous* (Ômura *et al.*, 2016). *Atlides* scents based on Robbins *et al.* (2012).

[7] Nieberding *et al.* (2008) identified three components included in the sex pheromone of *Bicyclus anynana* males: (Z)-9-tetradecen-1-ol (MSP1), hexadecanal (MSP2) and 6,10,14-tri-methylpentadecan-2-ol (MSP3). Some of the pheromone compounds of *B. anynana* are structurally identical to female moth sex pheromones (Yuvaraj *et al.*, 2018).

[8] The local co-occurrence of closely related *Bicyclus* species correlates with differences in the composition and quantity of their major male volatile compounds (Bacquet *et al.*, 2015).

[9] Bacquet *et al.* (2015, 2016), Dion *et al.* (2017), Visser *et al.* (2018). Courting males of different ages display relatively large changes in the concentration of hexadecanal, and MSP composition remained a reliable indicator of male individuality over several weeks (Nieberding *et al.*, 2012). However, Kehl *et al.* (2015c) concluded that sex pheromones do not function as honest signals of male quality, in young male *B. anynana.*

[10] Mann *et al.* (2017) identified 68 different compounds in the wing androconia of 17 *Heliconius* species and subspecies. Seven putative MSPs had previously been identified by Mérot *et al.* (2015), in both *H. melpomene* and *H. timareta.* According to Darragh *et al.* (2017), five putative MSPs – (Z)-9-octadecenal, octadecanal, (Z)-11-icosenal, icosanal and (Z)-13-docosenal – are thought to be produced and stored in the hindwing androconia of mature male *Heliconius melpomene*. A more recent study confirmed the most abundant compounds in the androconia of *H. melpomene* as: syringaldehyde, octadecanal, octadecan-1-ol, (Z)-11-icosenal and icosanal (Darragh *et al.*, 2019a). Octadecanal was found exclusively in *H. melpomene*, amongst seven tested *Heliconius* species (Darragh *et al.*, 2019b).

[11] See Boppré and Vane-Wright (1989) for a review of androconial systems – abdominal and alar organs – in danaine butterflies. For a comprehensive survey of alar, abdominal and appendage androconial organs in the Riodinidae, see Hall and Harvey (2002). For 3D imaging of concealed androconial scales in metalmark butterflies (Riodinidae), see: Montejo-Kovacevich *et al.* (2017). Photos of secondary sexual organs on the wings of New World hairstreak butterflies (Lycaenidae, Theclinae) are presented by Martins *et al.* (2018).

[12] Bálint *et al.* (2017), Robbins *et al.* (2012), Martins *et al.* (2016, 2018). See fig. 14 in Robbins *et al.* (2015): *Thereus oppia* showing the erect androconia attached to the inner margin of ventral forewing of *Thereus oppia* (also in *T. brocki*), which occur in no other Eumaeini.

[13] Rauser and Rutowski (2003). The alar androconial organ of the Actorion owlet, *Bia actorion* (Brassolini, Morphinae) is described by Vane-Wright and Boppré (2004).

[14] Eisner and Meinwald (2003). Two dihydropyrrolizines may be produced independently, in different organs in *Ideopsis similis*: hydroxydanaidal from the abdominal hairpencils and danaidone

from the hindwings (Honda *et al.*,1995). Danaidone acts as an arrestant pheromone on females, for *Danaus gilippus* and *Danaus chrysippus*, and is necessary for courtship success (Pliske and Eisner, 1969). Dihydropyrrolizine derivatives, such as danaidone (2,3-dihydro-7-methyl-1H-pyrrolizin-1-one), danaidal (6,7-dihydro-5H-pyrrolizine-1-carboxaldehyde) and hydroxydanaidal are common components of androconial secretions in many danaines (Honda *et al.*, 2016).

[15] Guillaumin (1963), Pivnick *et al.* (1992), Warren *et al.* (2009), Braby and Zwick (2015) and Hernández-Roldán *et al.* (2014). See Kristensen and Simonsen (2012) for further details on scent scales. Osmophores are a pheromone-dispersing mechanism (Wuest, 1998) delivered to the female antennae via direct contact in some, but not all, species (Sellier, 1972).

[16] Vetter and Rutowski (1978), Rutowski (1977a, 1980b) and Yoshida *et al.* (2000).

[17] Condamin (1973), Costanzo and Monteiro (2007), Nieberding *et al.* (2008) and Dion *et al.* (2016). Androconia are closely linked to reproductive isolation between biological species (Wakeham-Dawson *et al.*, 2007).

[18] Putative androconia are specialized brush-like scales within a shiny grey region of the male hindwing (Darragh *et al.*, 2017).

[19] Most of the 83 species in the tribe Danaini have androconia that produce cuticular PTPs, except for the eight species in the genera *Ideopsis* (Vane-Wright and Boppré, 1990). The sex brands of six danaids from Japan and Taiwan were described by Komae *et al.* (1983). Different forms of the alar glands were described by Brower and Jones (1965). The pheromone components produced by male *Idea leuconoe* include: geranyl methyl thioether, viridifloric β-lactone and 6-hydroxy-4-dodecanolide (Schulz and Nishida, 1996).

[20] Myers and Brower (1969), Komae *et al.* (1983), Boppré (1993), Nishida *et al.* (1996), Vane-Wright *et al.* (2002), Brower *et al.* (2010) and Honda *et al.* (2018). For studies on queen butterflies, *Danaus gilippus berenice*, see: Pliske and Eisner (1969), Pliske and Salpeter (1971), Schneider *et al.* (1975), and Eisner and Meinwald (2003).

[21] Boppré (1984, 1993, 1995) and Boppré and Vane-Wright (1989).

[22] Brower *et al.* (1965), Boppré *et al.* (1978), Schulz *et al.* (1993). Meinwald *et al.* (1969) discovered the sex pheromones produced by male, *Danaus gilippus berenice*.

[23] Hashimoto and Yara (2007, 2008) compared the microstructures of PTPs from nine male *Tirumala* (Danainae) species. Vane-Wright and Boppré (1990) describe the evolution of PTP production in the Danaini. PTPs in *Amauris ochlea* are described by Petty *et al.* (1977). Scales in alar organs are far more densely packed per unit area than normal wing-covering scales (Vane-Wright *et al.*, 2002).

[24] Pliske (1975b) and Boppré (1993).

[25] Honda (2006), Honda *et al.* (2005, 2016) and Schulz (1998). For a description of the androconial organs of *Parantica sita*, see Vane-Wright *et al.* (2002). The extrusion of the hairpencils, and their insertion into the alar wing pockets, by male *Danaus gilippus xanthippus* is described by Brower and Jones (1965); and for *Danaus chrysippus* by Boppré *et al.* (1978). Two compounds – 9,10-epoxytetrahydroedulan (ET) and viridiflorine b-lactone (VL) – were identified as major components from the hairpencils of males of *E. mulciber*.

[26] Edgar *et al.* (1976a, b), Boppré (1986), Dussourd *et al.* (1989), Trigo and Motta (1990), Nishida *et al.* (1996) and Nishida (2017). Some Ithomiinae disseminate aphrodisiacs during the later phases of the courtship flight, others mainly use these glands during male–female pursuits (Pliske, 1975a).

[27] Various plants, such as *Heliotropium indicum* L. (Boraginaceae) – including dead stems, seeds – containing PAs are visited by nymphalid butterflies in the subfamilies Ithomiinae and Danainae (Pliske, 1975b, Pliske *et al.*, 1976). For a detailed description of leaf-scratching behaviour by danaine butterflies, see Boppré (1983). A new PA derivative was present in the androconial organs of *Prittwitzia hymenaea, Mechanitis isthmia veritabilis* and *Tithorea harmónia fúria* (Ithomiinae): methyl hydroxydanaidoate (Schulz *et al.*, 1988). Edgar *et al.* (1973) observed danaid butterflies feeding on the carcasses of dead males.

[28] Forsberg and Wiklund (1989), Andersson *et al.* (2004), Mellström and Wiklund (2009) and Larsdotter-Mellström *et al.* (2016a). Male *Heliconius erato* transfer antiaphrodisiac pheromones from abdominal clasper scent glands to females at mating (Gilbert, 1976).
[29] Andersson *et al.* (2003) considered that antiaphrodisiacs are honest signals, conveying the information that the female has already mated. Estrada *et al.* (2011) however, argued that they were not always honest signals of her receptivity to mate; rather, they are 'an instrument of male control over female remating'.
[30] Based on Schulz *et al.* (2007, 2008), Estrada (2009) and Estrada *et al.* (2011). Scent glands and clubs likely play a role in receiving, storing and disseminating antiaphrodisiac pheromones (Jiggins, 2016). The stink clubs are on the posterior margin of the eighth sternum in a passion vine butterfly, *Neruda metis* sp. n. (Heliconiine), from Brazil (Moreira and Mielke, 2010); and are attached to the plate of sternum 7 in the Gulf fritillary, *Agraulis vanillae* (Heliconiine) (Scott, 1986a).
[31] Hayashi *et al.* (1985), Ockenfels *et al.* (1998) and Schulz *et al.* (2008). Plants can respond to insect feeding damage by releasing volatiles – including (E)-β-ocimene (Paré and Tumlinson, 1999).
[32] Virgin female *Pieris napi* are rendered strongly unattractive immediately after mating (Andersson *et al.*, 2000). The time spent *in copula* interferes with egg laying (Andersson *et al.*, 2003).
[33] Andersson *et al.* (2004) and Huigens *et al.* (2010). Indole is an aromatic heterocyclic organic compound with the formula: C_8H_7N.

10 Wing Colours and Iridescence

The Different Functions of Butterfly Wing Surfaces

The upper (dorsal) wing surfaces of butterflies are often brightly coloured and visually highly apparent, whilst the under (ventral) sides are usually dull and inconspicuous. Clearly, this is a gross generalization, as there are many butterflies with very drab, or even matt black upper wing surfaces, which is an adaptation to enable thermoregulation. Nevertheless, in general, there is a marked difference between the two wing surfaces, which reflects their contrasting functions. Upper wing surfaces are often used to signal messages – warnings or startling flashes aimed at predators, or sexual signals between males and females – whereas undersurfaces, visible in butterflies that rest with their wings held vertically, are generally a form of camouflage. In effect, the colours and patterns seen on butterfly wings have evolved as a result of three main forces of selection: thermoregulation, communication with members of the same species and reducing the risk of predation. The unique appearance of each species is, therefore, a product of the evolutionary interplay of these forces of natural and sexual selection.[1]

There is a spectacular variety of mechanisms by which butterflies have evolved ways of achieving such contrasting functional ends, including light-absorbing pigments, light-reflecting scales and so-called photonic crystals, which manipulate light to produce high levels of reflectance. Photonic crystals are three-dimensional materials composed of just two types of matter – air and insect cuticle – with different ways of bending light rays, i.e. different refractive indices. Only artificially fabricated by man as recently as 1987, photonic crystals are widespread in nature, from opals to butterfly wings, and have been described as a unique partnership between light and matter. They are called crystals because

© Raymond J.C. Cannon 2020. *Courtship and Mating in Butterflies: Reproduction, Mating Behaviour and Sexual Conflicts* (R.J.C. Cannon)

of their periodic or repeating structure, plus the fact that they grow or self-assemble during the pupal stage of the butterfly's development.[2]

Structural Colours and Nanoarchitecture

The dazzling iridescent wings and highly reflective wing patches found on some butterflies, such as the Menelaus blue morpho, *Morpho menelaus huebneri*, male (Fig. 10.1), are the result of an optical phenomena – predominantly light interference – produced by nanoscale structures, together with selective absorption by pigments. The tiny – micron- and submicron-sized – photonic structures are made up of multiple layers of chitin in the form of ridges; simply layers of cuticle interspersed with air spaces. The ridges run parallel to the long axis of the butterfly wing scale (Fig. 10.2) and scatter the incident light falling on it. The submicron-sized ridges, and three-dimensional cuticular lattices, are of a size comparable with that of the wavelength of light landing on them and produce what is called physical or structural colours. These so-called nanostructures are widespread in the animal kingdom – from crabs to weevils, butterflies to birds – and produce colours via interference, diffraction or scattering of light.[3]

The colours that are generated by optical interference and other such mechanisms via these physical nanostructures are far brighter, richer and more vivid than those produced by chemical pigmentation. They have been exploited by male butterflies to attract mates and/or threaten or intimidate rivals, i.e. males of the same species, often over considerable distances. The advantages of structural colours over chemical colours (pigments) is that they enable butterflies to adapt the hue, intensity and angle dependence of a colour; they can also reflect a large proportion of the light in select regions of the spectrum. Structural colours – usually blue and green hues – are produced by interactions between incident light and interference and reflections caused by the periodic surface structures

Fig. 10.1. Menelaus blue morpho (*Morpho menelaus huebneri*) male, Pará, Brazil. (Photo by Didier Descouens – own work, CC BY-SA 4.0, https://commons.wikimedia.org/w/index.php?curid=15789967)

Fig. 10.2. Birdwing butterfly wing scales showing longitudinal ridges. (a) *Ornithoptera Croesus*, (b) *O. tithonus*, (c) *O. priamus*, (d) *O. urvillianus*. Column I: picture of the animal, II: scale lattice of the colour patch on the dorsal wing surface, III: TEM cross-section images of structural coloured wing scales, IV: scatterograms of single wing scales. (From Wilts *et al*., 2015.)

on the scales. The effects range from simple light scattering to more complex multilayer interference, although all structural colours produced by butterfly scales are, as a result of what is termed, coherent scattering. That is to say, *constructive interference* between light waves scattered coherently by the nanostructures on the surfaces of the scales. In summary, structural colours are produced via a complex combination of optical effects, including interference, diffraction, absorption and scattering.[4]

Iridescence

Structural colours can be iridescent or non-iridescent. Iridescence is when surfaces – usually the wings in butterflies – appear to change brightness and colour as the angle of view or the angle of illumination changes. This results in highly conspicuous flashing – called *flicker contrast* – of the iridescent surface when butterflies like blue morphos flap their wings during flight, but flickering can also be produced by the butterfly when at rest, by adjusting the position of its wings. The orientations of both signaller (relative to the light source, usually the sun) and receiver (relative to the signaller) are clearly critical in terms of the behavioural utilization of iridescence. Iridescence is usually only visible within a narrow range. For example, the brilliant violet-blue iridescent colours produced on the dorsal wing surfaces of male purple emperors, *Apatura iris* (Fig. 10.3), are only visible within an 18° angular range. Similarly, the UV reflecting patches on the wings of flying males of the orange sulphur butterfly, *Colias eurytheme* (Fig. 2.15), are only visible over an angle of c. 14°, and are brightest when seen from above. The rapid wing beats of this species, *C. eurytheme* – approx. eight beats/s – mean that the brilliant UV flickering will pulse at about 16 Hz – a highly conspicuous signal, which has evolved to both repel other males and to be attractive to females.[5]

Black light-traps and solar collectors

Butterfly scales are not only adapted to reflecting light, they may also be extremely efficient at absorbing it. Rajah Brooke's birdwing, *Trogonoptera brookiana* (Fig. 10.4a, b), has large areas covered in black scales, which serve to highlight the green, iridescent leaflet-shaped features on the upper surfaces of both fore- and hindwings (Fig. 10.4b). The reflectance of the black scales is extremely low – never greater than 8%, and below 2–3% for

Fig. 10.3. Purple emperor (*Apatura iris*) male Forêt de Sagnebaude, Arfons, Tarn, France. (Photo by Bernard DUPONT. Attribution-ShareAlike 2.0 Generic (CC BY-SA 2.0).)

Fig. 10.4. Rajah Brooke's birdwing (*Trogonoptera brookiana*), KL butterfly park, Malaysia.

Fig. 10.5. Golden birdwing (*Troides aeacus thomsoni*) male, butterfly park, Bangkok, Thailand. (Photo by Steve from Bangkok, Thailand [CC BY-SA 2.0 (https://creativecommons.org/licenses/by-sa/2.0)].)

most of the range, throughout the spectrum from 400–900 nm – compared with the more reflective, iridescent green scales. In other words, the black areas are super-efficient light-traps, as a result of a microscopic honeycomb-like structure between the tiny ridges (1.6 µm apart) on the surfaces of the black scales. A similar arrangement of nanoscale ridges and holes is also responsible for the 'anti-reflection' properties of the black scales on the forewings – particularly the males – of the golden birdwing, *Troides aeacus* (Fig. 10.5). The light-absorbing properties of these black wing scales – which are natural 'solar collectors' – may also help with thermoregulation in species with extensive areas of black on their wings.[6]

Hydrophobicity and self-cleaning wings

The ability to shed water and prevent overloading of the wing is an important adaptation in many butterflies. Low wettability of wing surfaces is due to the property of super-hydrophobicity, which occurs when the contact angle between a water droplet and the wing surface is in excess of 150°. Butterfly wings have a nanostructured roughness – microscopic (<1 µm) protuberances – which means that droplets roll, or slide, across the wing surface with ease, and in the process sweep away any contaminating particles, such as dirt, dust and fungal spores. Another useful property arises from the fact that butterfly wing scales, and the nanostructures along them, are aligned in the same direction, so that water droplets roll off of the wing in one direction.[7]

Scales and Ridges

Butterfly wings are covered on both sides by rows of tiny – e.g. 50–100 µm wide and 150–200 µm long – overlapping scales, a bit like very thin, flat roof tiles or shingles (Fig. 10.6). Or as one researcher, Dr Bodo Wilts, eloquently put it: 'a tapestry of numerous small scales imbricates the wings of butterflies, often causing strikingly colourful patterns'. Scales are

Fig. 10.6. Butterfly wing scales, overlapping like very thin, flat roof tiles. Scales of a tropical swallowtail butterfly. (Created by Alex Wild.)

modified bristles, highly flattened, with hollow interiors, like an empty sac. They develop rapidly from single cells during the pupal stage, expanding and then dying to leave a skeleton of chitin filled with air and pigments. The scales usually point away from the body of the butterfly and are attached to the wing membrane via a tiny shaft – which inserts into a socket cell (Fig. 10.7) – and are slightly angled at 10–20° relative to the wing surface. The wings are in effect just two cells thick. Wing scales are easily detached, which might be an adaptation to escape from spider's webs: scales that become attached to the viscid threads of a spider's web can be sacrificed to allow the butterfly to regain its freedom (but not all escape!). The colours seen on a butterfly wing are produced by the combined layers of scales on either side of the wing, together with the wing membrane itself. The overall effect of these individually coloured scales has been described as like that of a pointillist painting, although a single scale can sometimes produce more than one colour, i.e. reflectance peak.[8]

Most species have two layers of wing scales – a base or ground layer, and an upper layer of larger cover scales – on both dorsal and ventral sides of their wings. In some species, such as the colourful Paris peacock swallowtail butterfly, *Papilio paris* (Fig. 10.8), iridescent cover scales are laid on top of black base scales, which enhance the effect by absorbing any stray light. The scales are very delicate, typically 1–2 μm in thickness, and are denuded by wear and tear as butterflies age. Cover scales are visible on the intact wing, but basal or ground scales are almost invisible, as they are obscured by the cover scales. The wing membrane itself is also very thin. Together, these five elements – scale, scale, wing membrane, scale, scale – with their optical properties resulting from nanostructures and embedded pigments, determine the colours of a butterfly wing. Scales can vary markedly in size and shape but, depending on the species, there are of the order of 200–600 scales/mm^2 of wing. Each scale is usually a unique colour, although some scales appear as a mosaic of different hues – a tessellated pattern – under polarized light; an effect caused by the fused domains of individual gyroid crystals, of which more below. Scales are not

Fig. 10.7. Wing scales of the painted lady butterfly, *Vanessa cardui*. (a) *V. cardui* is a common lepidopteran species found throughout the world. Scale bar 1 cm. (b) The wing is covered in overlapping layers of scale cells as seen in this reflected light image in the region of one of the ventral hindwing eyespots. The scales themselves are composed of chitin, but in the adult are no longer alive, rather just the remains of once-living cells. Scale bar 2 mm. (c) Higher magnification image of wing scales. The scales are arranged like overlapping tiles. Note that each scale is just a single colour. The wing scales all point towards the proximal end of the wing (away from the body), which in this image is oriented down. Scale bar 150 μm. (d) SEM image showing finger-like projection (green arrow) at the end of scale cells and the socket (blue arrow) where scales are attached to the wing epithelium. Scale bar 65 μm. (e) SEM of base of scale cell where the scale shaft (red arrow) inserts into the socket cell (blue arrow). Scale bar 20 μm. (f) High magnification SEM showing that scale cell surfaces are ornately patterned with longitudinal ridges (yellow arrow) and cross-ribs between the ridges. Scale bar 2 μm. (g) 6-day pupal wing scale stained with fluorescent Wheat Germ Agglutinin (WGA), a lectin that binds to chitin. Green arrow indicates finger like projections, and red arrow indicates the shaft of the scale (scale has been pulled from the socket). By this stage of development, pupal scales already have the overall size and morphology characteristic of adult scales. Scale bar 20 μm. (From Dinwiddie *et al.*, 2014, fig. 1.).

always flat: red wing scales – which produce blue iridescence – on the wings of the Lysander cattleheart, *Parides lysander* (Fig. 10.9), are highly curved in a gutter-like shape. Similarly, scales from the iridescent patches on the satyrid butterfly, *Pierella luna*, are of three types: flat, curled upwards and tubular.[9]

Each scale consists of two layers, or lamellae, held together by a series of tiny pillars, called trabeculae. The lower layer of the scale is flat and smooth – between 100 and 200 nm in thickness – whilst the upper

Fig. 10.8. Paris peacock (*Papilio paris*) iridescent wing scales, Thailand.

Fig. 10.9. Lysander cattleheart (*Parides lysander*), dorsal view, Cacao, Crique Baguot, French Guiana. (Photo by Muséum de Toulouse [CC BY-SA 4.0 (https://creativecommons.org/licenses/by-sa/4.0)].)

layer consists of a series of longitudinal ridges, or striae – about 1–2 µm apart – and transverse cross-ribs, which create a three-dimensional lattice, or honeycomb structure. These elaborate three-dimensional nanostructures – so-called *perforated multilayers* – between the lamellae, cause the observed colours, like iridescence. Different types of these regular photonic structures – often called *nanoarchitecture* – have been identified based on the shapes of the multilayers and the 3-D honeycomb of chitin and air spaces. Fine differences in the structure of the nanoarchitecture – i.e. the size and separation of the holes within the honeycomb – in the scales of different species produce different colours on the wings of butterflies. The scales can also act as thin-film reflectors, or diffraction gratings, which physically manipulate the incoming light, selectively reflecting certain wavelengths and blocking others.[10]

Pigments

Butterfly scales are pigmented, and this, together with their structural properties, determines their final coloration. If no pigments were present,

Fig. 10.10. Ulysses butterfly (*Papilio ulysses ambiguus*). Ambon, Maluku, Indonesia. Photo by Bernord Dupont. Flickr Creative Commons.

scales would be white. Melanin is the most commonly occurring pigment in butterfly wings, responsible for brown and black colours, although nanostructures also contribute to the black appearance in some species, such as Ulysses butterfly, *Papilio ulysses* (Fig. 10.10). Some butterfly families have evolved their own characteristic pigments, such as: *papiliochromes* (responsible for the yellow and cream colours of papilionids like the Japanese yellow swallowtail, *Papilio xuthus*, Fig. 2.2a, b), *ommochromes* (which produce red and yellow colours in nymphalids) and *pterins* (contained in the scales of pierids). The ultraviolet-absorbing pigment, *leucopterin*, occurs in white, *Pieris* species, and the violet-absorbing *xanthopterin* and blue-absorbing *erythropterin* pigments occur in yellow- and orange-coloured Coliadinae. Pigments are usually found in scales that do not have nanostructures, and they selectively absorb light of a certain colour, i.e. in a restricted wavelength range. There are, however, some species where pigments are combined with photonic structures. For example, the hindwings of male Magellan birdwing butterflies, *Troides magellanus*, contain fluorescent molecules – papiliochrome II – embedded in 3-D photonic crystals, which together produce iridescent glints and a bright yellow colour in daylight. In other words, this colour is produced by a combination of chemical coloration – restricted light absorbance – and structural coloration: specific light reflection.[11]

Blue Morpho Wings

Butterflies in the subfamily Morphinae, morphos and owl butterflies (Fig. 10.11) include many species where the males possess shiny, blue

Fig. 10.11. Owl butterfly (*Caligo atreus*) and blue morphos (*Morpho peleides*) feeding on oranges, Artis Royal Zoo in Amsterdam, NL.

metallic wings; e.g. *Morpho rhetenor*, *Morpho menelaus* (Fig. 10.1), *Morpho adonis*, *Morpho didius*, *Morpho epistrophus*, *Morpho helenor* and *Morpho cypris*. The wings of the females are generally duller than the males, or completely non-reflective. The undersides of morpho wings are usually dark brown (Fig. 10.15b), although the eyespots in some species are remarkably colourful, as shown in these photographs of *Morpho peleides* (Fig. 10.12a, b). There is, however, considerable diversity in terms of the size, shape and blue-toned structural colours of the wings of butterflies in the genus *Morpho*, of which there are about 30 species and 150 subspecies. Some, like the giant blue morpho, *M. didius*, have dull, cobalt blue wings, whereas others, like Sulkowsky's morpho, *Morpho sulkowskyi* (Fig. 10.13), have a more whitish-blue or pearly colouring. In other words, some species have more glossy wings than others, but they all exhibit very high levels of reflectivity – typically about 75% – and these uniform blue reflections are spread across a broad angular range. Differences in the intensity of reflections from different *Morpho* species are related to nanostructures: the more intense blue reflections are produced when the scale ridges have tall stacks of lamellae, acting as highly reflective multilayers (Fig. 10.14). The intense blue iridescence is primarily used for in-flight signalling between morpho males and can be seen from an incredible distance: blue flashing morphos have been reported as visible from aeroplanes![12]

A remarkable feature of morpho butterflies is that the wing colour changes to some extent with the angle of observation. This effect can be seen when the wing is tilted vertically or rotated horizontally, so butterflies like *M. menelaus* (Fig. 10.1) have somewhat different iridescent colours depending on their position relative to the observer. These vibrant iridescent colours are predominantly used for 'in-flight signalling' between males of the same species. Some species – such as *M. rhetenor* (Fig. 10.15a, b), *M. cypris* and *M. didius* – have just one layer of iridescent ground scales on the dorsal surfaces of their wings; the cover scales have atrophied. Other species – such as *M. didius* and *M. helenor* – have

Fig. 10.12. Blue morphos (*Morpho peleides*) feeding on oranges, Artis Royal zoo, Amsterdam.

Fig. 10.13. Sulkowsky's morpho (*Morpho sulkowskyi*), male and female. (Photo by Anaxibia [CC BY-SA 4.0 (https://creativecommons.org/licenses/by-sa/4.0)].)

two layers: comprising the iridescent ground scales plus a covering of near-transparent, glass scales, which cause strong diffraction; further spreading out the angle over which incident light is reflected. Both cover and ground scales have microscopic ridges, typically about 1,300 thin, parallel ridges per millimetre running down their lengths; and tiny

Fig. 10.14. Iridescent colours (*Morpho sulkowskyi*) caused by tree-like nanostructures on its wings (Photo by Anaxibia [CC BY-SA 4.0 (https://creativecommons.org/licenses/by-sa/4.0)].)

Fig. 10.15. (a) Rhetenor blue morpho (*Morpho rhetenor rhetenor*) male dorsal side. (b) Ventral side. (Photo by Didier Descouens – own work, CC BY-SA 4.0, https://commons.wikimedia.org/w/index.php?curid=22824176)

differences in nanoarchitecture – i.e. the thickness of the ridges and their periodic spacing – between species determine the structural colours we see on the whole wing. Small differences in the colour of the ground scales – probably due to the concentration of melanin pigments – also affect the overall colour.[13]

The fine ridging on the ground scales creates a periodic or repeating structure – called a diffraction grating – in the horizontal scale plane. The ridges are constructed of a series of overlapping lamellae – the so-called multilayer system – each about 90–100 nm thick (depending on species), which, in cross-section, appear as Christmas (or pine) tree-like geometric shapes running perpendicular to the longitudinal plane of the scale (Fig. 10.14). These 10–12 stacked multilayers of cuticle sticking out on either side of the thousands of ridges running the length of the scales produce interference effects in the blue part of the visible spectrum. It is the interference in light between these layers of air and cuticle (with their different refractive indices) that produces the iridescent blue and violet colours in morpho butterflies. Cuticle material accounts for only about 30% of the volume within these ridged nanostructures, which are slightly tilted or bent over on the scale plane. Together with the ridged, glass-like, transparent cover scales, these minute, Christmas tree-like structures on the surface of the basal scales produce a diffraction grating, which can reflect up to 70–80% of the blue light that falls on the wings. Man-made diffraction gratings – nowhere near as beautiful as a morpho's wing – include metallic-coloured holograms on some credit cards.[14]

Lycaenid Wings

Like the morphos, the wings of lycaenid butterflies are also clad in cover and ground (basal) scales. The cover scales are ridged and unmelanized, and function as photonic crystals, usually reflecting short wavelength (UV/blue) light. Differences in hues are correlated with variations in the shape and pattern of these nanostructures, whilst the ground scales usually contain the light-absorbing pigment, melanin. Male lycaenids typically have bright, iridescent colours – e.g. blue, green, purple, silver and gold – usually, but not exclusively, on the dorsal surfaces of their wings. For example, the yellow tinsel, *Catapaecilma subochrea* (Fig. 10.16), has beautiful, highly reflective, silver scales on its ventral wing surfaces.

The iridescence seen on the upper wing surfaces of many lycaenids is usually only present on the males, like the common gem (*Poritia hewitsoni tavoyana*) male in Fig. 10.17a, whereas the cryptic patterns found on ventral wing surfaces are common to both sexes (Fig. 10.17b). Some very similar *Cyanophrys* species – which occur together in the same tropical environment – appear almost identical to human eyes but have subtly different structural colours. The males have iridescent blue colour patterns on their dorsal wings, which are species-specific, and probably function as a sexual signal during courtship and/or mate recognition. Both males and females have the same green structural colour on their ventral hindwings however, which presumably helps them hide in green foliage. These butterflies have also evolved vision that enables them to discriminate very subtle differences in iridescent blue colours, e.g. between 480 and 490 nm, and have effectively separated themselves using these fine colour gradations, whilst living within the same habitat. Subtle shades of iridescent blue define each species![15]

Fig. 10.16. Yellow tinsel (*Catapaecilma subochrea*), Thailand.

Fig. 10.17. (a) Common gem (*Poritia hewitsoni tavoyana*) male upper side showing iridescence, Thailand. (b) Common gem (*Poritia hewitsoni tavoyana*) male underside showing cryptic coloration, Thailand.

Polyommatine blues found in Europe also have subtly different nanostructures that produce species-specific colours. Male Adonis blues, *Polyommatus bellargus* (Fig. 10.18), for example, generate conspicuously bright, colourful signals, as they flap their wings in flight. These flashing signals help them to find and attract mates, as well as advertising their presence to potential competitors. Different species also space themselves apart in time to avoid conflict: in a given location, their mating flights occur at different times of the year, so there is less need for them to differentiate themselves in terms of their coloration. A variety of different nanostructures have been found to generate blue colours in lycaenid butterflies, although many have so-called multilayers – alternating layers of chitin and air – within the scales that produce the hue of the reflected light. These nanostructures have a pepperpot structure and look a bit like microscopic Swiss cheese (Fig. 10.19) and have evolved to produce light of a certain colour, i.e. blue to near UV.[16]

Intrasexual communication, i.e. competition between males, is thought to be the major selective agent responsible for producing the brilliant

Fig. 10.18. Adonis blue (*Polyommatus bellargus*) male, Asturias, Spain.

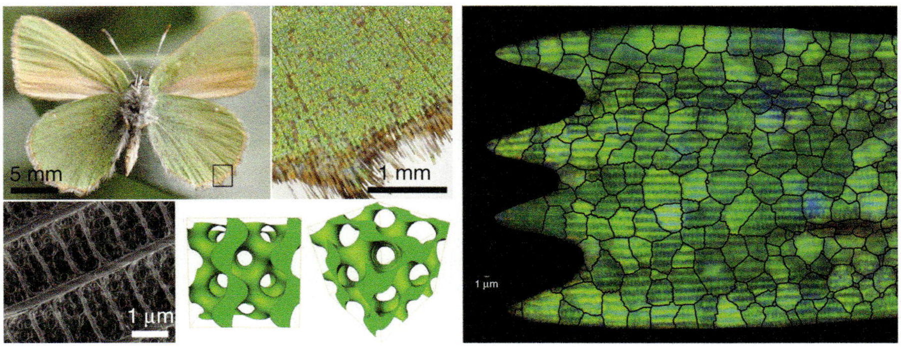

Fig. 10.19. Butterfly Callophrys rubi wing scale microstructure. Four photos on the left from Winter *et al.*, 2015; Clockwise from top left: (a) ventral side of C. rubi, (b) hundreds of shimmering green scales, (c) models of the left and right-handed gyroid structures, (d) SEM image of the ventral scale surface showing the periodic gyroid structure. Photo on right from Corkery and Tyrode, 2017; C. rubi wing scale showing domain boundaries of biophotonic gyroid nanostructures.

coloration in many species, including lycaenids. However, monophyletic groups – species that have all descended from a common ancestor – have similar nanoarchitecture. Nanostructures such as photonic crystals vary tremendously in size and shape, depending on the species and their position on the wing. Single photonic crystals on the shiny blue dorsal scales of the male Brazilian butterfly, *Cyanophrys remus* (Eumaeini), are relatively large (50 × 12 μm²), whilst those on the matte-green undersides are small, blue, green and yellow crystal grains, c. 3–10 μm in diameter.[17]

The reflected iridescence produced by light scattering from the dorsal wing scales of many lycaenids is highly directional, i.e. only observable from a narrow angular window. The green hairstreak, *Callophrys rubi* (Figs 10.19 and 10.20) – an Old World species with a broad Palaearctic distribution – however, generates a bright matte-green colour on the ventral (underside) of its wings – like that of *Cyanophrys remus*, which can be seen from all directions. This cryptic coloration provides the butterfly with camouflage when resting in a position with wings folded, revealing its undersides, amongst bright green foliage. The reflected green light is generated via a biophotonic crystal – a *gyroid* structure – within the scales on the undersides of both wings, whose 3-D lattice scatters the light in all directions (Fig. 10.19). Several lycaenid butterflies in the genera *Callophrys*, *Cyanophrys* and *Thecla* have evolved these biophotonic gyroid nanostructures within their wings. What is remarkable is that these crystal lattices are predominantly filled with air: the microstructural space occupied by solid cuticle varies between 17 and 40%, depending on species. Another type of hairstreak with bright green wings on their undersides is *Thecla opisena*, a New World species found in Mexico. The cover scales of this butterfly are iridescent green, reflecting light in a narrow band, with a peak wavelength of around 540 nm. The cover scales have an upper lamina comprising a grating of ridges and cross-ribs, which overlie a composite patchwork of faceted, gyroid-shaped crystallites, loosely connected to the lower lamina of each scale.[18]

Lycaenids are most sensitive to light between violet and UV, and to some extent, the spectral sensitivity in different species correlates with their wing colour, at least in *monomorphic* species – where the two sexes are approximately the same colour. In other words, they have evolved photoreceptors in their eyes that see their own wing colours best. For example, orange-coloured species – such as *Japonica lutea* and *Japonica saepestriata* – were significantly more sensitive to red light, whereas blue species – such as *Favonius ultramarinus* and *Favonius saphirinus* –

Fig. 10.20. Green hairstreak (*Callophrys rubi*), Aston Upthorpe, Oxfordshire. (Photo by Charles J. Sharp [CC BY-SA 4.0 (https://creativecommons.org/licenses/by-sa/4.0)].)

were more sensitive to UV. These differences were most pronounced in the males of these Japanese butterflies, perhaps in order to recognize prospective mates.[19]

In lycaenid species where the sexes are markedly different in colour (*dimorphic*), no marked differences in spectral sensitivities between species were found, at least those that were studied by a Japanese researcher Michio Imafuku of Kyoto University, Japan. The dorsal wing surfaces of the males of four Japanese lycaenids in the genus *Chrysozephyrus* – *Chrysozephyrus smaragdinus*, *Chrysozephyrus brillantinus*, *Chrysozephyrus hisamatsusanus* and *Chrysozephyrus ataxus* – all appear metallic green to the human eye, but they look very different to the butterflies themselves, which are able to see reflected UV light. There were differences between the species in terms of the amount of UV (315–350 nm) and green light (530–550 nm) they reflected. These brightly coloured, iridescent butterflies rely on structural coloration on their wings to produce the iridescence that they use in male-to-male interactions or in attracting females. Evolution does not appear to have modified the basic pattern of spectral sensitivity to colour, probably because the females are dark brown and mostly lacking in structural colours. As always in nature, there are exceptions, and in some species, such as *C. ataxus*, the females do have some areas of iridescence produced by structural coloration.[20]

Another example of sexual dichromatism is provided by the angled sunbeam, *Curetis acuta* (Lycaenidae) (Fig. 10.21). Males and females have a similar basic pattern, with central patches of colour on a brown forewing, but in the male the patch is a vibrant orange colour, whilst in the female it is white. The vivid forewing colours are largely hidden when *Curetis* butterflies rest with their wings folded, showing their silvery-whitish undersides (Fig. 10.22). The different colours on the forewings arise from a mixture of tiny light-reflecting scales – orange-red and black in the male, silvery-white and brown/black in the female – whose colours are created by both pigments and nanostructures. In contrast, the

Fig. 10.21. Angled sunbeam (*Curetis acuta*). (Photo by Nihal Jabin [CC BY-SA 4.0 (https://creativecommons.org/licenses/by-sa/4.0)].).

Fig. 10.22. Bright sunbeam (*Curetis bulis*), Thailand.

undersides of both male and female wings of *C. acuta* – and other congeneric species – are covered in silvery scales, which have the remarkable property of reflecting *direct* light but polarizing *reflected* light. This enables them to use their silvery-white undersides in ecologically different ways: they can reflect the colour of green foliage and remain camouflaged when resting on leaves, but they can also create bright silvery flashes whilst flying, in order to signal to each other. In other words, they can use their special silvery wing scales to both conceal or reveal their presence in their habitats by positioning themselves and controlling their orientation to the sun. This physical adaptation has, therefore, enabled them to evolve behaviours associated with two different requirements: avoiding natural enemies, yet being highly visible to each other.[21]

Partitioning the Function (Different Wings for Different Things)

The pipevine swallowtail butterfly, *Battus philenor* (Fig. 2.18), displays brilliant blue iridescence on both dorsal and ventral wing surfaces. Large pigmented orange spots located within a field of iridescence create an aposematic colour pattern on the ventral hindwings of both sexes. This combination of short wavelength-absorbing orange spots, together with long wavelength-reflecting iridescent patches, produces a highly effective signal, which will in effect flicker as the butterflies flap their wings and alter the angle of observation of the iridescence. In contrast to the undersides, the dorsal iridescent area on the hindwing in *B. philenor* is much brighter in males, suggesting that it is a secondary sexual character, used by females in selecting a mate.[22]

These examples illustrate how different butterfly species have evolved ways of separating out divergent and sometimes conflicting functions – such as, on the one hand (or wing!), highly conspicuous signals aimed at mates and rivals, and on the other, cryptic coloration evolved to escape detection by predators – by allocating them to different wing surfaces. Biologists refer to such conflicting requirements – in this case between maximizing the signals sent to intended recipients (mates and rivals) and minimizing their visibility to potential predators (charmingly called 'eavesdroppers') – as trade-offs. Where warning colours are concerned, the predator is the intended receiver, but the signals aimed in this direction can also be separated from those aimed at mates and rivals. The different patterns and colours found on different wing surfaces can be further refined in terms of communication by appropriate behavioural manipulation. Iridescence offers an opportunity to minimize this trade-off by enhancing the efficiency of the communication. In other words, brilliant flashes of colour can be aimed accurately at the intended receiver by a combination of restricted viewing angle and clever positioning by the signaller, thereby minimizing the likelihood of being seen by unwanted observers on the lookout for prey.[23]

This property of being able to hide in some situations, whilst being conspicuous and highly apparent in others is at the heart of butterfly biology, and they have evolved remarkable ways of achieving these somewhat conflicting objectives. Signal partitioning in this way is a solution to conflicting evolutionary pressures. Namely, forewing traits concerned with communication between the sexes – which are sexually selected and rapidly evolving – and hindwing patterns providing camouflage, which are more fixed, or stabilized as a result of natural selection.[24]

Pierid Wings

The pigments found in the wings of pierid butterflies give these species their characteristic white (e.g. Pierinae), and yellow and orange (e.g. Coliadinae) colours. The colours we see on these butterflies are what remains after the selective absorption of part of the visible spectrum by pigment-loaded granules within wing scales. Additional structural colours are produced by a combination of light scattering and iridescence. The spectrum of colours reflected by pierid wings is highly diverse and unique to each species, but almost all have peaks of short wavelength (<400 nm) UV light reflectance.[25]

In some male pierids – such as the orange sulphur, *Colias eurytheme* (Fig. 2.15), the African clouded yellow, *Colias electo*, and the large grass yellow, *Eurema hecabe* (Fig. 2.9), all in the subfamily Coliadinae – both forewing and hindwing dorsal surfaces exhibit an iridescent ultraviolet (UV) reflectance. In many other pierids, however – such as the clouded yellow, *Colias croceus* (Fig. 2.24a, b), the Cleopatra, *Gonepteryx cleopatra*, the common brimstone, *Gonepteryx rhamni* (Fig. 3.2) and the great orange tip, *Hebomoia glaucippe* (Fig. 9.22), to name just a few – iridescence is restricted to the tips of the forewings. Females often have differently coloured wings and usually lack the full iridescent wings, or iridescent wing tips, seen on the males (refer to Chapter 2, this volume). The pigments responsible for the colours in pierid wings belong to a class of molecules called pterins. The main pterin pigments are leucopterin, xanthopterin and erythropterin. Leucopterin – which is present in the wings of large and small white butterflies, *Pieris brassicae* (Fig. 9.23) and *Pieris rapae* (Fig. 7.11) – absorbs ultraviolet light. We humans cannot see UV light so the wings of these pierid butterflies appear white to us, but butterfly vision extends into the ultraviolet (see Chapter 4, this volume), so their wings appear brightly coloured to them. The xanthopterins absorb violet light as well as UV, so the wings look yellow; erythropterin absorbs blue light, so orange or red colours result. In many pierids, including *Pieris napi*, the wings of males are UV-absorbing, whilst those of females are more strongly UV reflecting (see Chapter 2, this volume). In general, pierid males have brighter, more colourful, pterin-based wing colours than females.[26]

In the wing scales of pierid butterflies, the pterins are in spheroidal, bead-like pigment granules called nanobeads, which both absorb and scatter light in an incoherent manner. Incoherent in this case means that the light reflections are randomly scattered: a result of the fact that the tiny beads are non-periodically – that is to say, randomly – arranged within the lumen of the scales. The elliptical pterin granules essentially function to increase the reflection of light at those wavelengths that are not absorbed by pigments. In other words, they enhance the scattering of light from the scales in which they are contained. The overall colour of pierid butterflies is therefore, produced by a combination of light absorption – by pterin pigments contained within the beads – and light scattering. In butterflies like the orange sulphur, *C. eurytheme* (Fig. 2.15), the pterins amplify the colours and intensity of the iridescence in males as they flap their wings during flight. The shape and arrangement of the beads within the scales results in particularly bright colours characteristic of pierids.[27]

The density of the beads often differs markedly between the sexes and determines the sexual dichromatism – 'his and her coloration' – of pierids, where the males of a species usually have highly reflective and often colourful dorsal wing surfaces. Males exhibit UV iridescence, produced by coherent scattering of light by specialized scales, located on the dorsal wing surfaces. For example, in the orange sulphur butterfly, *C. eurytheme*, the UV reflecting cover scales on the upper wing surfaces are composed of nanostructures that function as multilayer interference reflectors. These overlie yellow/orange pterin granules within the lumen of the scale, which creates a highly spectrally pure, UV reflectance. The situation is reversed in the Japanese subspecies of the small white butterfly, *Pieris rapae crucivora* (Fig. 7.11). In this subspecies – unlike the European subspecies, *P. rapae rapae* – the female wings are powerful UV reflectors, a feature the males use for mate recognition. The beads contain a UV-absorbing pterin pigment, so their absence in the females of this subspecies results in higher UV reflectance, compared with the males whose wing scales do contain these unique nanobeads. So male and female *P. rapae crucivora* have distinctly different colours; unfortunately, both sexes look the same to us![28]

In the small white, *P. rapae*, there are two layers of incompletely overlapping cover and ground scales on each side of the wing. These two scales give a wing reflectance of about 0.6: i.e. 60% of incident light is reflected. In the large white, *P. brassicae*, the wing reflectance is higher (up to 0.8), as a result of having stacks of three to four overlapping scales. The black spots on the forewings of *P. rapae* (Fig. 2.4a), and presumably other pierids such as *P. brassicae* (Fig. 10.23), are produced by the absence of elongated pterin beads; the light is, therefore, not reflected or scattered, but absorbed. The colours of each pierid species, therefore, is determined by the type of pigment(s), the pattern and arrangement of light-scattering pigment beads within the nanoarchitecture, and the shape of the individual wing scales. These iridescent butterflies radiate flashes of short-wavelength (blue or UV) light when they fly on sunny days, but exactly how this is received and perceived by would-be mates, competing males or predators is still not fully understood.[29]

Colour Badges as Honest Signals

The UV iridescence produced by male wings of butterflies such as *Colias eurytheme*, or blue morphos, appears to be a secondary sexual character that females evaluate when choosing males with which to mate. As such, these 'structural colour badges' are thought to be honest signals, or reliable information, of the phenotypic condition of the males. Evidence supporting the idea that structural colours function as sexual signals of phenotypic quality comes from the fact that they are proportionally more affected by stressful situations – such as poor diet or unfavourable temperatures – than many other biological traits. In other words, features like UV iridescence are highly sensitive to contingent factors and are thus good indicators of the condition of the male and the environment in which he has developed. Males with good genes, which have developed under favourable environmental conditions, will be bright and showy, and females will choose them on the basis that they are more likely to

Fig. 10.23. Large white (*Pieris brassicae*), Galicia, Spain.

be vigorous, and provide a large and fertile spermatophore. Presumably because they are costly or difficult to produce, the scales producing the effect are lost or worn down as the male butterflies age (Fig. 7.14). Anything that is difficult to obtain or produce – perhaps including some pigments that must be obtained from the environment – could function as an indicator of quality, since the individual needs to have the necessary ability (or genotype) and also needs to spend time and energy obtaining or synthesizing the commodity, or property. If this hypothesis of honest signalling is correct, then iridescence is in effect an elaborate ornamentation that allows would-be rivals or prospective mates to gauge the quality of the individual and to decide whether to engage in contests or mating, respectively. That said, not all male butterflies engage in contests, since not all species are territorial. There is, however, evidence from several studies that UV coloration has an inhibiting effect and deters other males of the same species from approaching and engaging in courting behaviour.[30]

Interestingly, colours produced by pigments are not generally thought to be involved in mate choice, in the same way as structural colours, perhaps because they are not so bright or deeply saturated and are easier to produce. The reality is as always, however, more complex, since pigments can accentuate the intensity of structural colours. For example, pteridine pigments accentuate the magnitude of the iridescence in the orange sulphur butterfly, *C. eurytheme*. Although there is relatively little definitive evidence that female butterflies choose between males on the basis of the quality of the intensity, hue or saturation of their reflective colours, this is largely because experiments of female mate choice are so difficult and logistically demanding to carry out, e.g. requiring large numbers of virgin females under experimental conditions. Nevertheless, the available evidence supports the idea that brilliant male, structurally coloured, sexual ornamentation is indicative of a range of male qualities, and evolves as a result of sexual selection (see Chapter 2, this volume).[31]

Papilionoid Wings

The large, colourful wings of swallowtail butterflies have colours produced by both pigmentary and structural means; often by a combination of the two. For example, in the Japanese yellow swallowtail butterfly, *P. xuthus* (Fig. 2.2a, b), the upper wing patterns are formed by arrangements of yellow, orange, black and blue scales. In the cream, orange and black scales, there are papiliochrome pigments located in the upper lamina of the scale. The lower thin lamina of each scale functions as a thin-film interference reflector of the light passing through the scale. As it is reflected back by the mirror-like lamina, the light is filtered – selectively absorbed – by the pigments to produce a reflectance spectrum that is 'tuned' – meaning adapted by evolution – to suit the tetrachromic vision (see Chapter 4, this volume) of these butterflies. The unpigmented blue scales in the iridescent crescents on the upper hindwing of *P. xuthus* produce a distinctive structural blue colour by a combination of nanostructures in the upper and lower laminae of the scales.[32]

Birdwings

In birdwing butterflies of the genera *Trogonoptera*, *Troides* and *Ornithoptera* the males are often brightly coloured – e.g. a combination of vivid iridescent colours such as green, blue and yellow, against a black background – and include a number of iconic species such as Rajah Brookes birdwing (Fig. 10.4), Queen Alexandra's birdwing (*Ornithoptera alexandrae*, the world's largest butterfly) and the Goliath birdwing (second largest in the world, see Fig. 2.27). In contrast to morphos, colours are produced almost exclusively by semi-transparent cover scales in *Ornithoptera* species, and ground scales simply provide a dark background. Some colours are generated via pigments – notably papiliochrome (yellow and cream) and melanin (black) – whilst nanostructures, tilted multilayers of cuticle and air, create iridescent colours when viewed from certain angles. In the Magellan birdwing, *Troides magellanus* – from the Philippines and Taiwan – a blue-green sheen of reflected iridescence is visible when the hindwings are positioned at a very low angle – the so-called, near-grazing incidence – relative to the viewer. In other words, the light is modified by nanostructures on the yellow scales to produce flashes of blue-green iridescence, when the hindwings are at a very low angle, relative to the horizontal. The golden birdwing, *Trioides aeacus* (Fig. 10.5) has bright yellow patches on its dorsal hindwings, which are produced by absorption and scattering of incident light by the pigment, papiliochrome. The yellow scales do not have a multilayered cuticular structure, hence there is no backscattering effect and no iridescence.[33]

Four papilionid species – *Papilio bromius*, *Papilio epiphorbas*, *Papilio nireus* and *Papilio oribazus* in the so-called *nireus* species group, a 'clade' with 13 members – have wings whose upper sides exhibit prominent blue-green bands set against a black wing background. These colours are generated by a dense lattice of structured, blue-green cover scales, which reflect mainly the blue-green component of incident light, rather like a mirror. These scales also contain a UV-violet-absorbing pigment, papiliochrome II, in the upper laminas of the scales, which causes them to fluoresce slightly. This pigment acts as a spectral filter that absorbs violet light, thus limiting the reflected light to the blue-green part of the spectrum, and this is the colour we see. The pale, yellow scales of swallowtails, such as *Papilio xuthus*, *Papilio demoleus*, *Papilio protenor*, *Papilio helenus*, *Papilio castor*, *Papilio polytes* and *Papilio dardanus*, also contain papiliochromes as the major pigments.[34]

Mixing it

Swallowtail butterflies, like *Papilio ulysses* (Fig. 10.10) and *Papilio blumei* (10.24), have a series of tiny depressions, concavities and ridges on their wing scales. These regular arrays of flat-bottomed depressions and ridges produce different colours depending on which surface – e.g. the flat bottom or the inclined sides – of the concavities the incident light is reflected off (Fig. 10.25). The observed iridescent blue and green colours are the result of a mixing together of individual colours – e.g. yellow and blue to make green – generated by the complex microscopic surfaces. The bright green iridescent bands on the upper wing surfaces of male emerald swallowtail butterflies, *Papilio palinurus* (Fig. 10.26), are also generated by a combination of blue and yellow colours. The yellow colour is produced by reflections off the bottom of the tiny (4–10 μm) diameter bowls or concavities, whilst the blue colour is generated from two 45° reflections off the sides of the bowl.[35]

Fig. 10.24. The peacock or green swallowtail (*Papilio blumei blumei*), male, North Sulawesi. (Photo by Alan Cassidy - own work, CC BY-SA 4.0, https://commons.wikimedia.org/w/index.php?curid=45401772)

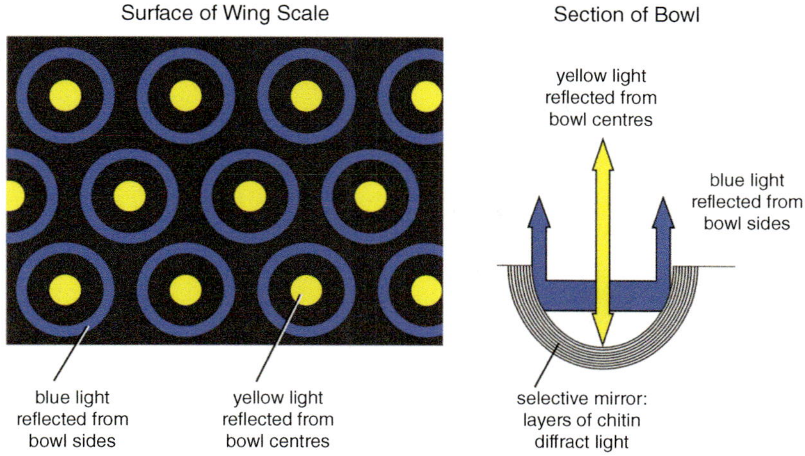

Fig. 10.25. Emerald swallowtail (*Papilio palinurus*) scale structure diagram. (From https://upload.wikimedia.org/wikipedia/commons/6/61/Emerald_Swallowtail_Papilio_palinurus_scale_structure_diagrams.svg)

Fig. 10.26. The emerald swallowtail (*Papilio palinurus daedalus*). (Photo by D. Gordon E. Robertson [CC BY-SA 3.0 (https://creative-commons.org/licenses/by-sa/3.0)].)

A characteristic property of these *Papilio* species – such as *P. blumei*, *P. peranthus* (Fig. 10.27) and *P. ulysses* (Fig. 10.10) – is that the observed colours of the iridescent green bands shift towards the blue part of the spectrum as the viewing angle increases. The differences in the hues of their iridescences – i.e. green, yellowish green or bluish green – are related to differences in the optical thickness and 'periodicities' of the air/cuticle bilayer stacks, of their respective wing scales. In other words, small differences in nanostructural details produce the observed differences in iridescent coloration.[36]

Males of the meliboeus swordtail, *Ancyluris meliboeus* – a South American riodinid – have striking metallic iridescence on the *undersides* of both fore- and hindwings. The upper wing surfaces are black with a prominent red stripe, which give it its alternative name: the redline doctor. The undersides vary from deep blue through to orange, depending on the angle of observation. Called a living jewel, the bright iridescence is generated by an unusual arrangement of highly tilted, multilayers of cuticle and air within discrete scale ridges, similar to that of morpho butterflies. The fact that these nanostructures are so highly tilted – by about 30° to the horizontal – means that the butterfly can produce vivid reflections within a relatively narrow window, or horizon, of iridescence. There is also a dark zone where no iridescence occurs; the butterfly thus possesses the ability to produce a strong flickering effect without much movement of the wings, by exposing and hiding the iridescent patches. Wing movements of just a few degrees generate ultra-high-contrast colour flickering.[37]

The Australasian butterfly *P. peranthus* (Fig. 10.27) has wings that are predominantly grey-black, with an area of iridescent green on the inner (or proximal) parts of the upper wings, i.e. closest to the body. The black areas that surround the bright green regions are produced by a unique light-absorbing nanostructure which increases the optical path of incident light, such that it is trapped and adsorbed, i.e. a solar trap (see above). The ridge structure also results in interference effects, which enhance the light trapping effect. The rare and beautiful, Kaiser-i-Hind, or Emperor of India butterfly, *Teinopalpus imperialis* (Fig. 10.28) is covered in iridescent green wing scales that contain photonic crystals: specifically, a network-like nanostructure, known as a *single gyroid*.[38]

Cattleheart butterflies (Papilionidae), such as the emerald-patched cattleheart, *Parides sesostris*, have iridescent, emerald green patches set against a black background on the upper surfaces of their forewings (Fig. 10.29), and crimson patches on the undersides. These colours are

Fig. 10.27. *Papilio peranthus* male, Kangean Islands, Indonesia. Male dorsal (left) and ventral (right) sides. (Photo by Muséum de Toulouse [CC BY-SA 4.0 (https://creativecommons.org/licenses/by-sa/4.0)].)

Fig. 10.28. Kaiser-i-Hind (*Teinopalpus imperialis*) male, ventral side. (Photo by Anaxibia [CC BY-SA 3.0 (https://creativecommons.org/licenses/by-sa/3.0)].)

Fig. 10.29. Emerald-patched cattleheart (*Parides sesostris*) male, dorsal side. (Photo by Muséum de Toulouse [CC BY-SA 4.0 (https://creativecommons.org/licenses/by-sa/4.0)].)

also produced by gyroid-type photonic crystals with circular air holes in the cuticular network – which occupy 60% of the volume – in square and triangular shaped lattices. In the cover scales, the c. 5-µm honeycomb structure lies directly on top of a similarly 5-µm layer of gyroid-type photonic crystals and contains a pigment that acts as a filter – absorbing UV and blue-wavelength light – so that the reflectance spectrum of the underlying photonic crystal becomes green.[39]

Nymphalid Wings

The upper (dorsal) wing surfaces of many nymphalid species are strikingly colourful, with highly reflective spots and vivid bands, formed by scales with different pigment-based colours, which also utilize structural colours.

The colours of nymphaline wings – seen on butterflies like the peacock, *Aglais io* (Fig. 3.27), painted lady, *Vanessa cardui* (Fig. 3.10), red admiral, *Vanessa atalanta* (Figs 3.22 and 3.23) and small tortoiseshell, *Aglais urticae* (Fig. 3.28) – are due to a complex combination of chemical and structural colours: thin films, pigments and stacked scales. There are individual scales of many colours – blue, black, red, orange, white and yellow – which are largely determined by internal pigments (blue and white being the exceptions). The yellow, orange and red scales contain pigments such as 3-OH-kynurenine and ommochrome (xanthommatin), and the black scales contain high concentrations of melanin. The lower laminas of these scales act as thin-film reflectors and the reflected light is partially absorbed by pigments within the scales. The blue cover scales, however – prominent in species such as the peacock, *A. io*, and small tortoiseshell (Fig. 10.30) – are virtually unpigmented and the diffuse colour is caused by nanostructures: light-scattering ridges and cross-ribs. The pigments responsible for the beautiful colours in *Heliconius* species (Figs 2.6 and 2.7) include: melanin (black), xanthommatin (orange), dihydroxanthommatin (red) and 3-hydroxykynurenine (3-OHK, yellow). Some *Heliconius* species – seven so-called iridescent specialists, including *Heliconius sara* (Fig. 7.15), exhibit angle-dependent iridescent blue or green structural colours, e.g. large patches of metallic blue on their dorsal wing surfaces. Iridescence is also present in a few other *Heliconius* species, including *Heliconius cydno* (Fig. 2.8), which has an iridescent blue colour.[40]

Male purple emperors, *Apatura iris* (Fig. 10.3), and lesser purple emperors, *A. Ilia* (apaturinae), have patches on their wings that generate intensive purple-blue iridescence in the UV region. Nanoarchitecture on the cover scales is responsible for the blue iridescence seen in both species, but the iridescence is only visible in a narrow angle, so during flight the colour of the dorsal side of the wings changes from violet-blue to brownish. In the great purple emperor, *Sasakia charonda* (Fig. 10.31), males have six different types of scales, ranging in colour from brown, yellow, red, iridescent purple-blue and iridescent pearly white. Colours are again formed by a combination of pigments and structural colours, whilst the iridescence is produced by a 3-D optical diffraction grating.[41]

Fig. 10.30. Small tortoiseshells (*Aglais urticae*) dorsal hindwings showing iridescent blue wing scales, UK.

Fig. 10.31. Japanese emperor or great purple emperor (*Sasakia charonda formosana*). (Photo by Peellden [CC BY-SA 4.0 (https://creativecommons.org/licenses/by-sa/4.0)].)

Clearwings and Nanopillars

Clearwing (ithomiine) butterflies of the genus *Greta* have large, clear, transparent areas of the wing, which make it difficult for predatory birds (and entomologists!) to track the butterfly whilst it is flying. To be transparent, the wing must neither absorb nor scatter light. These bare patches are less than 1 μm thick in the Andromica clearwing, *Greta andromica* (Fig. 10.32), and thin-film optics play a role in causing the reflectance spectra of these bare wing membranes. Light is reflected by both the upper and lower boundaries of the thin films and, depending on the thickness of the film, certain wavelengths are intensified or attenuated. Thus, white light is split into various component colours and interference patterns result, like the beautiful colours of the rainbow we see in soapy bubbles! Clearwing butterflies, like *Greta oto* (Fig. 10.33), also have tiny, nanoscale protuberances, called nanopillars (Fig. 10.34). The random height distributions of these submicroscopic protuberances causes the so-called, omnidirectional, anti-reflection properties of the transparent wings. The difference between the anti-reflection, corneal nipple arrays found on the eyes of some moths and butterflies (see Chapter 4, this volume) – which have a constant height and aspect ratio of 1:1 – is that the glasswing nanopillars are randomly sized, with much higher aspect ratios (up to 10), so they are much more like tiny, round-topped pillars (or cuticular microvilli it seems to me) than nipples.[42]

The swordtail butterfly, *Graphium sarpedon* (Fig. 10.35), also has bare wing patches, free of scales, but these bands – which run down the middle of both fore- and hindwings – contain high concentrations of pigments that produce a blue-green colour. Most of the underside of the blue-green patches are covered with plate-like, glass scales, which also have the optical properties of a thin film, such that reflected light is iridescent and strongly polarized. The overall colour of these semi-transparent, blue-green areas depends on incident light illuminating the wing from both sides, filtered and reflected by the pigments and glass scales, respectively.[43]

Fig. 10.32. Andromica clearwing (*Greta andromica*), Nymphalidae, Danainae, Ithomiini. (Photo by yakovlev.alexey from Moscow, Russia [CC BY-SA 2.0 (https://creativecommons.org/licenses/by-sa/2.0)].)

Fig. 10.33. Glasswing butterfly (*Greta oto*). (Photo by David Tiller [CC BY-SA 3.0 (https://creativecommons.org/licenses/by-sa/3.0)].)

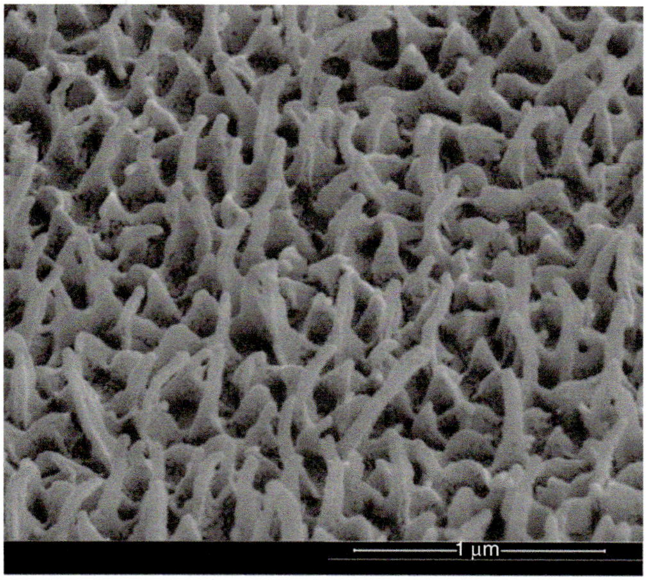

Fig. 10.34. Scanning electron micrograph (SEM) of random nanopillars on a glasswing butterfly (*Greta oto*) wing. (Photo by Radwanul H. Siddique, KIT [CC BY-SA 3.0 (https://creativecommons.org/licenses/by-sa/3.0)].)

Fig. 10.35. Common bluebottle (*Graphium sarpedon*), Thailand.

Concluding Remarks

Structural colours are found throughout the animal kingdom and probably evolved very early on in the fossil record, during the Cambrian period. Butterfly wings can be considered at a hierarchy of different scales: from the colours and patterns on the wings themselves; to the different types of scales; into the internal nanostructures on or within the scales; and finally, to the chemical components, including light-absorbing pigments. The end result of this complexity, the colourful effects seen by an observer, is therefore produced by the combination of a number of different factors, including: the interference, diffraction or absorption of the incident light on and in the wing, the nature and characteristics of the light source, the viewing angle, the brightness and colours of the surroundings, and so on. Structural colours can, therefore, be generated in a variety of different ways using a combination of optical effects. For example, iridescent green is generated by different types of nanostructures in different species, e.g. by sculpted multilayers, perforated multilayers and gyroid 3-D photonic crystals. Butterflies have evolved these different nanostructures – which exaggerate or suppress optical effects – to generate structural colours, which have been adapted to different purposes, such as 'in-flight signalling' and 'at rest' flashing.[44]

Whilst the colours and patterns of butterfly wings amaze and enthral us, we must remember that we only see a limited selection of the palette observable to the butterflies themselves. The splendour of UV colours eludes us, and we are largely blind to visual signals created by the controlled reflections of polarized light. Some animals, butterflies included, can perceive polarized light and create a visual perception akin to colour vision, which has been called 'false' colours. In other words, butterflies have a visual system that is capable of utilizing both the *colour* (spectral sensitivity) and *polarization* of light, at least in some circumstances, such as when making decisions about where to lay eggs.[45]

Most organisms have visual systems that analyse the colour, intensity and polarization of light separately, but some researchers think that such polarization-dependent colour vision used by butterflies might be more common than we realize. Butterflies whose wings reflect polarized light tend to occur in habitats where light levels are very low, or at best highly variable, such as the understorey levels of tropical forests. Here, they can utilize visual signals containing polarized light in order to enhance their visibility to potential mates, because it is less dependent on the intensity and spectrum of the ambient light. These intraspecific polarized signals are also less visible to predators, so butterflies inhabiting these gloomy forests are beautifully adapted to both reflecting and detecting polarized light and using it to their advantage. What a wonderful world it must be from a butterfly's perspective![46]

Notes

[1] Kingsolver (1985), Oliver *et al.* (2009) and De Keyser *et al.* (2015).
[2] Kertész *et al.* (2006), Lipson and Lu (2009) and Yoshioka *et al.* (2014b). Photonic crystals are usually made up of many smaller domains with different crystal orientations.
[3] Vukusic *et al.* (2001b). N.B. nanostructures can only be visualized using scanning electron microscopes (SEM) and transmission electron microscopy (TEM). SEM micrographs give an impression of the three-dimensional character of the structure, whilst TEM micrographs provide a two-dimensional view (Michielsen and Stevanga, 2008).
[4] Ingram and Parker (2008), Kemp (2006a), Prum *et al.* (1998, 2006), Wickham *et al.* (2006) and Niu *et al.* (2015).
[5] Rutowski *et al.* (2007b), Ćurčić *et al.* (2012) and Sun *et al.* (2013). Doucet and Meadows (2009) highlight the unique features of iridescent colours, including their capacity to be highly directional, remarkably bright and saturated.
[6] Zhao *et al.* (2011) and Han *et al.* (2015).
[7] Ding *et al.* (2008), Sun *et al.* (2009) and Webb *et al.* (2014).
[8] Eisner *et al.* (1964), Nijhout (1991), Ding *et al.* (2009), Kemp and Rutowski (2011), Wilts *et al.* (2013), Dinwiddie *et al.* (2014) and Stavenga *et al.* (2015).
[9] Lawrence *et al.* (2002), Stavenga *et al.* (2006b), Giraldo and Stavenga (2008), Kolle *et al.* (2010), Vigneron and Simonis (2010), Kemp and Rutowski (2011), Saba *et al.* (2014), Yoshida *et al.* (2014b) and Wilts *et al.* (2014). For the tessellated pattern, see fig. 1 in Saba *et al.* (2014) and fig. 3 in Corkery and Tyrode (2017).
[10] Vukusic *et al.* (1999), Wilts *et al.* (2008), Bálint *et al.* (2012) and Stavenga (2014).
[11] Vukusic *et al.* (2004), Wijnen *et al.* (2007), Wilts *et al.* (2011), Van Hooijdonk *et al.* (2012) and Stavenga *et al.* (2014a). Papiliochromes are pigments derived from a precursor molecule called kynurenine.
[12] Blandin and Purser (2013), Chazot *et al.* (2016), Giraldo and Stavenga (2016). Figure 14 from Potyrailo *et al.* (2015).
[13] Ghiradella (1994), Berthier *et al.* (2003), Niu *et al.* (2015, 2016), Vukusic *et al.* (1999, 2002), and Giraldo and Stavenga (2016). The inherent characteristics of the morpho wing were summarized by Kinoshita *et al.* (2002b) as: (i) a very high reflectivity in a particular wavelength range, (ii) uniform blue reflection in a broad angular range, (iii) glitter-like laser speckles and (iv) variations of colours amongst species.
[14] Vukusic *et al.* (1999) Vukusic and Sambles (2003), Berthier *et al.* (2003), Plattner (2004), Yoshioka and Kinoshita (2004), Ding *et al.* (2008), Poladian *et al.* (2009), Sun *et al.* (2013) and Han *et al.* (2014).
[15] Vértesy *et al.* (2006), Imafuku (2013), Saba *et al.* (2014), De Keyser *et al.* (2015). See Figs 3–11 in Bálint *et al.* (2009). The nanostructures responsible for the blue coloration in *Cyanophrys* species are probably under the influence of sexual selection, whereas those causing the green colour are affected by natural selection (Bálint *et al.*, 2009).
[16] Vértesy *et al.* (2006), Wilts *et al.* (2008) and Bálint *et al.* (2012). The pepperpot structure acts as a natural photonic band gap (PBG) material causing increased reflectance in the spectral range from blue to near UV (Biró *et al.*, 2003).
[17] Vukusic *et al.* (1999, 2009), Vukusic and Sambles (2003) and Kertész *et al.* (2008).
[18] Stavenga *et al.* (2009), Michielsen and Stevanga (2008), Michielson *et al.* (2009), Schröder-Turk *et al.* (2011), and Corkery and Tyrode (2017). N.B. A gyroid is a particular type of photonic structure with a repeating three-dimensional lattice, only discovered in 1970. See Wilts *et al.* (2017a) figs 4 and 5, for images of these so-called faceted gyroids.
[19] Imafuku (2013) and Imafuku *et al.* (2007).

[20] Imafuku *et al.* (2002a) and Imafuku (2013). Males of *Chrysozephyrus smaragdinus* and *Neozephyrus japonicus* can discriminate between the sexes of their own species using wing colour (Imafuku and Kitamura, 2015).
[21] Wilts *et al.* (2013).
[22] Rutowski *et al.* (2010) and Rutowski and Rajyaguru (2013).
[23] White *et al.* (2014).
[24] According to Oliver *et al.* (2009), both dorsal and forewing characters are more labile through evolutionary time than those on the ventral surface and hindwing.
[25] Stavenga and Leertouwer (2007), fig 5a in Pirih *et al.* (2011) and Wilts *et al.* (2017b).
[26] Yagi (1954), Meyer-Rochow and Järvilehto (1997), Morehouse *et al.* (2006), Stavenga and Leertouwer (2007), Wijnen *et al.* (2007), Pirih *et al.* (2011) and Wilts *et al.* (2011, 2012). See fig. 1 in White *et al.* (2012) and fig. 1 in Kemp (2007a).
[27] Rutowski *et al.* (2005), Morehouse *et al.* (2006), Luke *et al.* (2009) and Wilts *et al.* (2017b).
[28] Obara and Hidaka (1968), Obara and Majerus (2000), Kemp *et al.* (2005), Giraldo and Stavenga (2007) and Rutowski *et al.* (2007b), Fukano *et al.* (2012). Male *P. rapae* typically have a wing scale bead-array density of 21 ± 2 μm^{-2}, whereas females have a bead-array density of 14.5 ± 0.5 μm^{-2} (Luke *et al.*, 2009).
[29] See fig. 4 in Stavenga *et al.* (2006b), and fig. 11 in Matin *et al.* (2010).
[30] Kemp *et al.* (2006), Kemp *et al.* (2005), Kemp and Rutowski (2007), Doucet and Meadows (2009), Luke *et al.* (2009) and Pecháček *et al.* (2014).
[31] Kemp (2007a), Kemp *et al.* (2005, 2006, 2011) and Kemp and Rutowski (2007).
[32] Stavenga *et al.* (2015).
[33] Lawrence *et al.* (2002), Dechkrong *et al.* (2011) and Kazama *et al.* (2017).
[34] Umebachi (1977), Trzeciak *et al.* (2012), Wilts *et al.* (2012) and Kerkut (2013).
[35] Vukusic *et al.* (2000), Crne *et al.* (2011), and Diao and Liu (2011).
[36] Liu *et al.* (2010), Stavenga *et al.* (2012, 2014b, 2015), Wilts *et al.* (2012) and Tam *et al.* (2013), see fig. 5 in Tam *et al.* (2013).
[37] Vukusic *et al.* (2001b, 2002).
[38] Han *et al.* (2013), Argyros *et al.* (2002) and Saba *et al.* (2014).
[39] Pouya (2012) and Wilts *et al.* (2012, 2014).
[40] Gilbert *et al.* (1988), Stavenga *et al.* (2014b), Wasik *et al.* (2014) and Parnell *et al.* (2018). So-called 'iridescent specialists' include: *Heliconius antiochus*, *Heliconius leuchadia*, *Heliconius sara*, *Heliconius hewitsoni*, *Heliconius sapho*, *Heliconius congener* and *Heliconius eleuchia*.
[41] Pantelić *et al.* (2011), Ćurčić *et al.* (2012) and Matějková-Plšková *et al.* (2009, 2011).
[42] Binetti *et al.* (2009) and Siddique *et al.* (2015).
[43] Stavenga *et al.* (2010, 2012, 2014b).
[44] Parker (2000), Gonzato and Pont (2004), Wickham *et al.* (2006) and Wilts *et al.* (2012).
[45] Sweeney *et al.* (2003) and Marshall *et al.* (2014).
[46] Kelber (1999), Kelber *et al.* (2001), Cronin *et al.* (2003), Douglas *et al.* (2007) and Vukusic (2011).

Summing Up

I began this journey, of trying to understand as much as possible about butterfly behaviour as I could, as a relative novice. The more I researched the topic, the more I realized what a gargantuan task it was. For each topic I strayed into, I discovered scientists who have dedicated their lives trying to push back the boundaries of knowledge, in one area after another. As I mentioned at the outset, perhaps I was foolhardy to attempt to communicate all these areas by myself. Nevertheless, I think it has been worthwhile trying to understand a broad field and attempting to communicate it to a wider audience. There are some scientists who are great communicators but, for many, communicating their research to a wider audience is not a high priority, although in the Internet age this is changing. I have greatly enjoyed my journey of discovery, represented by this book, and I hope that readers will have found it equally rewarding reading it. Like the BBC, I have tried to be entertaining as well as educative, and I hope I have not slipped up too many times on the way.

I have already rounded off each individual chapter as best I could. My goal is not to find and present a grand synthesis. I have tried to draw some conclusions, but it is fair to say that there are many gaps, gaping holes even, in the literature. We don't know or fully understand many aspects of courtship and mating behaviour of some our own common, sometimes sadly declining, UK species. Many aspects of their behaviour remain unknown, or undescribed, in even the best-studied species. There are, no doubt, some naturalists who know a great deal about some species, but who have not recorded their knowledge in the form of a publication. The Internet is a growing resource of photographs and videos, but it is extremely difficult, if not impossible, to synthesize these observations, beyond referring to a few good examples.

© Raymond J.C. Cannon 2020. *Courtship and Mating in Butterflies: Reproduction, Mating Behaviour and Sexual Conflicts* (R.J.C. Cannon)

Regarding attempting to make generalizations, Professor Rutowski, Arizona State University – 75 of whose research papers I refer to in this book – advises us to interpret what we see in terms of butterfly behaviour with care: the function of certain kinds of behaviour – he refers to ascending flights, for example – will vary according to species and context. The nature of interactions between butterflies in general will depend on a host of factors, including some or all, of the following: species, sex, age, experience, habitat, context, phenotype and genotype. Chapter 5 has, I hope, sketched out the varieties of courtship behaviour displayed by a representative range of different butterflies. Whilst there are many common elements, there is also a high degree of variability.[1]

The extent of courtship usually depends on the receptivity of the female. Eager females – it was males who were supposed to be eager in the Old Days (see Chapter 2, this volume); females were supposed to be coy! – often actively solicit courtships from males, and their willingness can sometimes shorten the sequence of courtship behaviours. Nevertheless, like all good stories, certain elements of the repertoire are essential – usually the transfer of a chemical signal, the brush of a hairpencil or the contact between scent brand and antenna – in order to reach a happy ending. However, the evaluation of these signals is also a key element in the process, and choices play a fundamental role both in individual predilections and as drivers of evolution (Chapter 2). Both sexes can be choosy and mutual mate choices occur.[2]

Persistence certainly seems to pay off for the males of some species, as evidenced by the rock grayling studies of Pinzari and Sbordoni, who found that males can repeat certain behavioural sequences up to 25 times, before the female 'decides' or 'chooses' to engage in copulation with him.[3] In times of mate scarcity, such as the extreme female bias that can occur as a result of male-killing *Wolbachia* bacteria, roles can become reversed, and the rare males can have their pick of the pack! There are extraordinary reports that individual males in such situations can mate more than 50 times in their lifetimes.[4]

The old Tinbergen–Lorenz ideas of stimulus and release do, I think, still hold some water. The process of courtship leading to mating needs to proceed via a two-way exchange, where an action by one sex releases a reciprocal action in the other sex. Anyone who has engaged in their own courtship attempts knows how this works (!), and we also know that, like butterflies, the process can break down at any stage. As stated above, the transfer of a male pheromone – and for all we know, maybe a corresponding female scent – appears to be an essential element in most (but not all – those strange monarch butterflies again! See Chapter 9) courtships. However, the role of female pheromones is mostly inferred or assumed, and there may be exciting discoveries of the future in this area.

Detailed studies, some using video recorders, have broken down butterfly behaviours into a series of units, or key elements. These are also commonly clustered into different phases – air, air–ground, ground – and so on. Whilst there are a host of common, basic behavioural elements – e.g.

wing flickering, wing flapping, nudging, bowing, antennal palpation, proboscis waving, and so on – almost like a tool kit of potentially utilizable actions with specific functions, how these are assembled into a species-specific repertoire is probably infinite in scope. This is because each of these actions brings different, and often unique elements – different chemicals, different visual signals, different smells, different tastes and different colours – to the attention of the receiver. So, whilst wing flickering might look superficially similar in different species, the frequency of the vibration, the amplitude of the wing stroke, the intensity of the reflection and the spectrum of the colours might all be different from one species to another. Finally, any one signal – be it physical, chemical, auditory or visual – can have different effects on different receivers (e.g. intra- and intersexually), and all of them can interact to produce a combined effect. How butterflies receive and evaluate these messages is clearly still something of a mystery, but perhaps that's a good thing.[5]

In reviewing this subject, I was struck by how some areas of research seemed to have stopped or at least tailed off in recent years. There are still some terrific studies being carried out on butterfly behaviour, but I have the feeling that some of these, if not firmly attached to a molecular approach, would probably fail to receive funding in some countries, these days. One thing that has impressed me, though, is the diversity of researchers, from all around the world – including many from tropical countries – who are carrying out exciting behavioural research – often linked to conservation – on local butterflies. There is now a kaleidoscope of information on different species, but still a gamut of possible things to investigate in the future.

Notes

[1] Rutowski (1997a).
[2] Rutowski (1997a).
[3] Rutowski (1997a).
[4] Pinzari and Sbordoni (2013).
[5] Boppré (1984) and Vane-Wright and Boppré (1993).

Glossary

Androconia. Scent scales found mainly on the wings of male butterflies, but sometimes in other locations on the body, where pheromones are produced and disseminated.
Andromorph. A female that resembles a male. Andromorphs benefit from avoiding male harassment.
Anisogamy. Sexual reproduction by the fusion of two gametes that differ in size and shape. Males can produce large numbers of small sperm, but females produce a much more limited number of large eggs.
Apposition eye. Compound eye found in diurnal insects, such as butterflies, in which light entering each optic unit reaches the retina of each ommatidium as a single spot and an image is made up of all the spots.
Apyrene sperm. Small anucleate spermatazooans (sperm).
Assortative mating. Where individuals with similar phenotypes mate with one another more frequently than would be expected, if mating patterns were random.
Bateman's principle. Variability in reproductive success is greater in males than in females. Mating is usually more costly to females, as they invest more in each offspring, and as a result they should be choosy (Bateman, 1948).
Batesian mimicry. A form of mimicry where one harmless species that is palatable to a predator (the mimic) mimics the appearance of a harmful or noxious species (the model).
Brightness. The perceived intensity of a stimulus, independent of hue and saturation.
Capital breeders. Females of species whose reproductive output is largely determined by the quantity of (mostly protein) resources they obtained during feeding as larvae.
Chroma. The purity or intensity of colour. Or, the perceived strength of colour, in terms of the difference from a neutral grey, similarly illuminated.

© Raymond J.C. Cannon 2020. *Courtship and Mating in Butterflies: Reproduction, Mating Behaviour and Sexual Conflicts* (R.J.C. Cannon)

Chromophore. An atom or part of a molecule whose presence is responsible for its colour.
Colour constancy. The ability to consistently classify or perceive colour stimuli as largely invariant, despite widely varying illumination.
Conspecific species. Different species that are members of the same genus.
Contest competition. A situation where available resources, such as mates, are monopolized by one, or a few individuals, thus preventing reproduction by other individuals.
Coremata. Glandular, pheromone disseminating structures in butterflies and moths, which are everted during courtship, to form the elaborate, scent-tufts – also called hairpencils – at the end of the abdomen.
Costs of reproduction. A reduction in parental fitness caused by the allocation of resources to reproduction. Usually, a reduction in longevity or a diminished capacity to reproduce in the future.
Cryptic female choice. In species where females mate more than once, the choice of which male's sperm to use for fertilization continues after insemination.
Cuticular hydrocarbons (CHCs). Thin waxy layers on the insect's cuticle, consisting of a mix of saturated and unsaturated, straight chain and branched hydrocarbons. Involved in species recognition and social communication during courtship and mating.
Cyanogenic glucosides (CGs). Cyanogenic glucosides are so-called 'phytoanticipins', compounds that protect a plant from any pathogen or herbivore due to their bitter taste and the release of toxic hydrogen cyanide when the tissues are damaged, e.g. by consumption. Many butterflies and moths are able to *de novo* synthesize CGs, but some sequester them from their host plants, as well.
Danaidone. The heterocyclic ketone (2,3-dihydro-7-methyl-1H-pyrrolizin-1-one) is a common major component of the hairpencil odour in many butterfly species in the genera: *Danaus*, *Amauris* and *Lycorea*. Derived from **pyrrolizidine alkaloids** found in plants.
Darwin–Bateman effect. Male fitness increases steeply with the number of matings with different females, due ultimately to anisogamy.
Dihydropyrrolizines. Volatile compounds – such as **danaidone** and hydroxydanaidal – found in the alar scent organs (sex brands) and abdominal hairpencils of male danaine butterflies. Chemically related to the pyrrolizidine alkaloids.
Dimorphic. Species where the two sexes are different shapes and colours.
Eclosion. The emergence of the adult butterfly from the pupal case. Also, the emergence of a larva from an egg.
Eupyrene sperm. Large nucleated spermatozoa.
Eye shine. Light reflected back by the **tapetum**, which is not absorbed by the rhabdom.
Flavonoids. Naturally occurring substances found in fruits, vegetables, grains, bark, roots, stems, flowers, etc., which absorb UV radiation and protect plants from damage.

Genotype. The entire set of genes in an organism; its complete heritable genetic identity, which gives rise to the **phenotype**.
Gynandromorph. An individual organism that contains both male and female characteristics. In butterflies, gynandromorphs may be bilateral (male on one side, female on the other), or mosaic (random distributions of male and female elements).
Gynomorph. A male that resembles a female.
Gyroid. A particular type of photonic structure with a repeating three-dimensional lattice; only discovered in 1970. In butterfly scales, a porous material consisting of two components – i.e. air and solid chitin – each of which has a network-like shape, e.g. one air-filled and the other one composed of solid chitin.
Hairpencils. A pair of eversible secondary sexual organs located within the male abdomen of danaine butterflies and everted during courtship displays (Chapter 9). See **coremata**.
Hue. The category of colour, e.g. red, green, blue, etc.
Income breeders. Females of species whose reproductive output is determined by a combination of larval feeding and resources obtained – often in the form of protein contributions from males – during the adult stage.
Lek. A territory in which males gather to display, fight or compete with each other in the presence of females. Derived from the Swedish word for 'play', meaning an aggregation of males engaging in competitive displays, with the aim of attracting females to mate with them.
Lek polygyny. When groups of males aggregate at a lek, or lekking site, which females visit to choose mates.
Life history. All of the factors that contribute directly to the production and survival of offspring. Including: age at first reproduction, frequency of mating, number of offspring produced per reproductive event and parental investment in each offspring.
Luminance contrast. The difference in colour and brightness between objects in the same field of view.
Mate choice. The evaluation of qualities in a potential mate and the selection of a given individual.
Mating plug. A gelatinous secretion produced by the male during mating, in an attempt to seal the genital opening of the female, which later hardens into a plug.
Milkweed butterflies. A tribe (Danaini) of about 160 large, often brightly coloured species, found in tropical and temperate habitats throughout the world. Includes the monarch (*Danaus plexippus*), the African queen or plain tiger (*Danaus chrysippus*), the tree-nymph (*Idea leuconoe*) and various crows (*Euploea* spp.).
Monandry. Where females mate only once.
Monomorphic. Species where the two sexes are approximately the same shape or colour.
Müllerian mimicry. A form of mimicry, or biological resemblance, where two or more unrelated, noxious or distasteful species have come to mimic each other's warning signals, to their mutual benefit.

Nanostructures. Forms of matter characterized by, usually repeating, spatial patterns of the order of a few, to a few hundred, nanometres (1 nm = 1×10^{-9} m).
Nuptial gifts. Nutritional gifts – e.g. nutrients and proteins in the form of a **spermatophore** – given by one partner during sexual reproduction.
Ommatidia (singular: **ommatidium**). The individual optical units – typically hexagonal in cross-section – that collectively form the compound eyes of insects.
Ommochrome. A major class of orange, red and yellow colour pigments in butterflies, derived from the amino acid tryptophan.
Operational sex ratio (OSR). The number of males and females in a population willing to mate at any given time.
Opsins. Light-active proteins such as **rhodopsin**, which enable vision, via the conversion of photons of light into electrochemical signals.
Optical diffraction grating. An optical component with a periodic (i.e. regular) structure that splits and diffracts light into several beams travelling in different directions
Papiliochromes. Cream- and yellow-coloured pigments that occur in the wings and bodies of butterflies in the family Papilionidae, derived from a precursor molecule called kynurenine.
Phenotype. The observable physical properties of an organism, including its appearance, development and behaviour, resulting from the interaction of its **genotype** with the environment.
Pheromone. A chemical substance produced and released into the environment by an animal, which affects the behaviour or physiology of another individual, usually a member of the opposite sex, of its species.
Pheromone-transfer particles (PTPs). Tiny (1–2 μm) cuticular particles, containing male specific **pheromones**, produced by some butterflies in the subtribes Danaina and Amaurina.
Photonic band gap. Incident light is reflected rather than transmitted, so no photons pass through a material.
Photonic crystals. Periodic **nanostructures** which have a band gap that allows some wavelengths of light to pass, but not others.
Photoreceptor. Cells containing the photopigment **rhodopsin** that respond to light.
Polyandry. Where females mate multiple times (usually with different partners).
Polygynandrous species. Where both sexes are promiscuous and mate more than once.
Protandry. In entomology is usually taken to mean a situation where one sex (usually the male) develops faster than the other, and emerges (via eclosion) as an adult before the other.
Pterins. A class of pigments of different colours found in butterfly wings of pierid species, which absorb ultraviolet light. Pterins are concentrated in pigment granules, or beads, in wing scales.
Pyrrolizidine alkaloids (PAs). Naturally occurring toxic compounds found in many plants throughout the world, including the genera *Senecio*, *Crotalaria*, *Cynoglossum*, *Amsinckia*, *Heliotropium* and *Echium*. Potent

liver toxins that may cause poisoning of livestock, wildlife and humans. Sequestered by many butterflies as precursors for the manufacture of pheromones, and as defensive compounds, toxic or unpalatable to predators.

Rayleigh scattering. A scattering process caused by particles or **nanostructures**, much smaller than the wavelength of light.

Realized sex ratio. The ratio of males to females in the same generation in a population.

Refractory period. In female butterflies, a period of time, after mating, when they reject the advances of males.

Releasers or releasing stimuli. External stimuli – such as movements, shapes or colours – that produce neural and motor responses in the signal receiver, resulting in specific behaviours.

Resource holding potential (RHP). The ability of an animal to win a contest, or fight, if one were to take place (Parker, 1974), and derived from factors such as its strength, size, experience and prior occupancy of a territory.

Rhabdom. The transparent, crystalline rod-like structure that lies beneath the cornea and occurs in the central part of each **ommatidium** in compound eyes.

Rhodopsin. Light-sensitive protein molecule that changes shape when it absorbs a photon of light, becoming straighter.

Saturation (colour). The deepness or spectral purity of colour, e.g. pink is a less saturated version of red or purple.

Scramble competition. A situation in which available resources, such as mates, are accessible to all competitors.

Sexual antagonistic coevolution (SAC). An evolutionary arms race between the sexes, involving changes in sexual morphology over time to counteract the impact of traits in the opposite sex, in order to achieve the maximum reproductive success.

Sexual dimorphism. The morphological, physiological and behavioural differences that exist between males and females of the same species.

Sexual selection. The selection of traits associated with competition for mates (Shuker, 2010), or differences in survivorship or offspring production, as a result of phenotypic differences between individuals in traits that affect access to mates.

Spectral tuning. Factors affecting the spectral absorbance of a visual pigment or the spectral sensitivity of a **photoreceptor** cell.

Spermatophore. A protein capsule containing spermatozoa, and other ejaculate material, transferred during mating.

Supernormal stimuli. Extreme stimuli never encountered in nature.

Supposition eye. Compound eyes that form a single erect image. Most nocturnal insects (such as moths) possess refracting superposition eyes, where the lenses and photoreceptors are separated by a wide optically clear region.

Synapomorphy. A character or property present in an ancestral species, which is shared exclusively by its evolutionary descendants.

Tapetum. In butterflies, a layer at the base of the **ommatidium** composed of air-filled cells that act as mirrors, and bounce any unabsorbed light back up through the rhabdom.
Tetrachromatic. Organisms like butterflies that possess four types of light receptors.
Thanatosis. Feigning death or 'playing possum'.
Tonic immobility (TI). The adoption of a motionless posture, triggered by physical contact or very close proximity – e.g. harassment by males of females – in a state of reduced responsiveness to external stimulation. This state of 'playing possum', or **thanatosis**, is maintained for a time even after release by the predator, or departure of the harasser, but the individual can recover, if uninjured (after Humphreys and Ruxton, 2018).
Trade-off. Balancing an increase in currents costs – usually associated with a single activity – against an increase in future benefits.
Trichromatic. Organisms like humans whose retina contains three types of colour receptors (called cone cells in vertebrates).
Ultraviolet light. The region of the electromagnetic spectrum that has wavelengths between 100 and 400 nm, comprising UVA (315–400 nm), UVB (280–315 nm) and UVC (100–280 nm).
Zahavi handicap. A theory formulated by Amotz Zahavi (1928–2017), which proposes that certain structures or behaviours function as signals of the status of the organism, perhaps as an indication of good genes.

Annex 1

Internet Links

For a video of cryptic wood white (*Leptidea juvernica*) courtship display, by Peter Eeles, see: https://youtu.be/blpiQHdeCV4

'Not now, dear – female butterfly closes wings to avoid sex'. Japanese small copper, *Lycaena phlaeas daimio*: https://www.reuters.com/article/us-japan-butterfly/not-now-dear-female-butterfly-closes-wings-to-avoid-sex-idUSTRE75012V20110601

Hoskins, Adrian; Learn About Butterflies: https://www.learnaboutbutterflies.com

© Raymond J.C. Cannon 2020. *Courtship and Mating in Butterflies: Reproduction, Mating Behaviour and Sexual Conflicts* (R.J.C. Cannon)

References

Ackery, P.R. (1988) Hostplants and classification: a review of nymphalid butterflies. *Biological Journal of the Linnean Society*, 33(2), 95–203.

Ackery, P.R., de Jong, R. and Vane-Wright, R.I. (1999) The butterflies: Hedyloidea, Hesperioidea and Papilionoidea. In: Kristensen, N.P. (ed.), *Lepidoptera, Moths and Butterflies. 1. Evolution, Systematics, and Biogeography. Handbook of Zoology*, 4(35) Lepidoptera. De Gruyter, Berlin, pp. 263–300.

Adamski, P. (2004) Sex ratio of apollo butterfly *Parnassius apollo* (Lepidoptera: Papilionidae) – facts and artifacts. *European Journal of Entomology*, 101, 341–344.

Adamski, P. and Witkowski, Z. (2006) Male patrolling modes in apollo butterfly *Parnassius apollo* (L.): simulation of optimal choice (Lepidoptera: Papilionidae). *Nature Conservation*, 62, 27–36.

Ah-King, M. (2011) Female sexual selection in light of the Darwin–Bateman paradigm. *Behavioral Ecology*, 22(6), 1142–1143.

Ah-King, M., Barron, A.B. and Herberstein, M.E. (2014) Genital evolution: why are females still understudied? *PLoS Biology*, 12(5), e1001851.

Alcock, J. (1983) Territoriality by hilltopping males of the great purple hairstreak, *Atlides halesus* (Lepidoptera, Lycaenidae): convergent evolution with a pompilid wasp. *Behavioral Ecology and Sociobiology*, 13(1), 57–62.

Alcock, J. (1984) Hilltopping in the nymphalid butterfly *Chlosyne californica* (Lepidoptera). *American Midland Naturalist*, 113, 69–75.

Alcock, J. (1987) Leks and hilltopping in insects. *Journal of Natural History*, 21(2), 319–328.

Alcock, J. (1988) The mating system of three territorial butterflies in Costa Rica. *Journal of Research on the Lepidoptera*, 26, 89–97.

Alcock, J. (1994) Alternative mate-locating tactics in the butterfly *Chlosyne californica* (Lepidoptera, Nymphalidae). *Ethology*, 97, 103–118.

Alcock, J. (2010) The hilltopping mating system of the duskywing skipper *Erynnis tristis* (Lepidoptera: Hesperiidae). *Journal of Research on the Lepidoptera*, 43, 1–5.

Alcock, J. and Dodson, G. (2008) The diverse mating systems of hilltopping insects. *American Entomologist*, 54(2), 80–87.

Alcock, J. and Gwynne, D. (1987) The mating system of *Vanessa kershawi*: males defend landmark territories as mate encounter sites. *Journal of Research on the Lepidoptera*, 26(1), 116–124.

Alcock, J. and O'Neill, K.M. (1986) Density-dependent mating tactics in the grey hairstreak, *Strymon melinus* (Lepidoptera: Lycaenidae). *Journal of Zoology*, 209, 105–113.

Alexander, R.D., Marshall, D.C. and Cooley, J.R. (1997) Evolutionary perspectives on insect mating. In: Choe, J.C. and Crespi, B.J. (eds) *The Evolution of Mating Systems in Insects and Arachnids*. Cambridge University Press, Cambridge, UK, pp. 4–31.

Almbro, M. and Kullberg, C. (2009) The downfall of mating: the effect of mate-carrying and flight muscle ratio on the escape ability of a pierid butterfly. *Behavioral Ecology and Sociobiology*, 63(3), 413.

Andersson, J., Borg-Karlson, A.K. and Wiklund, C. (2000) Sexual cooperation and conflict in butterflies: a male-transferred anti-aphrodisiac reduces harassment of recently mated females. *Proceedings of the Royal Society of London B: Biological Sciences*, 267(1450), 1271–1275.

Andersson, J., Borg-Karlson, A.K. and Wiklund, C. (2003) Antiaphrodisiacs in pierid butterflies: a theme with variation! *Journal of Chemical Ecology*, 29(6), 1489–1499.

Andersson, J., Borg-Karlson, A.K. and Wiklund, C. (2004) Sexual conflict and anti-aphrodisiac titre in a polyandrous butterfly: male ejaculate tailoring and absence of female control. *Proceedings of the Royal Society of London B: Biological Sciences*, 271(1550), 1765–1770.

Andersson, J., Borg-Karlson, A.K., Vongvanich, N. and Wiklund, C. (2007) Male sex pheromone release and female mate choice in a butterfly. *Journal of Experimental Biology*, 210(6), 964–970.

Andersson, M.B. (1994) *Sexual Selection*. Princeton University Press, Princeton, NJ.

Andersson, M. and Simmons, L.W. (2006) Sexual selection and mate choice. *Trends in Ecology and Evolution*, 21(6), 296–302.

Argyros, A., Manos, S., Large, M.C.J., McKenzie, D.R., Cox, G.C. and Dwarte, D.M. (2002) Electron tomography and computer visualisation of a three-dimensional 'photonic' crystal in a butterfly wing-scale. *Micron*, 33(5), 483–487.

Arikawa, K. (2001) Hindsight of butterflies: the *Papilio* butterfly has light sensitivity in the genitalia, which appears to be crucial for reproductive behavior. *BioScience*, 51(3), 219–225.

Arikawa, K. and Aoki, K. (1982) Response characteristics and occurrence of extraocular photoreceptors on lepidopteran genitalia. *Journal of Comparative Physiology*, 148(4), 483–489.

Arikawa, K., Eguchi, E., Yoshida, A. and Aoki, K. (1980) Multiple extraocular photoreceptive areas on genitalia of butterfly *Papilio xuthus*. *Nature*, 288, 700–702.

Arikawa, K., Mizuno, S., Kinoshita, M. and Stavenga, D.G. (2003) Coexpression of two visual pigments in a photoreceptor causes an abnormally broad spectral sensitivity in the eye of the butterfly *Papilio xuthus*. *The Journal of Neuroscience*, 23(11), 4527–4532.

Arikawa, K., Kinoshita, M. and Stavenga, D.G. (2004) Color vision and retinal organization in butterflies. In: Prete, F.R. (ed.), *Complex Worlds from Simpler Nervous Systems*. MIT, Cambridge, MA, pp. 193–219.

Arikawa, K., Wakakuwa, M., Qiu, X., Kurasawa, M. and Stavenga, D.G. (2005) Sexual dimorphism of short-wavelength photoreceptors in the small white butterfly, *Pieris rapae crucivora*. *Journal of Neuroscience*, 25(25), 5935–5942.

Arnqvist, G. (1998) Comparative evidence for the evolution of genitalia by sexual selection. *Nature*, 393, 784–786.

Arnqvist, G. and Nilsson, T. (2000) The evolution of polyandry: multiple mating and female fitness in insects. *Animal Behavior*, 60, 145–164.

Arnqvist, G. and Rowe, L. (1995) Sexual conflict and arms races between the sexes: a morphological adaptation for control of mating in a female insect. *Proceedings of the Royal Society of London B: Biological Sciences*, 261, 123–127.

Arnqvist, G. and Rowe, L. (2002) Antagonistic coevolution between the sexes in a group of insects. *Nature*, 415(6873), 787.

Arnqvist, G. and Rowe, L. (2005) *Sexual Conflict*. Princeton University Press, Princeton, NJ.

Austad, S.N., Jones, W.T. and Waser, P.M. (1979) Territorial defence in speckled wood butterflies: why does the resident always win? *Animal Behaviour*, 27, 960–961.

Awata, H., Matsushita, A., Wakakuwa, M. and Arikawa, K. (2010) Eyes with basic dorsal and specific ventral regions in the glacial apollo, *Parnassius glacialis* (Papilionidae). *Journal of Experimental Biology*, 213(23), 4023–4029.

Bacquet, P.M.B., Brattström, O., Wang, H.L., Allen, C.E., Löfstedt, C., Brakefield, P.M. and Nieberding, C.M. (2015) Selection on male sex pheromone composition contributes to butterfly reproductive isolation. *Proceedings of the Royal Society of London B: Biological Sciences*, 282(1804), 20142734.

Bacquet, P.M., de Jong, M.A., Brattström, O., Wang, H.L., Molleman, F., Heuskin, S., Lognay, G., Löfstedt, C., Brakefield, P.M., Vanderpoorten, A. and Nieberding, C.M. (2016) Differentiation in putative male sex pheromone components across and within populations of the African butterfly *Bicyclus anynana* as a potential driver of reproductive isolation. *Ecology and Evolution*, 6(17), 6064–6084.

Baguette, M., Vansteenwegen, C., Convi, I. and Neve, G. (1998) Sex-biased density-dependent migration in a metapopulation of the butterfly *Proclossiana eunomia*. *Acta Oecologica*, 19(1), 17–24.

Baidya, S., Basu, D.N., Roy, S. and Basu Roy, A. (2018) Occurrence of interspecific mating between two species of *Danaus* Kluk, 1780 (Lepidoptera: Nymphalidae) in nature. *Psyche*. https://doi.org/10.1155/2018/3059017

Baker, R.R. (1972) Territorial behaviour of the nymphalid butterflies, *Aglais urticae* (L.) and *Inachis io* (L.). *The Journal of Animal Ecology*, 41, 453–469.

Bálint, Z., Moser, A., Kertesz, K., Biró, L.P. and Parker, A.R. (2009) A supposition: structural colours resulting from both natural and sexual selection on an individual wing in the butterfly genus *Cyanophrys* (Lepidptera: Lycaenidae). *Annales Historico-Naturales Musei Nationalis Hungarici*, 101, 63–79.

Bálint, Z., Kertész, K., Piszter, G., Vértesy, Z. and Biró, L.P. (2012) The well-tuned blues: the role of structural colours as optical signals in the species recognition of a local butterfly fauna (Lepidoptera: Lycaenidae: Polyommatinae). *Journal of the Royal Society Interface*, 9(73), 1745–1756.

Bálint, Z., Heath, A., Katona, G., Kertész, K. and Sáfián, S. (2017) Male secondary sexual characters in Aphnaeinae wings (Lepidoptera: Lycaenidae). *Opuscula Zoologia (Budapest)* 48(1), 27–34.

Barbero, F. (2016) Cuticular lipids as a cross-talk among ants, plants and butterflies. *International Journal of Molecular Sciences*, 17(12), 1966.

Bastock, M. (1967) *Courtship: a Zoological Study*. Heinemann Educational Publishers, London.

Bateman, A.J. (1948) Intra-sexual selection in Drosophila. *Heredity* 2(3), 349–368.

Bates, H.W. (1863) *The Naturalist on the River Amazons*. 2 vols. Murray, London.

Baughman, J.F., Murphy, D.D. and Ehrlich, P.R. (1988) Population structure of a hilltopping butterfly. *Oecologia*, 75(4), 593–600.

Bear, A. and Monteiro, A. (2013) Male courtship rate plasticity in the butterfly *Bicyclus anynana* is controlled by temperature experienced during the pupal and adult stages. *PLoS ONE*, 8(5), e64061.

Bear, A., Prudic, K.L. and Monteiro, A. (2017) Steroid hormone signaling during development has a latent effect on adult male sexual behavior in the butterfly *Bicyclus anynana*. *PLoS ONE*, 12(3), e0174403.

Beck, J., Mühlenberg, E. and Fiedler, K. (1999) Mud-puddling behavior in tropical butterflies: in search of proteins or minerals? *Oecologia*, 119(1), 140–148.

Beltrán, M., Jiggins, C.D., Brower, A.V., Bermingham, E. and Mallet, J. (2007) Do pollen feeding, pupal-mating and larval gregariousness have a single origin in *Heliconius* butterflies?

Inferences from multilocus DNA sequence data. *Biological Journal of the Linnean Society*, 92(2), 221–239.

Beltrán, M., Brower, A.V.Z. and Jiggins, C. (2010) *Heliconius* Kluk 1780 TS = *Papilio charithonia* L., 1767 (designated by Hemming, 1933) longwings or passion-flower butterflies. Version 21 July 2010. Available at: http://tolweb.org/Heliconius/72231/2010.07.21 in the Tree of Life Web Project, http://tolweb.org

Benítez, H.A., Parra, L.E., Sepulveda, E. and Sanzana, M.J. (2011) Geometric perspectives of sexual dimorphism in the wing shape of Lepidoptera: the case of *Synneuria* sp. (Lepidoptera: Geometridae). *Journal of the Entomological Research Society*, 13(1), 53.

Bennett, V.J., Smith, W.P. and Betts, M.G. (2012) Evidence for mate guarding behavior in the Taylor's checkerspot butterfly. *Journal of Insect Behavior*, 25(2), 183–196.

Bennett, V.J., Pack, S.M., Smith, W.P. and Betts, M.G. (2013) Sex-biased dispersal in a rare butterfly and the implications for its conservation. *Journal of Insect Conservation*, 17(5), 949–958.

Benson, W.W., Haddad, C.F.B. and Zikan, M. (1989) Territorial behavior and dominance in some Heliconiine butterflies (Nymphalidae). *Journal of the Lepidopterists' Society*, 43(1), 33–49.

Benyamini, D. (1992) New courtship posture in females of two Chilean butterflies: rejective or receptive? *The Journal of Research on the Lepidoptera*, 31(3–4), 287–292.

Berger, D., Olofsson, M., Gotthard, K., Wiklund, C. and Friberg, M. (2012) Ecological constraints on female fitness in a phytophagous insect. *The American Naturalist*, 180(4), 464–480.

Bergman, M. (2011) *The Evolution of Territoriality in Butterflies*. Doctoral dissertation, Department of Zoology, Stockholm University, Sweden.

Bergman, M. and Rutowski, R.L. (2016) Eye morphology and visual acuity in the pipevine swallowtail (*Battus philenor*) studied with a new method of measuring interommatidial angles. *Biological Journal of the Linnean Society*, 117(3), 646–654.

Bergman M. and Wiklund C. (2009a) Differences in mate location behaviours between residents and nonresidents in a territorial butterfly. *Animal Behaviour*, 78, 1161–1167.

Bergman M. and Wiklund C. (2009b) Visual mate detection and mate flight pursuit in relation to sunspot size in a woodland territorial butterfly. *Animal Behaviour*, 78, 17–23.

Bergman, M., Gotthard, K., Berger, D., Olofsson, M., Kemp, D.J. and Wiklund, C. (2007) Mating success of resident versus non-resident males in a territorial butterfly. *Proceedings of the Royal Society B: Biological Sciences*, 274(1618), 1659–1665.

Bergman, M., Gotthard, K. and Wiklund, C. (2011) Mate acquisition by females in a butterfly: the effects of mating status and age on female mate-locating behaviour. *Animal Behaviour*, 81(1), 225–229.

Bergman, M., Lessios, N., Seymoure, B.M. and Rutowski, R.L. (2015) Mate detection in a territorial butterfly – the effect of background and luminance contrast. *Behavioral Ecology*, 26(3), 851–860.

Bergström, J. and Wiklund, C. (1992) Effects of size and nuptial gifts on butterfly reproduction: can females compensate for a smaller size through male-derived nutrients? *Behavioral Ecology* and *Sociobiology*, 52, 296–302.

Bergström, J. and Wiklund, C. (2005) No effect of male courtship intensity on female remating in the butterfly *Pieris napi*. *Journal of Insect Behaviour*, 18, 479–489.

Bernard, G.D. and Remington, C.L. (1991) Color vision in *Lycaena* butterflies: spectral tuning of receptor arrays in relation to behavioral ecology. *Proceedings of the National Academy of Sciences*, 88(7), 2783–2787.

Berthier, S. (2007) *Iridescences: the Physical Colors of Insects*. Springer Science & Business Media, New York.

Berthier, S., Charron, E. and Da Silva, A. (2003) Determination of the cuticle index of the scales of the iridescent butterfly *Morpho menelaus*. *Optics Communications*, 228(4–6), 349–356.

Berwaerts, K., Van Dyck, H. and Aerts, P. (2002) Does flight morphology relate to flight performance? An experimental test with the butterfly, *Pararge aegeria*. *Functional Ecology*, 16(4), 484–491.

Bhakare, M. and Smetacek, P. (2010) Two instances of inter-generic mating by Lycaenidae (Lepidoptera) in Maharashtra, India. *Journal of Research on the Lepidoptera*, 43, 23–25.

Bhardwaj, S., Prudic, K.L., Bear, A., Dasgupta, M., Wasik, B.R., Tong, X., Cheong, W.F., Wenk, M.R. and Monteiro, A. (2017) Sex differences in 20-hydroxyecdysone hormone levels control sexual dimorphism in *Bicyclus anynana* wing patterns. *Molecular Biology and Evolution*, 35(2), 465–472.

Binetti, V.R., Schiffman, J.D., Leaffer, O.D., Spanier, J.E. and Schauer, C.L. (2009) The natural transparency and piezoelectric response of the *Greta oto* butterfly wing. *Integrative Biology*, 1(4), 324–329.

Birkhead, T.R. and Pizzari, T. (2002) Evolution of sex: postcopulatory sexual selection. *Nature Reviews Genetics*, 3(4), 262.

Biró, L.P. and Vigneron, J.P. (2011) Photonic nanoarchitectures in butterflies and beetles: valuable sources for bioinspiration. *Laser and Photonics Reviews*, 5(1), 27–51.

Biró, L.P., Bálint, Z., Kertész, K., Vértesy, Z., Márk, G.I., Horváth, Z.E., Balázs, J., Méhn, D., Kiricsi, I., Lousse, V. and Vigneron, J.P. (2003) Role of photonic-crystal-type structures in the thermal regulation of a Lycaenid butterfly sister species pair. *Physical Review E*, 67(2), 021907.

Bissoondath, C.J. and Wiklund, C. (1995) Protein content of spermatophores in relation to monandry/polyandry in butterflies. *Behavioral Ecology and Sociobiology*, 37(6), 365–371.

Bissoondath, C.J. and Wiklund, C. (1996a) Effect of male mating history and body size on ejaculate size and quality in two polyandrous butterflies, *Pieris napi* and *Pieris rapae* (Lepidoptera: Pieridae). *Functional Ecology*, 10(4), 457–464.

Bissoondath, C.J. and Wiklund, C. (1996b) Male butterfly investment in successive ejaculates in relation to mating system. *Behavioral Ecology and Sociobiology*, 39(5), 285–292.

Bitzer, R.J. (1995) Territorial behavior of the red admiral butterfly, *Vanessa atalanta* (L.) (Lepidoptera: Nymphalidae). Iowa State University Retrospective Theses and Dissertations. 10881. Available at: http://lib.dr.iastate.edu/rtd/10881 (accessed 24 July 2019).

Bittzer R.J and Shaw K.C. (1979) Territorial behavior of the red admiral, *Vanessa atalanta* (L.) (Lepidoptera: Nymphalidae). *Journal of Research on the Lepidoptera*, 18, 36–49.

Blackiston, D., Briscoe, A.D. and Weiss, M.R. (2011) Color vision and learning in the monarch butterfly, *Danaus plexippus* (Nymphalidae). *Journal of Experimental Biology*, 214(3), 509–520.

Blandin, P. and Purser, B. (2013) Evolution and diversification of Neotropical butterflies: insights from the biogeography and phylogeny of the genus *Morpho fabricius*, 1807 (Nymphalidae: Morphinae), with a review of the geodynamics of South America. *Tropical Lepidoptera Research*, 23(2), 62–85.

Blut, C., Wilbrandt, J., Fels, D., Girgel, E.I. and Lunau, K. (2012) The 'sparkle' in fake eyes – the protective effect of mimic eyespots in Lepidoptera. *Entomologia Experimentalis et Applicata*, 143(3), 231–244.

Boggs, C.L. (1981) Selection pressures affecting male nutrient investment at mating in heliconiine butterflies. *Evolution*, 35(5), 931–940.

Boggs, C.L. (1990) A general model of the role of male-donated nutrients in female insects' reproduction. *The American Naturalist*, 136(5), 598–617.

Boggs, C.L. and Gilbert, LE. (1979) Male contribution to egg production in butterflies: evidence for transfer of nutrients at mating. *Science*, 206, 83–84.

Boggs, C.L. and Jackson, L.A. (1991) Mud puddling by butterflies is not a simple matter. *Ecological Entomology*, 16(1), 123–127.

Boggs, C.L. and Nieminen, M. (2004) Checkerspot reproductive biology. In: Ehrlich P.R. and Hanski I. (eds), *On the Wings of Checkerspot: a Model System for Population Biology*.Oxford University Press, Oxford, UK, pp. 92–111.

Bonduriansky, R. (2001) The evolution of male mate choice in insects: a synthesis of ideas and evidence. *Biological Reviews*, 76, 305–339.

Bonduriansky, R. (2009) Reappraising sexual coevolution and the sex roles. *PLoS Biology*, 7(12), e1000255.

Bonte, D. and Van Dyck, H. (2009) Mate-locating behaviour, habitat-use, and flight morphology relative to rainforest disturbance in an Afrotropical butterfly. *Biological Journal of the Linnean Society*, 96(4), 830–839.

Boppré, M. (1978) Chemical communication, plant relationships, and mimicry in the evolution of danaid butterflies. *Entomologia Experimentalis et Applicata*, 24(3), 264–277.

Boppré, M. (1983) Leaf-scratching – a specialized behaviour of danaine butterflies (Lepidoptera) for gathering secondary plant substances. *Oecologia*, 59, 414–416.

Boppré, M. (1984) Chemically mediated interactions between butterflies In: Vane-Wright, R.I., Ackery, P.R. (eds), *The Biology of Butterflies* (Symposia of the Royal Entomological Society of London, 11). Academic Press, London, pp. 259–275.

Boppré, M. (1986) Insects pharmacophagously utilizing defensive plant chemicals (pyrrolizidine alkaloids). *Naturwissenschaften*, 73, 17–26.

Boppré, M. (1993) The American monarch: courtship and chemical communication of a peculiar danaine butterfly. *Biology and Conservation of the Monarch Butterfly*, (38), 29.

Boppré, M. (1995) Pharmacophagy: sex and drugs and butterflies. *Biologie in Unserer Zeit*, 25, 8–17.

Boppré, M. and Vane-Wright, R.I. (1989) Androconial systems in Danainae (Lepidoptera): functional morphology of *Amauris, Tirumala, Danaus* and *Euploea*. *Zoological Journal of the Linnean Society*, 97, 101–133.

Boppré, M., Petty, R.L., Schneider, D. and Meinwald, J. (1978) Behaviorally mediated contacts between scent organs: another prerequisite for pheromone production in *Danaus chrysippus* males (Lepidoptera). *Journal of Comparative Physiology*, 126(2), 97–103.

Braby, M.F. (1996) Mating frequency in bush-brown butterflies (Nymphalidae: Satyrinae). *Journal of the Lepidopterists Society*, 50, 80–86.

Braby, M. and Nishida, K. (2011) Mate-location behavior in Pereute Herrich-Schaffer butterflies (Lepidoptera: Pieridae), with a review of male behavior at encounter sites in the subtribe Aporiina. *The Journal of Research on the Lepidoptera*, 44, 51–60.

Braby, M.F. and Zwick, A. (2015) Taxonomic revision of the *Taractrocera ilia* (Waterhouse) complex (Lepidoptera: Hesperiidae) from north-western Australia and mainland New Guinea based on morphological and molecular data. *Invertebrate Systematics*, 29(5), 487–509.

Brackenbury, J. (2019) Tandem flight in butterflies – who does the pedalling? *Atropos*, 62, 16–17.

Brakefield, P.M. (1987) Tropical dry and wet season polyphenism in the butterfly *Melanitis leda* (Satvrinae): phenotypic plasticity and climatic correlates. *Biological Journal of the Linnean Society*, 31, 175–191.

Brakefield, P.M., El Filali, E., Van der Laan, R., Breuker, C.J., Saccheri, I.J. and Zwaan, B. (2001) Effective population size, reproductive success and sperm precedence in the butterfly, *Bicyclus anynana*, in captivity. *Journal of Evolutionary Biology*, 14(1), 148–156.

Brandley, N.C., Speiser, D.I. and Johnsen, S. (2013) Eavesdropping on visual secrets. *Evolutionary Ecology*, 27(6), 1045–1068.

Breuker, C.J. and Brakefield, P.M. (2002) Female choice depends on size but not symmetry of dorsal eyespots in the butterfly *Bicyclus anynana*. *Proceedings of the Royal Society of London B: Biological Sciences* B, 269, 1233–1239.

Bridarolli, A., Atak, S. and Herm, C. (2016) Analysis of Wilhelm Ostwald's 'Colour Organ' with Raman microspectroscopy. *Applied Physics A*, 122(11), 982.

Briscoe, A.D. (2008) Reconstructing the ancestral butterfly eye: focus on the opsins. *Journal of Experimental Biology*, 211(11), 1805–1813.
Briscoe, A.D. and Bernard, G.D. (2005) Eyeshine and spectral tuning of long wavelength-sensitive rhodopsins: no evidence for red-sensitive photoreceptors among five *Nymphalini* butterfly species. *Journal of Experimental Biology*, 208(4), 687–696.
Briscoe, A.D. and Chittka, L. (2001) The evolution of color vision in insects. *Annual Review of Entomology*, 46(1), 471–510.
Briscoe, A.D., Bernard, G.D., Szeto, A.S., Nagy, L.M. and White, R.H. (2003) Not all butterfly eyes are created equal: Rhodopsin absorption spectra, molecular identification, and localization of ultraviolet-, blue-, and green-sensitive rhodopsin-encoding mRNAs in the retina of Vanessa cardui. *Journal of Comparative Neurology*, 458(4), 334–349.
Briscoe, A.D., Bybee, S.M., Bernard, G.D., Yuan, F., Sison-Mangus, M.P., Reed, R.D., Warren, A.D., Llorente-Bousquets, J. and Chiao, C.C. (2010) Positive selection of a duplicated UV-sensitive visual pigment coincides with wing pigment evolution in *Heliconius* butterflies. *Proceedings of the National Academy of Sciences*, 107(8), 3628–3633.
Brooks, D.R. and McLennan, D.A. (2012) *The Nature of Diversity: an Evolutionary Voyage of Discovery*. University of Chicago Press, Chicago, Illinois.
Brower, A.V.Z. (2013) Introgression of wing pattern alleles and speciation via homoploid hybridization in *Heliconius* butterflies: a review of evidence from the genome. *Proceedings of the Royal Society of London B: Biological Sciences*, 280, 20122302.
Brower, A.V., Wahlberg, N., Ogawa, J.R., Boppré, M. and Vane-Wright, R.I. (2010) Phylogenetic relationships among genera of danaine butterflies (Lepidoptera: Nymphalidae) as implied by morphology and DNA sequences. *Systematics and Biodiversity*, 8(1), 75–89.
Brower, L.P. and Jones, M.A. (1965) Precourtship interaction of wing and abdominal sex glands in male *Danaus* butterflies. *Proceedings of the Royal Entomological Society of London. Series A, General Entomology* 40(10–12), 147–151. Blackwell Publishing Ltd, Oxford, UK.
Brower L.P., Brower, J.V.Z. and Cranston, F.P. (1965) Courtship behavior of the queen butterfly, *Danaus gilippus berenice*. *Zoologica*, 50, I–39.
Brower, L.P., Oberhauser, K.S., Boppré, M., Brower, A.V. and Vane-Wright, R.I. (2007) Monarch sex: ancient rites, or recent wrongs. *Antenna*, 31(1), 12–18.
Brown Jr, K.S. (1984) Adult-obtained pyrrolizidine alkaloids defend ithomiine butterflies against a spider predator. *Nature*, 309(5970), 707.
Brown, R.M. (2006) A case of suspected coral snake (*Hemibungarus calligaster*) mimicry by lepidopteran larvae (*Bracca* sp.) from Luzon Island, Philippines. *Raffles Bulletin of Zoology*, 54(2), 225–227.
Brown, W.D. and Alcock, J. (1990) Hilltopping by the red admiral butterfly: mate searching alongside congeners. *Journal of Research on the Lepidoptera*, 29(1), 1–10.
Brown, W.D., Crespi, B.J., and Choe, J.C. (1997) Sexual conflict and the evolution of mating systems. In: Choe, J.C. and Crespi, B.J. (eds), *The Evolution of Mating Systems in Insects and Arachnids*. Cambridge University Press, Cambridge, UK, pp. 352–377.
Brunton, C.F. and Majerus, M.E. (1995) Ultraviolet colours in butterflies: intra- or inter-specific communication? *Proceedings: Biological Sciences*, 260(1358), 199–204.
Brunton, C.F.A., Baxter, J.D., Quartson, J.A.S. and Panchen, A.L. (1990) The valesina morph of the butterfly *Argynnis paphia* (L.) in Corsica, 1988. The *Entomologists' Record and Journal of Variation*, 102(1–2), 31–7.
Brussard, P.F. and Ehrlich, P.R. (1970) The population structure of *Erebia epipsodea* (Lepidoptera: Satyrinae). *Ecology*, 51(1), 119–129.
Burghardt, F., Knüttel, H., Becker, M. and Fiedler, K. (2000) Flavonoid wing pigments increase attractiveness of female common blue (*Polyommatus icarus*) butterflies to mate-searching males. *Naturwissenschaften*, 87(7), 304–307.

Burk, T. (1982) Evolutionary significance of predation on sexually signalling males. *The Florida Entomologist*, 65(1), 90–104.

Burns, J.M. (1997) Presidential address 1996: on the beauties, uses, variation, and handling of genitalia. *Journal of the Lepidopterists' Society*, 51(1), 1–8.

Buschbeck, E.K. and Friedrich, M. (2008) Evolution of insect eyes: tales of ancient heritage, deconstruction, reconstruction, remodeling, and recycling. *Evolution: Education and Outreach*, 1(4), 448–462.

Butenandt, A., Beckmann, R. and Stamm, D. (1961) Über den Sexuallockstoff des Seidenspinners, II. Konstitution und Konfiguration des Bombykols. *Hoppe-Seyler's Zeitschrift für physiologische Chemie*, 324(1), 84–87.

Bybee, S.M., Yuan, F., Ramstetter, M.D., Llorente-Bousquets, J., Reed, R.D., Osorio, D. and Briscoe, A.D. (2012) UV photoreceptors and UV-yellow wing pigments in *Heliconius* butterflies allow a color signal to serve both mimicry and intraspecific communication. *The American Naturalist*, 179(1), 38–51.

Caballero-Mendieta, N. and Cordero, C. (2012) Enigmatic liaisons in Lepidoptera: a review of same-sex courtship and copulation in butterflies and moths. *Journal of Insect Science*, 12, 138.

Caballero-Mendieta, N. and Cordero, C. (2013) Male mating costs in a butterfly that produces small ejaculates. *Physiological Entomology*, 38(4), 318–325.

Calabrese M.J., Ries L., Matter S.F., Debinski D.D., Auckland J.N., Roland J. and Fagan W.F. (2008) Reproductive asynchrony in natural butterfly populations and its consequences for female matelessness. *Journal of Animal Ecology*, 77, 746–756.

Calhoun, J.V. (1996) Conquering soldiers: the successful invasion of Florida by *Danaus eresimus* (Lepidoptera: Nymphalidae). *Holarctic Lepidoptera*, 3(1), 7–18.

Callaghan, C.J. (1983) Study of isolating mechanisms among Neotropical butterflies of the subfamily Riodininae. *Journal of Research on the Lepidoptera*, 21(3), 159–176.

Cardoso, M.Z. and Gilbert, L.E. (2007) A male gift to its partner? Cyanogenic glycosides in the spermatophore of longwing butterflies (*Heliconius*). *Naturwissenschaften*, 94, 39–42.

Cardoso, M.Z. and Silva, E.S. (2015) Spermatophore quality and production in two Heliconius butterflies with contrasting mating systems. *Journal of Insect Behavior*, 28(6), 693–703.

Cardoso, M.Z., Roper, J.J. and Gilbert, L.E. (2009) Prenuptial agreements: mating frequency predicts gift-giving in *Heliconius* species. *Entomologia Experimentalis et Applicata*, 131, 109–114.

Carvalho, A.P.S., Orr, A.G. and Kawahara, A.Y. (2017) A review of the occurrence and diversity of the sphragis in butterflies (Lepidoptera, Papilionoidea). *ZooKeys*, 694, 41–70. doi:10.3897/zookeys.694.13097

Carvalho, A.P.S., Mota, L.L. and Kawahara, A.Y. (2019) Intersexual 'arms race'and the evolution of the sphragis in Pteronymia butterflies. *Insect Systematics and Diversity*, 3(1), 3.

Carvalho, M.R.M., Peixoto, P.E.C. and Benson, W.W. (2016) Territorial clashes in the Neotropical butterfly *Actinote pellenea (*Acraeinae): do disputes differ when contests get physical? *Behavioral Ecology and Sociobiology*, 70, 199–207.

Cepero, L.C., Rosenwald, L.C. and Weiss, M.R. (2015) The relative importance of flower color and shape for the foraging monarch butterfly (Lepidoptera: Nymphalidae). *Journal of Insect Behavior*, 28(4), 499–511.

Chapman, T., Arnqvist, G., Bangham, J. and Rowe, L. (2003) Sexual conflict. *Trends in Ecology and Evolution*, 18(1), 41–47.

Charlat, S., Reuter, M., Dyson, E.A., Hornett, E.A., Duplouy, A., Davies, N., Roderick, G.K., Wedell, N. and Hurst, G.D. (2007) Male-killing bacteria trigger a cycle of increasing male fatigue and female promiscuity. *Current Biology*, 17(3), 273–277.

Chaves, G.W. (2004) Ecologia de agrupamentos e interações agonisticas na borboleta *Charis cadytis* (Riodinidae). Doctoral thesis. Universidade Estadual de Campinas, Brazil.

Chaves, G.W., Patto, C.E. and Benson, W.W. (2006) Complex non-aerial contests in the lekking butterfly *Charis cadytis* (Riodinidae). *Journal of Insect Behavior*, 19(2), 179–196.

Chazot, N., Panara, S., Zilbermann, N., Blandin, P., Le Poul, Y., Cornette, R., Elias, M. and Debat, V. (2016) *Morpho* morphometrics: shared ancestry and selection drive the evolution of wing size and shape in *Morpho* butterflies. *Evolution*, 70(1), 181–194.

Chen, P.J., Arikawa, K. and Yang, E.C. (2013) Diversity of the photoreceptors and spectral opponency in the compound eye of the golden birdwing, *Troides aeacus formosanus*. *PLoS ONE*, 8(4), e62240.

Chen, P.J., Awata, H., Matsushita, A., Yang, E.C. and Arikawa, K. (2016) Extreme spectral richness in the eye of the common bluebottle butterfly, *Graphium sarpedon*. *Frontiers in Ecology and Evolution*, 4, 8. doi:10.3389/fevo.2016.00018

Chittka, L. and Osorio, D. (2007) Cognitive dimensions of predator responses to imperfect mimicry. *PLoS Biology*, 5(12), e339.

Clark, A.H. (1926) Notes on the odors of some New England butterflies. *Psyche: A Journal of Entomology*, 33(1), 1–5.

Clark, D. (2011) An unusual pairing. *News of the Lepidopterists' Society*, 59, 36.

Clayton, D.L. and Petr, D. (1992) Sexual differences in habitat preference and behavior of *Oeneis chryxus* (Nymphalidae: Satyrinae). *Journal of the Lepidopterists Society*, 46(2), 110–118.

Clutton-Brock, T. (2009) Sexual selection in females. *Animal Behaviour*, 77(1), 3–11.

Codella Jr, S.G. and Lederhouse, R.C. (1989) Intersexual comparison of mimetic protection in the black swallowtail butterfly, *Papilio polyxenes*: experiments with captive blue jay predators. *Evolution*, 43(2), 410–420.

Condamin, M. (1973) Monographie du genre *Bicyclus* (Lepidoptera Satyridae). *Mémoires de l'Institut Fondamental d'Afrique Noire*, 88, 1–324.

Conner, W.E. and Corcoran, A.J. (2012) Sound strategies: the 65-million-year-old battle between bats and insects. *Annual Review of Entomology*, 57, 21–39.

Cook, P.A. and Wedell, N. (1996) Ejaculate dynamics in butterflies: a strategy for maximizing fertilization success? *Proceedings of the Royal Society of London B: Biological Sciences*, 263(1373), 1047–1051.

Cook, S.E., Vernon, J.G., Bateson, M. and Guilford, T. (1994) Mate choice in the polymorphic African swallowtail butterfly, *Papilio dardanus*: male-like females may avoid sexual harassment. *Animal Behaviour*, 47, 389–397.

Corbera, G., Escrivà, À. and Corbera, J. (2011) Hilltopping de les papallones diürnes al turó d'Onofre Arnau (Mataró, Maresme). *L'Atzavara*, 20, 59–68.

Cordero, C. (1993) The courtship behavior of *Callophrys xami* (Lycaenidae). *Journal of Research on the Lepidoptera*, 32, 99–106.

Cordero, C. (1997) Why do some male *Callophrys xami* (Lycaenidae) shift their territories? *Journal of the Lepidopterists' Society*, 51(4), 295–303.

Cordero, C. (1999) Is spermatophore number a good measure of mating frequency in female *Callophrys xami* (Lycaenidae)? *Journal of the Lepidopterists' Society*, 53(4), 169–170.

Cordero, C. (2000) Trade-off between fitness components in males of the polygynous butterfly *Callophrys xami* (Lycaenidae): the effect of multiple mating on longevity. *Behavioral Ecology and Sociobiology*, 48(6), 458–462.

Cordero, C. (2002) Matings without spermatophore transfer and with transfer of two spermatophores in *Callophrys xami* (Lycaenidae). *Journal of the Lepidopterists' Society*, 56(2), 106–108.

Cordero, C. (2005) The evolutionary origin of signa in female Lepidoptera: natural and sexual selection hypotheses. *Journal of Theoretical Biology*, 232(3), 443–449.

Cordero, C. (2010) On the function of cornuti, sclerotized structures of the endophallus of Lepidoptera. *Genetica*, 138, 27–35.

Cordero, C. and Baixeras, J. (2015) Sexual selection within the female genitalia in Lepidoptera. In: Peretti, A.V. and Aisenberg, A. (eds), *Cryptic Female Choice in Arthropods*. Springer International Publishing, Cham, Switzerland, pp. 325–350. doi:10.1007/978–3-319–17894–3_12

Cordero, C. and Miller, J.S. (2012) On the evolution and function of caltrop cornuti in Lepidoptera – potentially damaging male genital structures transferred to females during copulation. *Journal of Natural History*, 46(11–12), 701–715.

Cordero, C.R. and Soberón, J. (1990) Non-resource based territoriality in males of the butterfly *Xamia xami* (Lepidoptera: Lycaenidae). *Journal of Insect Behavior*, 3(6), 719–732.

Cordero, C., Macías, R. and Jiménez, G. (2000) The number of copulations of territorial males of the butterfly *Callophrys xami* (Lycaenidae). *Journal of Research on the Lepidoptera*, 35, 78–89.

Corkery, R. and Tyrode, E.C. (2017) On the colour of wing scales in butterflies: iridescence and preferred orientation of single gyroid photonic crystals. *Interface Focus*, 7, 20160154. http://dx.doi.org/10.1098/rsfs.2016.0154

Costanzo, K. and Monteiro, A. (2007) The use of chemical and visual cues in female choice in the butterfly *Bicyclus anynana*. *Proceedings of the Royal Society B: Biological Sciences*, 274, 845–851.

Courtney, S.P. (1980) Studies on the biology of the butterflies *Anthocaris cardamines* (L) and *Pieris napi* (L), in relation to speciation in Pierinae. Doctoral dissertation, Durham University, UK.

Coutsis, J.G. (2012) Re-examination of the female genitalia of N American *Icaricia lupini* and *I. acmon* and description of those of the closely allied *I. neurona* and *I. shasta* (Lepidoptera: Lycaenidae, Polyommatiti). *Phegea*, 40(2), 53–55.

Crane, J. (1954) Spectral reflectance characteristics of butterflies (Lepidoptera) from Trinidad, BWI. *Zoologica*, 39, 85–115.

Crane, J. (1955) Imaginal behavior of a Trinidad butterfly, *Heliconius erato hydara* Heiwitson, with special reference to the social use of color. *Zoologica (NY)*, 40, 167–196.

Crane, J. (1957) Imaginal behavior in butterflies of the family Heliconiidae: changing social patterns and irrelevant actions. *Zoologica*, 42, 135–145.

Crne, M., Sharma, V., Blair, J., Park, J.O., Summers, C.J. and Srinivasarao, M. (2011) Biomimicry of optical microstructures of *Papilio palinurus*. *EPL (Europhysics Letters)*, 93(1), 14001.

Cronin, T.W. and Bok, M.J. (2016) Photoreception and vision in the ultraviolet. *Journal of Experimental Biology*, 219(18), 2790–2801.

Cronin, T.W., Shashar, N., Caldwell, R.L., Marshall, J., Cheroske, A.G. and Chiou, T.H. (2003) Polarization vision and its role in biological signaling. *Integrative and Comparative Biology*, 43(4), 549–558.

Cronin, T.W., Johnsen, S., Marshall, N.J., and Warrant, E.J. (2014) *Visual Ecology*. Princeton University Press, Princeton, NJ.

Ćurčić, S.B., Pantelić, D.V., Ćurčić, B.P., Savić-Šević, S.N., Makarov, S.E., Lačković, V.B., Labudović-Borović, M.M., Ćurčić, N.B. and Stojanović, D.V. (2012) Micro- and nanostructures of iridescent wing scales in purple emperor butterflies (Lepidoptera: *Apatura ilia* and *A. iris*). *Microscopy Research and Technique*, 75(7), 968–976.

Da, X.W., Zhang, R., Chen, G.L., Ren, Q.M., Du and Bo, Y.F.L. (2016) Why do males of *Parnassius imperator* fight for bare rocks but not the nectar flower during mate selection? *Ethology*, 122, 552–560.

Daily, G.C., Ehrlich, P.R. and Wheye, D. (1991) Determinants of spatial distribution in a population of the subalpine butterfly *Oeneis chryxus*. *Oecologia*, 88(4), 587–596.

Dana, R.P. (1991) Conservation management of the prairie skippers *Hesperia dacotae* and *Hesperia ottoe*. Station Bulletin 594–1991. Minnesota Agricultural Experiment Station, University of Minnesota, St. Paul, Minnesota.

Daniels, J.C. (2007) Courtship solicitation by females of the barred sulphur butterfly (*Eurema daira*) (Lepidoptera: Pieridae). *Journal of Animal Behaviour*, 20, 129–135.

Dapporto, L., Hardy, P.B. and Dennis, R.L. (2018) Evidence for adaptive constraints on size of marginal wing spots in the grayling butterfly, *Hipparchia semele*. *Biological Journal of the Linnean Society*, 126(1), 131–145.

Darragh, K., Vanjari, S., Mann, F., Gonzalez-Rojas, M.F., Morrison, C.R., Salazar, C., Pardo-Diaz, C., Merrill, R.M., McMillan, W.O., Schulz, S. and Jiggins, C.D. (2017) Male sex pheromone components in *Heliconius* butterflies released by the androconia affect female choice. *PeerJ*, 5, p.e3953.

Darragh, K., Byers, K.J., Merrill, R.M., McMillan, W.O., Schulz, S. and Jiggins, C.D. (2019a) Male pheromone composition depends on larval but not adult diet in *Heliconius melpomene*: diet affects *Heliconius* male pheromones. *Ecological Entomology*, 44(3), 397–405. doi:10.1111/een.12716

Darragh, K., Montejo-Kovacevich, G., Kozak, K.M., Morrison, C.R., Figueiredo, C.M., Ready, J.S., Salazar, C., Linares, M., Byers, K.J., Merrill, R.M. and McMillan, W.O. (2019b) Species specificity and intraspecific variation in the chemical profiles of *Heliconius* butterflies across a large geographic range. *bioRxiv*, 573469. doi:https://doi.org/10.1101/573469

Davies, N.B. (1978) Territorial defence in the speckled wood butterfly (*Pararge aegeria*): the resident always wins. *Animal Behaviour* 26, 138–147.

Davis, A.K. and Rendón-Salinas, E. (2010) Are female monarch butterflies declining in eastern North America? Evidence of a 30-year change in sex ratios at Mexican overwintering sites. *Biology Letters*, 6(1), 45–47.

Davis, A.K., Cope, N., Smith, A. and Solensky, M.J. (2007). Wing color predicts future mating success in male monarch butterflies. *Annals of the Entomological Society of America*, 100(2), 339–344.

De Bona, S., Valkonen, J. K., López-Sepulcre, A. and Mappes, J. (2015) Predator mimicry, not conspicuousness, explains the efficacy of butterfly eyespots. *Proceedings of the Royal Society B: Biological Sciences*, 282(1806), 20150202.

Dechkrong, P., Jiwajinda, S., Dokchan, P., Kongtungmon, M., Chomsaeng, N., Chairuangsri, T., Yu, C.C., Hsiao, C.N. and Shiojiri, M. (2011). Fine structure of wing scales of butterflies, *Euploea mulciber* and *Troides aeacus*. *Journal of Structural Biology*, 176(1), 75–82.

Deinert, E.I. (2003) Sexual selection in *Heliconius hewitsoni*, a pupal mating butterfly. In: Boggs L.C., Watt, W.B. and Ehrlich, P.R. (eds), *Butterflies: Ecology and Evolution Taking Flight*. University of Chicago Press, Chicago, Illinois, pp. 91–108.

Deinert, E.I., Longino, J.T. and Gilbert, L.E. (1994) Mate competition in butterflies. *Nature*, 370(6484), 23–24.

de Ibarra, N.H., Langridge, K.V. and Vorobyev, M. (2015) More than colour attraction: behavioural functions of flower patterns. *Current Opinion in Insect Science*, 12, 64–70.

De Jong, R. (2016) Reconstructing a 55-million-year-old butterfly (Lepidoptera: Hesperiidae). *European Journal of Entomology*, 113, 423–428. doi:10.14411/eje.2016.055

De Keyser, R., Breuker, C.J., Hails, R.S., Dennis, R.L. and Shreeve, T.G. (2015) Why small is beautiful: wing colour is free from thermoregulatory constraint in the small lycaenid butterfly, *Polyommatus icarus*. *PloS ONE*, 10(4), e0122623.

De Nardin, J. and De Araújo, A.M. (2011) Kin recognition in immatures of *Heliconius erato phyllis* (Lepidoptera; Nymphalidae). *Journal of Ethology*, 29(3), 499–503.

Dennis, R.L.H. (1982a) Mate location strategies in the wall brown butterfly, *Lasiommata megera* (L.) (Lepidoptera: Satyridae): wait or seek? *Entomologist's Record and Journal of Variation*, 94, 209–214 and 95, 7–10.

Dennis, R.L.H. (1982b) Patrolling behaviour in orange tip butterflies within the Bollin valley in north Cheshire, and a comparison with other pierids. *Vasculum*, 67, 17–25.

Dennis, R.L.H. (1983) Spatial patterning in butterfly oviposition. Density influences and the 'edge effect' in *Anthocharis cardamines* (L.) (Lep., Pieridae). *Bulletin of the Amateur Entomologists' Society*, 42, 55–60.

Dennis, R.L.H. (1984) Brimstone butterflies (*Gonepteryx rhamni)* (L.) playing possum. *Entomologist's Gazette*, 35, 6–7.

Dennis, R.L.H. (1998) Thanatosis in *Inachis io* (L.) Nymphalidae. *Entomologist's Record and Journal of Variation*, 110, 115–116.

Dennis, R.L.H. (2003) Playing possum as an alternative to mate-refusal posture in *Pararge aegeria* (L.), Satyrinae. *Entomologist's Record and Journal of Variation*, 115, 293.

Dennis, R.H. (2004) Landform resources for territorial nettle–feeding Nymphalid butterflies: biases at different spatial scales. *Animal Biodiversity and Conservation*, 27(2), 37–45.

Dennis, R.L.H. and Dennis, M.P. (2008) Territorial hill topping in British Butterflies: high summits present a special case. *Entomologist's Gazette*, 59, 227–232.

Dennis, R.L.H. and Shreeve, T.G. (1988) Hostplant-habitat structure and the evolution of butterfly mate-locating behaviour. *Zoological Journal of the Linnean Society*, 94(4), 301–318.

Dennis, R.L. and Sparks, T.H. (2005) Landscape resources for the territorial nymphalid butterfly *Inachis io*: microsite landform selection and behavioral responses to environmental conditions. *Journal of Insect Behavior*, 18(5), 725–742.

Dennis, R.L.H. and Williams, W.R. (1987) Mate location behavior of the large skipper butterfly *Ochlodes venata*: flexible strategies and spatial components. *Journal of the Lepidopterists' Society*, 41, 45–64.

Dewsbury, D.A. (1982) Ejaculate cost and male choice. *The American Naturalist*, 119(5), 601–610.

Dewsbury, D.A. (2005) The Darwin–Bateman paradigm in historical context. *Integrative and Comparative Biology*, 45(5), 831–837.

Diao, Y.Y. and Liu, X.Y. (2011) Mysterious coloring: structural origin of color mixing for two breeds of *Papilio* butterflies. *Optics Express*, 19(10), 9232–9241.

Dickinson, J.L. and Rutowski, R.L. (1989) The function of the mating plug in the chalcedon checkerspot butterfly. *Animal Behaviour*, 38, 154–162.

Dincă, V., Lukhtanov, V.A., Talavera, G. and Vila, R. (2011) Unexpected layers of cryptic diversity in wood white Leptidea butterflies. *Nature Communications*, 2, 324.

Dincă, V., Wiklund, C., Lukhtanov, V.A., Kodandaramaiah, U., Norén, K., Dapporto, L., Wahlberg, N., Vila, R. and Friberg, M. (2013) Reproductive isolation and patterns of genetic differentiation in a cryptic butterfly species complex. *Journal of Evolutionary Biology*, 26(10), 2095–2106.

Dinesh, A.S. and Venkatesha, M.G. (2013) Analysis of the territorial, courtship and coupling behavior of the hemipterophagous butterfly, *Spalgis epius* (Westwood) (Lepidoptera: Lycaenidae). *Journal of Insect Behaviour*, 26(2), 149–164.

Dinesh, A.S., Venkatesha, M.G. and Ramakrishna, S. (2010) Development, life history characteristics and behaviour of mealybug predator, *Spalgis epius* (Westwood) (Lepidoptera: Lycaenidae) on *Planococcus citri* (Risso) (Homoptera: Pseudococcidae). *Journal of Pest Science*, 83, 339–345.

Ding, Y., Xu, S., Zhang, Y., Wang, A.C., Wang, M.H., Xiu, Y., Wong, C.P. and Wang, Z.L. (2008). Modifying the anti-wetting property of butterfly wings and water strider legs by atomic layer deposition coating: surface materials versus geometry. *Nanotechnology*, 19(35), 355708 (7pp.).

Ding, Y., Xu, S. and Wang, Z.L. (2009) Structural colors from *Morpho peleides* butterfly wing scales. *Journal of Applied Physics*, 106(7), 074702.

Dinwiddie, A., Null, R., Pizzano, M., Chuong, L., Krup, A.L., Tan, H.E. and Patel, N.H. (2014). Dynamics of F-actin prefigure the structure of butterfly wing scales. *Developmental Biology*, 392(2), 404–418.

Dion, E., Monteiro, A. and Yew, J.Y. (2016) Phenotypic plasticity in sex pheromone production in *Bicyclus anynana* butterflies. *Scientific Reports*, 6, 39002.

Dion, E., Pui, L.X. and Monteiro, A. (2017) Early-exposure to new sex pheromone blend alters mate preference in female butterflies and in their offspring. bioRxiv, 214635.

Doucet, S.M. and Meadows, M.G. (2009) Iridescence: a functional perspective. *Journal of the Royal Society Interface*, 6(Suppl 2), S115–S132.

Douglas, J.M., Cronin, T.W., Chiou, T.H. and Dominy, N.J. (2007). Light habitats and the role of polarized iridescence in the sensory ecology of Neotropical nymphalid butterflies (Lepidoptera: Nymphalidae). *Journal of Experimental Biology*, 210(5), 788–799.

Douwes, P. (1975) Territorial behaviour in *Heodes virgaureae* L. (Lep., Lycaenidae) with particular reference to visual stimuli. *Norwegian Journal of Entomology*, 22, 143–154.

Downey, J.C. (1962) Inter-specific pairing in Lycaenidae. *Journal of the Lepidopterists' Society*, 16, 235–237.

Dreisig, H. (1995) Thermoregulation and flight activity in territorial male graylings, *Hipparchia semele* (Satyridae), and large skippers, *Ochlodes venata* (Hesperiidae). *Oecologia*, 101(2), 169–176.

Drummond, B.A. (1984) Multiple mating and sperm competition in the Lepidoptera. In: Smith, R.I. (ed.), *Sperm Competition and the Evolution of Animal Mating Systems*. Academic Press, London, pp. 291–370.

Dudley, R. (1990) Thanatosis in the Neotropical butterfly *Caligo illioneus* (Nymphalidae: Brassolinae). *The Journal of Research on the Lepidoptera (USA)*, 28(1/2), 125–126.

Duplouy, A., Woestmann, L., Gallego Zamorano, J. and Saastamoinen, M. (2018) Impact of male condition on his spermatophore and consequences for female reproductive performance in the Glanville fritillary butterfly. *Insect Science*, 25(2), 284–296.

Dussourd, D.E., Harvis, C.A., Meinwald, J. and Eisner, T. (1989) Paternal allocation of sequestered plant pyrrolizidine alkaloid to eggs in the danaine butterfly, *Danaus gilippus*. *Experientia*, 45(9), 896–898.

Dyson, E.A. and Hurst, G.D. (2004) Persistence of an extreme sex-ratio bias in a natural population. *Proceedings of the National Academy of Sciences*, 101(17), 6520–6523.

Eberhard, W.G. (1985) *Sexual Selection and Animal Genitalia*. Harvard University Press, Cambridge, Massachusetts.

Eberhard, W.G. (1991) Copulatory courtship and cryptic female choice in insects. *Biological Reviews*, 66(1), 1–31.

Eberhard, W.G. (1994) Evidence for widespread courtship during copulation in 131 species of insects and spiders, and implications for cryptic female choice. *Evolution*, 48(3), 711–733.

Eberhard, W.G. (1996) *Female Control: Sexual Selection by Cryptic Female Choice*. Princeton University Press, Princeton, New Jersey.

Eberhard, W.G. (2010) Evolution of genitalia: theories, evidence, and new directions. *Genetica*, 138, 5–18.

Edgar, J.A., Culvenor, C.C.J. and Robinson, G.S. (1973) Hairpencil dihydropyrrolizines of Danainae from the New Hebrides. *Australian Journal of Entomology*, 12(2), 144–150.

Edgar, J.A., Cockrum, P.A. and Frahn, J.L. (1976a) Pyrrolizidine alkaloids in *Danaus plexippus* L. and *Danaus chrysippus* L. *Experientia*, 32(12), 1535–1537.

Edgar, J.A., Culvenor, C.C. and Pliske, T.E. (1976b) Isolation of a lactone, structurally related to the esterifying acids of pyrrolizidine alkaloids, from the costal fringes of male Ithomiinae. *Journal of Chemical Ecology*, 2(3), 263–270.

Edmunds, M. (1969) Evidence for sexual selection in the mimetic butterfly *Hypolimnas misippus* L. *Nature*, 221(5179), 488.

Edward, D.A., Stockley, P. and Hosken, D.J. (2015) Sexual conflict and sperm competition. *Cold Spring Harbor Perspectives in Biology*, 7(4), a017707.

Eff, D. (1962) A little about the little-known *Papilio indra minori*. *Journal of the Lepidopterists' Society*, 16, 137–142.

Ehrlich, A.H. and Ehrlich, P.R. (1978) Reproductive strategies in the butterflies: I. Mating frequency, plugging, and egg number. *Journal of the Kansas Entomological Society*, 51, 666–697.

Ehrlich, P.R. and Hanski, I. (eds) (2004) *On the Wings of Checkerspots: a Model System for Population Biology*. Oxford University Press, Oxford, UK.

Ehrlich, P.R. and Raven, P.H. (1964) Butterflies and plants: a study in coevolution. *Evolution*, 18(4), 586–608.

Ehrlich, P.R., Launer, A.E. and Murphy, D.D. (1984) Can sex ratio be defined or determined? The case of a population of checkerspot butterflies. *The American Naturalist*, 124(4), 527–539.

Eisner, T. (2003a) *For Love of Insects*. Harvard University Press, Harvard, CT.

Eisner, T. (2003b) Living fossils: on lampreys, Baronia, and the search for medicinals. *BioScience*, 53, 265–269.

Eisner, T. and Meinwald, J. (2003) Alkaloid-derived pheromones and sexual selection in Lepidoptera. In: Blomquist, G. and Vog, R.G. (eds), *Insect Pheromone Biochemistry and Molecular Biology, the Biosynthesis and Detection of Pheromones and Plant Volatiles*. Academic Press, London, pp. 341–368.

Eisner, T., Alsop, R. and Ettershank, G. (1964) Adhesiveness of spider silk. *Science*, 146(3647), 1058–1061.

Elgar, M.A. and Pierce, N.E. (1988) Mating success and fecundity in an ant-tended Lycaenid butterfly. In: Clutton-Brock, T.H. (ed.), *Reproductive Success*. University of Chicago Press, Chicago, Illinois, pp. 59–75.

Ellers, J. and Boggs, C.L. (2002) The evolution of wing color in *Colias* butterflies: heritability, sex linkage, and population divergence. *Evolution*, 56(4), 836–840.

Ellers, J. and Boggs, C.L. (2003a) The evolution of wing color: male mate choice opposes adaptive wing color divergence in *Colias* butterflies. *Evolution*, 57(5), 1100–1106.

Ellers, J. and Boggs, C.L. (2003b) Evolutionary genetics of dorsal wing colour in *Colias* butterflies. *Journal of Evolutionary Biology*, 17(4), 752–758.

Ellers, J. and Boggs, C.L. (2004) Functional ecological implications of intraspecific differences in wing melanization in *Colias* butterflies. *Biological Journal of the Linnean Society*, 82(1), 79–87.

Emlen, S.T. and Oring, L.W. (1977) Ecology, sexual selection, and the evolution of mating systems. *Science*, 197(4300), 215–223.

Emmel, T.C. (1972) Mate selection and balanced polymorphism in the tropical nymphalid butterfly, *Anartia fatima*. *Evolution*, 26(1), 96–107.

Emmel, T.C. (1973) On the nature of the polymorphism and mate selection phenomena in *Anartia fatima* (Lepidoptera: Nymphalidae). *Evolution*, 27(1), 164–165.

Emmel, T.C. and Garraway, E. (1990) Ecology and conservation biology of the *Homerus* swallowtail in Jamaica. *Tropical Lepidoptera*, 1(2), 63–76.

Environment and Climate Change Canada (2017) Management Plan for the Weidemeyer's Admiral (*Limenitis weidemeyerii*) in Canada [Proposed]. Species at Risk Act Management Plan Series. Environment and Climate Change Canada, Ottawa.

Epstein, M.E. (1987) Mating behavior of *Acraea andromacha andromacha* (Fabricius) (Nymphalidae) in New Caledonia. *Journal of the Lepidopterists' Society (USA)*, 41(2), 119–131.

Estrada, C. (2009) Sexual behavior, intraspecific signaling and the evolution of mimicry among closely related species. Doctoral dissertation, the University of Texas at Austin, May 2009.

Estrada, C., Yildizhan, S., Schulz, S. and Gilbert, L.E. (2010) Sex-specific chemical cues from immatures facilitate the evolution of mate guarding in *Heliconius* butterflies. *Proceedings of the Royal Society of London B: Biological Sciences*, 277, 407–413.

Estrada, C., Schulz, S., Yildizhan, S. and Gilbert, L.E. (2011) Sexual selection drives the evolution of antiaphrodisiac pheromones in butterflies. *Evolution*, 65(10), 2843–2854.

Etheridge, K. (2011) Maria Sibylla Merian and the metamorphosis of natural history. *Endeavour*, 35(1), 16–22.

Ferkau, C. and Fischer, K. (2006) Costs of reproduction in male *Bicyclus anynana* and *Pieris napi* butterflies: effects of mating history and food limitation. *Ethology*, 112(11), 1117–1127.

Ferreira, C.P., Cerqueira, L.P., Uezu, A., Oliveira, E.J. and Peixoto, P.E. (2016) Territory availability at different spatial scales does not affect fighting investment in males of the lycaenid butterfly *Strymon mulucha*. *Ethology*, 122(10), 798–806.

Finkbeiner, S.D., Briscoe, A.D. and Reed, R.D. (2014) Warning signals are seductive: relative contributions of color and pattern to predator avoidance and mate attraction in *Heliconius* butterflies. *Evolution*, 68, 3410–3420. doi:10.1111/evo.12524

Finkbeiner, S.D., Fishman, D.A., Osorio, D. and Briscoe, A.D. (2017) Ultraviolet and yellow reflectance but not fluorescence is important for visual discrimination of conspecifics by *Heliconius erato*. *Journal of Experimental Biology*, 220(7), 1267–1276.

Firman, R.C., Gasparini, C., Manier, M.K. and Pizzari, T. (2017) Postmating female control: 20 years of cryptic female choice. *Trends in Ecology and Evolution*, 32(5), 368–382.

Fischer, K. (2007) Control of reproduction and a survival cost to mating in female *Bicyclus anynana* butterflies. *Ecological Entomology*, 32(6), 674–681.

Fischer, K. and Fiedler, K. (2001) Resource-based territoriality in the butterfly *Lycaena hippothoe* and environmentally induced behavioural shifts. *Animal Behaviour*, 61(4), 723–732.

Fischer, K., Perlick, J. and Galetz, T. (2008) Residual reproductive value and male mating success: older males do better. *Proceedings of the Royal Society of London B: Biological Sciences*, 275(1642), 1517–1524.

Fischer, K., Karl, I., Heuskin, S., Janowitz, S. and Dötterl, S. (2015) Kin recognition and inbreeding avoidance in a butterfly. *Ethology*, 121(10), 977–984.

Fischer, K., Karl, I., Dublon, I.A. and Kehl, T. (2018) A reply to Nieberding and Holveck: beyond experimental design and proximate mechanisms-mate choice in the face of sexual conflict. *Frontiers in Zoology*, 15(1), 19.

Ford, E.B. (1945) *Butterflies*. Collins, London.

Fordyce, J.A., Nice, C.C., Forister, M.L. and Shapiro, A.M. (2002) The significance of wing pattern diversity in the Lycaenidae: mate discrimination by two recently diverged species. *Journal of Evolutionary Biology*, 15(5), 871–879.

Forsberg. J and Wiklund, C. (1989) Mating in the afternoon: time-saving in courtship and remating by females of a polyandrous butterfly *Pieris napi* L. *Behavioral Ecology and Sociobiology*, 25, 349–356.

Fosdick, M.C. (1973) A population study of the neotropical nymphalid butterfly *Anartia amathea* in Ecuador. *Journal of Research on the Lepidoptera*, 11, 65–80.

Francini, R.B., Freitas, A.V.L. and Brown, K.S. (2005) Rediscovery of *Actinote zikani* (D'Almeida) (Nymphalidae, Heliconiinae, Acraeini): natural history, population biology and conservation of an endangered butterfly in SE Brazil. *Journal of the Lepidopterists' Society*, 59(3), 134.

Francini, R.B., Souza, T.S., Ramos, R.R. and Cerávolo, M.S. (2015) Quantifying wing pattern sexual dimorphism in the butterfly *Anartia amathea roeselia* Eschscholtz (Lepidoptera: Nymphalidae). *International Journal of Fauna and Biological Studies*, 2(2), 43–52.

Frederiksen, R. and Warrant, E.J. (2008) Visual sensitivity in the crepuscular owl butterfly *Caligo memnon* and the diurnal blue morpho *Morpho peleides*: a clue to explain the evolution of nocturnal apposition eyes? *Journal of Experimental Biology*, 211(6), 844–851.

Freeman, A.L. (1998) Butterflies as Signal Receivers. Doctoral dissertation, University of Oxford, UK.

Freese, A. and Fiedler, K. (2002) Experimental evidence for specific distinctness of the two wood white butterfly taxa, *Leptidea sinapis* and *L. reali* (Pieridae) *Nota Lepidopterologica*, 25, 39–59.

Freitas, A.V., Benson, W.W., Marini-Filho, O.J. and De Carvalho, R.M. (1997) Territoriality by the dawn's early light: the Neotropical owl butterfly *Caligo idomenaeus* (Nymphalidae: Brassolinae). *Journal of Research on the Lepidoptera*, 34(1), 14–20.

Freitas, A.V.L., Barbosa, E.P., Willmott, K.R., Wahlberg, N. and Lamas, G. (2016) 'Species' from two different butterfly genera combined into one: description of a new genus of Euptychiina (Nymphalidae: Satyrinae) with unusually variable wing pattern. *Revista Brasileira de Entomologia*, 60(2), 157–165.

Freitas, A.V.L., Francini, R.B., Paluch, M. and Barbosa, E.P. (2018) A new species of Actinote Hübner (Nymphalidae: Heliconiinae: Acraeini) from southeast Brazil. *Revista Brasileira de Entomologia*, 62(2), 135–147.

Frentiu, F.D. and Briscoe, A.D. (2008) A butterfly eye's view of birds. *BioEssays*, 30(11–12), 1151–1162.

Frentiu, F.D., Bernard, G.D., Cuevas, C.I., Sison-Mangus, M.P., Prudic, K.L. and Briscoe, A.D. (2007) Adaptive evolution of color vision as seen through the eyes of butterflies. *Proceedings of the National Academy of Sciences*, 104(suppl 1), 8634–8640.

Frey, D. (1999) Resistance to mating by female monarch butterflies. *Biología de la mariposa monarca/Biology of the Monarch Butterfly*. 1997 North American Conference on the Monarch Butterfly, Commission for Environmental Cooperation, 1999, pp. 79–88.

Frey, D., Leong, K. L., Peffer, E., Smidt, R.K. and Oberhauser, K. (1998) Mate pairing patterns of monarch butterflies (*Danaus plexippus* L.) at a California overwintering site. *Journal of the Lepidopterists Society*, 52, 84–97.

Friberg, M. and Wiklund, C. (2007) Generation-dependent female choice: behavioral polyphenism in a bivoltine butterfly. *Behavioral Ecology*, 18(4), 758–763.

Friberg, M., Bergman, M., Kullberg, J., Wahlberg, N. and Wiklund, C. (2008a) Niche separation in space and time between two sympatric sister species – a case of ecological pleiotropy. *Evolutionary Ecology*, 22(1), 1–18.

Friberg, M., Vongvanich, N., Borg-Karlson, A.K., Kemp, D.J., Merilaita, S. and Wiklund, C. (2008b) Female mate choice determines reproductive isolation between sympatric butterflies. *Behavioral Ecology and Sociobiology*, 62(6), 873–886.

Friberg, M., Leimar, O. and Wiklund, C. (2013) Heterospecific courtship, minority effects and niche separation between cryptic butterfly species. *Journal of Evolutionary Biology*, 26(5), 971–979.

Fric, Z. and Konvička, M. (2000) Adult population structure and behaviour of two seasonal generations of the European map butterfly, *Araschnia levana*, species with seasonal polyphenism (Nymphalidae). *Nota Lepidopterologica*, 23, 2–25.

Fukano, Y., Satoh, T., Hirota, T., Nishide, Y. and Obara, Y. (2012) Geographic expansion of the cabbage butterfly (*Pieris rapae*) and the evolution of highly UV-reflecting females. *Insect Science*, 19(2), 239–246.

Fullard, J.H., Otero, L.D., Orellana, A. and Surlykke, A. (2000) Auditory sensitivity and diel flight activity in Neotropical Lepidoptera. *Annals of the Entomological Society of America*, 93(4), 956–965.

Gage, M.J. (1994) Associations between body size, mating pattern, testis size and sperm lengths across butterflies. *Proceedings of the Royal Society B: Biological Sciences*, 258(1353), 247–254.

Galicia, I., Sánchez, V. and Cordero, C. (2008) On the function of signa, a genital trait of female Lepidoptera. *Annals of the Entomological Society of America* 101(4), 786–793.

Gao, K., Li, X., Chen, F., Guo, Z. and Settele, J. (2016) Distribution and habitats of *Phengaris* (Maculinea) butterflies and population ecology of *Phengaris teleius* in China. *Journal of Insect Conservation*, 20(1), 1–10.

Gallo, F., Beretta, S., Salogni, G. and Bonato, L. (2015) Partial separation of climatic niche between two cryptic species of Leptidea (Lepidoptera: Pieridae) in the Alpine–Adriatic region. *Italian Journal of Zoology*, 82(4), 573–581.

García-Barros, E. (1989) Estudio comparativo de los caracteres biológicos de dos satirinos *Hipparchia statilinus* (Hufnagel, 1766) e *H. semele* (L., 1758) (Lepidoptera, Nymphalidae, Satyrinae). *Miscellània Zoològica*, 13, 85–96.

Garzón-Orduña, I.J. (2012) Phylogenetic evidence for loss of sound production and a shift in sexual recognition signals in *Hamadryas* butterflies (Nymphalidae: Biblidinae). *Systematic Entomology*, 37(1), 84–101.

Gavrilets, S. (2000) Rapid evolution of reproductive barriers driven by sexual conflict. *Nature*, 403(6772), 886.

Gavrilets, S., Arnqvist, G. and Friberg, U. (2001) The evolution of female mate choice by sexual conflict. *Proceedings of the Royal Society of London B: Biological Sciences*, 268(1466), 531–539.

Ghiradella, H. (1994) Structure of butterfly scales: patterning in an insect cuticle. *Microscopy Research and Technique*, 27(5), 429–438.

Ghiradella, H., Aneshansley, D., Eisner, T., Silberglied, R.E. and Hinton, H.E. (1972) Ultraviolet reflection of a male butterfly: interference color caused by thin-layer elaboration of wing scales. *Science*, 178(4066), 1214–1217.

Gilbert, L.E. (1972) Pollen feeding and reproductive biology of *Heliconius* butterflies. *Proceedings of the National Academy of Sciences*, 69(6), 1403–1407.

Gilbert, L.E. (1976) Postmating female odor in *Heliconius* butterflies: a male-contributed antiaphrodisiac? *Science*, 193, 419–420.

Gilbert, L.E., Forrest, H.S., Schultz, T.D. and Harvey, D.J. (1988) Correlations of ultrastructure and pigmentation suggest how genes control development of wing scales of *Heliconius* butterflies. *The Journal of Research on the Lepidoptera*, 26, 141–160.

Gilchrist, G.W. (1990) The consequences of sexual dimorphism in body size for butterfly flight and thermoregulation. *Functional Ecology*, 4(4), 475–487.

Gilchrist, G.W. and Rutowski, R.L. (1986) Adaptive and incidental consequences of the alba polymorphism in an agricultural population of *Colias* butterflies: female size, fecundity, and differential dispersion. *Oecologia*, 68(2), 235–240.

Giraldo, M.A. and Stavenga, D.G. (2007) Sexual dichroism and pigment localization in the wing scales of *Pieris rapae* butterflies. *Proceedings of the Royal Society of London B: Biological Sciences*, 274(1606), 97–102.

Giraldo, M. A. and Stavenga, D.G. (2008) Wing coloration and pigment gradients in scales of pierid butterflies. *Arthropod Structure and Development*, 37(2), 118–128.

Giraldo, M.A. and Stavenga, D.G. (2016) Brilliant iridescence of *Morpho* butterfly wing scales is due to both a thin film lower lamina and a multilayered upper lamina. *Journal of Comparative Physiology A*, 202(5), 381–388.

Glaeser, G. and Paulus, H.F. (2015) *The Evolution of the Eye*. Springer, Heidelberg, Germany.

Glassberg, J. (2005) Butterfly eye-dentification. *American Butterflies*, 13(2), 34–35. Summer 2005.

Gomez-Diaz, C. and Benton, R. (2013) The joy of sex pheromones. *EMBO Reports*, 14(10), 874–883.

Gonzato, J.C. and Pont, B. (2004) A phenomenological representation of iridescent colors in butterfly wings. WSCG '2004: Short Communications: the 12th International Conference in Central Europe on Computer Graphics, Visualization and Computer Vision 2004, 2–6 February 2004, Plzeň, pp. 79–86.

Gordon, I.J. (1987) Natural selection for rare and mimetic colour pattern combinations in wild populations of the diadem butterfly, *Hypolimnas misippus* L. *Biological Journal of the Linnean Society*, 31(1), 1–23.

Gorelick, G.A. (1971) Biosystematic study of two species of *Callophrys* (*Callophrys*) in California (Lycaenidae). *Journal of the Lepidopterists' Society*, 25(2), 1–42.

Gould, S.J.R. and Lewontin, C. (1979) The Spandrels of San Marco and the panglossian paradigm: A critique of the adaptationist programme. *Proceedings of the Royal Society of London B: Biological Sciences*, 205(1161), 581–598.

Grafen, A. (1990a) Biological signals as handicaps. *Journal of Theoretical Biology*, 144(4), 517–546.

Grafen, A. (1990b) Sexual selection unhandicapped by the Fisher process. *Journal of Theoretical Biology*, 144(4), 473–516.

Graham, S.M., Watt, W.B. and Gall, L.F. (1980) Metabolic resource allocation vs. mating attractiveness: adaptive pressures on the 'alba' polymorphism of *Colias* butterflies. *Proceedings of the National Academy of Sciences*, 77(6), 3615–3619.

Green, K.K. and Madjidian, J.A. (2011) Active males, reactive females: stereotypic sex roles in sexual conflict research? *Animal Behaviour*, 81(5), 901–907.

Grether, G.F., Kolluru, G.R. and Nersissian, K. (2004) Individual colour patches as multicomponent signals. *Biological Reviews*, 79(3), 583–610.

Grula, J.W., McChesney, J.D. and Taylor, O.R. (1980) Aphrodisiac pheromones of the sulfur butterflies *Colias eurytheme* and *C. philodice* (Lepidoptera, Pieridae). *Journal of Chemical Ecology*, 6(1), 241–256.

Guillaumin, M. (1963) Le complexe odoriférant thoraco-abdominal de *Pyrgus malvae* L. (Lep. Hesperiidae). *Bulletin de la Societe Entomologique de France*, 68, 128–136.

Gwynne, D.T. and Simmons, L.W. (1990) Experimental reversal of courtship roles in an insect. *Nature*, 346(6280), 172.

Hafernik, J.E. (1982) *Phenetics and Ecology of Hybridization in Buckeye Butterflies: Lepidoptera: Nymphalidae* (Vol. 93). University of California Press, Berkeley, CA.

Haikola, S. (2004) Inbreeding depression in the Glanville fritillary butterfly (*Melitaea cinxia*). *Annales Zoologici Fennici*, 40, 483–493.

Haikola, S., Fortelius, W., O'Hara, R.B., Kuussaari, M., Wahlberg, N., Saccheri, I.J., Singer, M.C. and Hanski, I. (2001) Inbreeding depression and the maintenance of genetic load in *Melitaea cinxia* metapopulations. *Conservation Genetics*, 2(4), 325–335.

Haikola, S., Singer, M.C. and Pen, I. (2004) Has inbreeding depression led to avoidance of sib mating in the Glanville fritillary butterfly (*Melitaea cinxia*)? *Evolutionary Ecology*, 18(2), 113–120.

Hall, J.P. and Harvey, D.J. (2002) A survey of androconial organs in the Riodinidae (Lepidoptera). *Zoological Journal of the Linnean Society*, 136(2), 171–197.

Hamilton, W.D. (1964) The genetical evolution of social behaviour. II. *Journal of Theoretical Biology*, 7(1), 17–52.

Han, Z., Niu, S., Zhang, L., Liu, Z. and Ren, L. (2013) Light trapping effect in wing scales of butterfly *Papilio peranthus* and its simulations. *Journal of Bionic Engineering*, 10(2), 162–169.

Han, Z., Niu, S., Yang, M., Mu, Z., Li, B., Zhang, J., Ye, J. and Ren, L. (2014) Unparalleled sensitivity of photonic structures in butterfly wings. *RSC Advances*, 4(85), 45214–45219.

Han, Z., Li, B., Mu, Z., Yang, M., Niu, S., Zhang, J. and Ren, L. (2015) An ingenious super light trapping surface templated from butterfly wing scales. *Nanoscale Research Letters*, 10(1), 344.

Hardy, I.C. (1998) Butterfly battles: on conventional contests and hot property. *Trends in Ecology and Evolution*, 13(10), 385–386.

Hardy, P.B. (2018) Inter-specific courtship and mating in butterflies of the genus *Pieris* Schrank, 1801 (Lepidoptera: Pieridae). *Entomologist's Gazette*, 69(3), 168–170.

Hart, N.S. (2001) The visual ecology of avian photoreceptors. *Progress in Retinal and Eye Research*, 20(5), 675–703.

Hasegawa, M. and Watanabe, M. (2008) Changes in mate preference of male *Colias erate poliographus* in relation to copulation experience of females. *Japan Journal of Entomology*, 11, 105–114.

Hashimoto, K. and Yara, O. (2007) Comparison of microstructure of pheromone transfer particles (PTPs) in the genus *Tirumala* (Lepidoptera, Nymphalidae, Danainae). *Transactions of the Lepidopterological Society of Japan*, 58(2), 177–182.

Hashimoto, K. and Yara, O. (2008) Comparative morphological study in the genus *Tirumala* (Lepidoptera, Nymphalidae, Danainae): sexual isolation and PTPs. *Lepidoptera Science*, 59(4), 305–311.

Häuser, C. (1993) Die inneren weiblichen Organe der Tagfalter (Rhopalocera): Vergleichende Morphologie und phylogenetische Interpretation (Insecta, Lepidoptera) *Zoologische Jahrbücher Syst.*, 120, 389–439.

Hayashi, N., Kuwahara, Y. and Komae, H. (1978) The scent scale substances of male *Pieris* butterflies (*Pieris melete* and *Pieris napi*). *Experientia*, 34(6), 684–685.

Hayashi, N., Nishi, A., Murakami, T., Maeshima, K., Komae, H. and Sakao, T. (1985) The scent substances of pierid butterflies (*Hebomoia glaucippe* Linnaeus) and the volatile components of their food plants (*Crataeva religiosa* Forst.). *Zeitschrift für Naturforschung C*, 40(1–2), 47–50.

Hazel, W.N. (1990) Sex-limited variability and mimicry in the swallowtail butterfly *Papilio polyxenes* Fabr. *Heredity*, 65, 109–114.

Heikkilä, M., Kaila, L., Mutanen, M., Pena, C. and Wahlberg, N. (2012) Cretaceous origin and repeated tertiary diversification of the redefined butterflies. *Proceedings of the Royal Society B: Biological Sciences*, 279(1731), 1093–1099.

Heikkilä, M., Mutanen, M., Wahlberg, N., Sihvonen, P. and Kaila, L. (2015) Elusive ditrysian phylogeny: an account of combining systematized morphology with molecular data (Lepidoptera). *BMC Evolutionary Biology*, 15(1), 260.

Heitzman R. (1964) The story of a 'mixed up' *Thorybes pylades* (Hesperiidae). *Journal of the Lepidopterists' Society*, 18, 169–170.

Henning, S.F. (1990) Courtship and mating in butterflies. *The Entomologists' Record and Journal of Variation*, 102, 205–213.

Hernandez, M.I.M. and Benson, W.W. (1998) Small-male advantage in the territorial tropical butterfly *Heliconius sara* (Nymphalidae): a paradoxical strategy? *Animal Behaviour*, 56, 533–540.

Hernández-Roldán, J.L., Bofill, R., Dapporto, L., Munguira, M.L. and Vila, R. (2014) Morphological and chemical analysis of male scent organs in the butterfly genus *Pyrgus* (Lepidoptera: Hesperiidae). *Organisms Diversity and Evolution*, 14(3), 269–278.

Herrell, J. and Hazel, W. (1995) Female-limited variability in mimicry in the swallowtail butterfly *Papilio polyxenes* Fabr. *Heredity*, 75(1), 106–110.

Heuskin, S., Vanderplanck, M., Bacquet, P., Holveck, M.J., Kaltenpoth, M., Engl, T., Pels, C., Taverne, C., Lognay, G. and Nieberding, C.M. (2014) The composition of cuticular compounds indicates body parts, sex and age in the model butterfly *Bicyclus anynana* (Lepidoptera). *Frontiers in Ecology and Evolution*, 2, 37. doi.org/10.3389/fevo.2014.00037

Hidaka, T. (1973) Logic of mating behavior of Lepidoptera. *Annals of the New York Academy of Sciences*, 223(1), 70–76.

Hidaka, T and Yamashita, K. (1975) Wing color pattern as the releaser of mating behavior in the swallowtail butterfly, *Papilio xuthus* L. (Lepidoptera: Papilionidae). *Applied Entomology and Zoology*, 10(4), 263–267.

Hill, G.E. (1994) Trait elaboration via adaptive mate choice: sexual conflict in the evolution of signals of male quality. *Ethology Ecology and Evolution*, 6, 351–370.

Hill Jr, H.F., Wenner, A.M. and Wells, P.H. (1976) Reproductive behavior in an overwintering aggregation of monarch butterflies. *American Midland Naturalist*, Jan 1, 10–19.

Hiroki, M. and Kato, Y. (1996) Age-related sexual receptivity change in virgin females of a butterfly, *Eurema hecabe*. *Applied Entomology and Zoology*, 31(3), 455–458.

Hiroki, M., Obara, Y. and Kato, Y. (1998) Changes in age-related reproductive tactics in the female of the butterfly, *Eurema hecabe*. *The Science of Nature*, 85(11), 551–552.

Hirota, T. and Obara, Y. (2000) The influence of air temperature and sunlight intensity on mate-locating behavior of *Pieris rapae crucivora*. *Zoological Science*, 17(8), 1081–1087.

Hirota, T., Hamano, K. and Obara, Y. (2001) The influence of female post-emergence behavior on the time schedule of male mate-locating in *Pieris rapae crucivora*. *Zoological Science*, 18(4), 475–482.

Hiroyoshi, S. and Reddy, G. (2018) Field and laboratory studies on the ecology, reproduction, and adult diapause of the Asian comma butterfly, *Polygonia c-aureum* L. (Lepidoptera: Nymphalidae). *Insects*, 9(4), 169. doi:10.3390/insects9040169

Honda, K. (1980) Odor of a papilionid butterfly. *Journal of Chemical Ecology*, 6(5), 867–873.

Honda, K., Tada, A. and Hayashil, N. (1995) Dihydropyrrolizines from the male scent-producing organs of a danaid butterfly, *Ideopsis similis* (Lepidoptera: Danaidae) and the morphology of alar scent organs. *Applied Entomology and Zoology*, 30(3), 471–477.

Honda, K., Honda, Y., Yamamoto, S. and Ômura, H. (2005) Differential utilization of pyrrolizidine alkaloids by males of a danaid butterfly, *Parantica sita*, for the production of danaidone in the alar scent organ. *Journal of Chemical Ecology*, 31(4), 959–964.

Honda, K., Honda, Y., Matsumoto, J., Tsuruta, Y., Yagi, W., Ômura, H. and Honda, H. (2016) Production and sex-pheromonal activity of alkaloid-derived androconial compounds in the danaine butterfly, *Parantica sita* (Lepidoptera: Nymphalidae: Danainae). *Biological Journal of the Linnean Society*, 119(4), 1036–1059.

Honda, K., Matsumoto, J., Sasaki, K., Tsuruta, Y. and Honda, Y. (2018) Uptake of plant-derived specific alkaloids allows males of a butterfly to copulate. *Scientific Reports*, 8(1), 5516.

Honda, Y. (2006) Major components in the hairpencil secretion of a butterfly, *Euploea mulciber* (Lepidoptera, Danaidae): their origins and male behavioral responses to pyrrolizidine alkaloids. *Journal of Insect Physiology*, 52, 1043–1053.

Hornett, E.A., Charlat, S., Duplouy, A.M., Davies, N., Roderick, G.K., Wedell, N. and Hurst, G.D. (2006) Evolution of male-killer suppression in a natural population. *PLoS Biology*, 4(9), e283.

Hosken, D.J. and Stockley, P. (2004) Sexual selection and genital evolution.*Trends in Ecology and Evolution*, 19(2), 87–93.

Hosken, D.J., Taylor, M.L., Hoyle, K. Higgins, S. and Wedell, N. (2008) Attractive males have greater success in sperm competition. *Current Biology*, 18, 553–554.

Hossie, T.J. and Sherratt, T.N. (2012) Eyespots interact with body colour to protect caterpillar-like prey from avian predators. *Animal Behaviour*, 84(1), 167–173.

Hossie, T.J. and Sherratt, T.N. (2014) Does defensive posture increase mimetic fidelity of caterpillars with eyespots to their putative snake models. *Current Zoology*, 60, 76–89.

Hossie, T.J., Skelhorn, J., Breinholt, J.W., Kawahara, A.Y. and Sherratt, T.N. (2015) Body size affects the evolution of eyespots in caterpillars. *Proceedings of the National Academy of Sciences*, 112(21), 6664–6669.

Hughes, L., Chang, B.S.W., Wagner, D. and Pierce, N.E. (2000) Effects of mating history on ejaculate size, fecundity, longevity, and copulation duration in the ant-tended lycaenid butterfly, *Jalmenus evagoras*. *Behavioral Ecology and Sociobiology*, 47(3), 119–128.

Huigens, M.E., Woelke, J.B., Pashalidou, F.G., Bukovinszky, T., Smid, H.M. and Fatouros, N.E. (2010) Chemical espionage on species-specific butterfly anti-aphrodisiacs by hitchhiking *Trichogramma* wasps. *Behavioral Ecology*, 21(3), 470–478.

Humphreys, R.K. and Ruxton, G.D. (2018) A review of thanatosis (death feigning) as an anti-predator behaviour. *Behavioral Ecology and Sociobiology*, 72(2), 22.

Huq, M., Bhardwaj, S. and Monteiro, A. (2019) Male *Bicyclus anynana* butterflies choose females on the basis of their ventral UV-reflective eyespot centers. *Journal of Insect Science*, 19(1), 25.

Ide, J.Y. (2010) Weather factors affecting the male mate-locating tactics of the small copper butterfly (Lepidoptera: Lycaenidae). *European Journal of Entomology*, 107(3), 369.

Ide, J.Y. (2011) Avoiding male harassment: wing-closing reactions to flying individuals by female small copper butterflies. *Ethology*, 117, 630–637.

Ide, J.Y. (2014) Age-related changes in the frequency of harassment-avoidance behaviour of virgin females of the small copper butterfly, *Lycaena phlaeas* (Lepidoptera: Lycaenidae). *European Journal of Entomology*, 111(3), 417.

Imafuku, M. (2013) Sexual differences in spectral sensitivity and wing colouration of 13 species of Japanese Thecline butterflies (Lepidoptera: Lycaenidae). *European Journal of Entomology*, 110(3), 435.

Imafuku, M. and Hirose, Y. (2016) Effect of bright wing color of males on other males in *Favonius taxila* (Lepidoptera: Lycaenidae) with sexual dimorphism in wing color. *Entomological Science*, 19(2), 138–141.

Imafuku, M. and Kitamura, T. (2015) Ability of males of two theclini species (Lepidoptera: Lycaenidae) to discriminate between sexes and different types of females based on the colour of their wings. *European Journal of Entomology*, 112(2): 328–333.

Imafuku, M. and Kitamura, T. (2018) Preference of virgin females for male wing color in a sexually dichromatic butterfly, *Pseudozizeeria maha* (Lycaenidae). *The Journal of the Lepidopterists' Society*, 72(3), 212–217.

Imafuku, M., Ohtani, T. and Takeuchi, T. (2000) Copulation of *Neozephyrus japonicus* (Lycaenidae) under captive conditions. *Lepidoptera Science*, 52(1), 1–10.

Imafuku M., Hirose Y. and Takeuchi T. (2002a) Wing colors of *Chrysozephyrus* butterflies (Lepidoptera; Lycaenidae): ultraviolet reflection by males. *Zoological Science*, 19, 175–183.

Imafuku, M., Gotoh, S. and Takeuchi, T. (2002b) Ultraviolet reflection by the male of *Narathura bazalus* (Lycaenidae). *Lepidoptera Science*, 53(4), 197–203.

Imafuku M., Shimizu I., Imai H. and Shichida Y. (2007) Sexual difference in color sense in a lycaenid butterfly, *Narathura japonica*. *Zoological Science*, 24, 611–613.

Imafuku M., Kubota H.Y. and Inouye K. (2012) Wing colors based on arrangement of the multilayer structure of wing scales in lycaenid butterflies (Insecta: Lepidoptera). *Journal of Entomological Science*, 15, 400–407.

Ingram, A.L. and Parker, A.R. (2008) A review of the diversity and evolution of photonic structures in butterflies, incorporating the work of John Huxley (the Natural History Museum, London from 1961 to 1990). *Philosophical Transactions of the Royal Society of London B: Biological Sciences*, 363(1502), 2465–2480.

Isaac, N.J., Mallet, J. and Mace, G.M. (2004) Taxonomic inflation: its influence on macroecology and conservation. *Trends in Ecology and Evolution*, 19(9), 464–469.

Jablonski, D. and Raup, D.M. (1995) Selectivity of end-Cretaceous marine bivalve extinctions. *Science*, 268(5209), 389–391.

Jain, A., Kunte, K. and Webb, E.L. (2016) Flower specialization of butterflies and impacts of non-native flower use in a transformed tropical landscape. *Biological Conservation*, 201,184–191.

Janowitz, S. A. and Fischer, K. (2010) Costing reproduction: effects of mating opportunity on mating success in male *Bicyclus anynana* butterflies. *Behavioral Ecology and Sociobiology*, 64(12), 1999–2006.

Janzen, D.H., Hallwachs, W. and Burns, J.M. (2010) A tropical horde of counterfeit predator eyes. *Proceedings of the National Academy of Sciences*, 107(26), 11659–11665.

Jiggins, C.D. (2016) *The Ecology and Evolution of Heliconius Butterflies*. Oxford University Press, Oxford, UK. Published to Oxford Scholarship Online, April 2017.

Jiggins, C.D., Naisbit, R.E., Coe, R.L. and Mallet, J. (2001) Reproductive isolation caused by colour pattern mimicry. *Nature*, 411, 302–305.

Jiggins, F.M., Hurst, G.D. and Majerus, M.E. (2000) Sex-ratio-distorting *Wolbachia* causes sex-role reversal in its butterfly host. *Proceedings of the Royal Society of London B: Biological Sciences*, 267(1438), 69–73.

Jiggins, F.M., Randerson, J.P., Hurst, G.D. and Majerus, M.E. (2002) How can sex ratio distorters reach extreme prevalences? Male-killing *Wolbachia* are not suppressed and have near-perfect vertical transmission efficiency in *Acraea encedon*. *Evolution*, 56(11), 2290–2295.

Jones, A.G. and Ratterman, N.L. (2009) Mate choice and sexual selection: what have we learned since Darwin?. *Proceedings of the National Academy of Sciences*, 106(Supplement 1), 10001–10008.

Jones, R.T., Poul, Y.L., Whibley, A.C., Mérot, C., ffrench-Constant, R.H. and Joron, M. (2013) Wing shape variation associated with mimicry in butterflies. *Evolution*, 67(8), 2323–2334.

Jones, R.T., Salazar, P.A., Jiggins, C.D. and Joron, M. (2012) Evolution of a mimicry supergene from a multilocus architecture. *Proceedings of the Royal Society of London B: Biological Sciences*, 279(1727), 316–325.

Joron, M. and Brakefield, P.M. (2003) Captivity masks inbreeding effects on male mating success in butterflies. *Nature*, 424(6945), 191.

Kaitala, A. and Wiklund, C. (1994a) Polyandrous female butterflies forage for matings. *Behavioral Ecology and Sociobiology*, 35(6), 385–388.

Kaitala, A. and Wiklund, C. (1994b) Female mate choice and mating costs in the polyandrous butterfly *Pieris napi* (Lepidoptera: Pieridae). *Journal of Insect Behavior*, 8(3), 355–363.

Kan, E. and Hidaka, T. (1997) Role of male scent in the mating behavior of *Pieris melete* Ménétriès (Lepidoptera: Pieridae). *Journal of Ethology*, 15(2), 87–93.

Karl, I. and Fischer, K. (2013) Old male mating advantage results from sexual conflict in a butterfly. *Animal Behaviour*, 85(1), 143–149.

Karl, I., Heuskin, S. and Fischer, K. (2013) Dissecting the mechanisms underlying old male mating advantage in a butterfly. *Behavioral Ecology and Sociobiology*, 67(5), 837–849.

Karlson, P. and Lüscher, M. (1959) 'Pheromones': a new term for a class of biologically active substances. *Nature*, 183(4653), 55.

Karlsson, B. (1995) Resource allocation and mating systems in butterflies. *Evolution* 49(5), 955–961.

Karlsson, B. (1996) Male reproductive reserves in relation to mating system in butterflies: a comparative study. *Proceedings of the Royal Society of London B: Biological Sciences*, 263(1367), 187–192.

Karlsson, B. (1998) Nuptial gifts, resource budgets, and reproductive output in a polyandrous butterfly. *Ecology* 79(8), 2931–2940.

Karlsson, B., Leimar, O. and Wiklund, C. (1997) Unpredictable environments, nuptial gifts and the evolution of sexual size dimorphism in insects: an experiment. *Proceedings of the Royal Society of London B: Biological Sciences*, 264(1381), 475–479.

Kassarov, L. (2003) Are birds the primary selective force leading to evolution of mimicry and aposematism in butterflies? An opposing point of view. *Behaviour*, 140(4), 433–451.

Kather, R. and Martin, S.J. (2012) Cuticular hydrocarbon profiles as a taxonomic tool: advantages, limitations and technical aspects. *Physiological Entomology*, 37(1), 25–32.

Kato, Y. and Nakane, T. (1989) Male approach to pupae in the yellow butterfly, *Eurema hecabe*. *Journal of Ethology*, 7(1), 59–61.

Kato, Y. and Yamada, H. (2001) Effect of light irradiation on blue-green coloration of larval integument and adult wing in the butterfly *Graphium sarpedon* (Lepidoptera, Papilionidae). *Lepidoptera Science*, 52(4), 279–287.

Kato, Y. and Yoshioka, Y. (2003) Visual stimuli affecting male mating behavior of *Graphium sarpedon* (Lepidoptera, Papilionidae). *Lepidoptera Science*, 54(4), 209–219.

Kawagoe, T., Suzuki, N. and Matsumoto, K. (2001) Multiple mating reduces longevity of females of the windmill butterfly *Atrophaneura alcinous*. *Ecological Entomology*, 26(3), 258–262.
Kawahara, A.Y., and Breinholt, J.W. (2014) Phylogenomics provides strong evidence for relationships of butterflies and moths. *Proceedings of the Royal Society B: Biological Sciences*, 281, 20140970.
Kawasaki, M., Kinoshita, M., Weckström, M. and Arikawa, K. (2015) Difference in dynamic properties of photoreceptors in a butterfly, *Papilio xuthus*: possible segregation of motion and color processing. *Journal of Comparative Physiology A* 201(12), 1115–1123.
Kazama, M., Ichinei, M., Endo, S., Iwata, M., Hino, A. and Otaki, J.M. (2017) Species-dependent microarchitectural traits of iridescent scales in the triad taxa of *Ornithoptera* birdwing butterflies. *Entomological Science*, 20, 255–269.
Kehl, T. (2015) Old male mating advantage in Bicyclus anynana butterflies-sexual conflict or cooperation? Doctoral dissertation, The University of Greifswald, Greifswald, Germany.
Kehl, T., Karl, I. and Fischer, K. (2013) Old-male paternity advantage is a function of accumulating sperm and last-male precedence in a butterfly. *Molecular Ecology*, 22(16), 4289–4297.
Kehl, T., Burmeister, M.F., Donke, E., Köhn, N.A., Metschke, K., Pfender, D., Karl, I. and Fischer, K. (2014) Pheromone blend does not explain old male mating advantage in a butterfly. *Ethology*, 120(11), 1137–1145.
Kehl, T., Beaulieu, M., Kehl, A. and Fischer, K. (2015a) Old male sex: large ejaculate, many sperm, but few offspring. *Behavioral Ecology and Sociobiology*, 69(9), 1543–1552.
Kehl, T., Bensch, J., Böhm, F., Kniepkamp, B.O., Leonhardt, V., Schwieger, S. and Fischer, K. (2015b) Fat and sassy: factors underlying male mating success in a butterfly. *Entomologia Experimentalis et Applicata*, 155(3), 257–265.
Kehl, T., Dublon, I.A. and Fischer, K. (2015c) Young male mating success is associated with sperm number but not with male sex pheromone titres. *Frontiers in Zoology*, 12(1), 31.
Kelber, A. (1999) Why 'false' colours are seen by butterflies. *Nature*, 402(6759), 251–251.
Kelber, A. and Osorio, D. (2010) From spectral information to animal colour vision: experiments and concepts. *Proceedings of the Royal Society of London B: Biological Sciences*, 277(1688), 1617–1625.
Kelber, A., Thunell, C. and Arikawa, K. (2001) Polarisation-dependent colour vision in *Papilio* butterflies. *Journal of Experimental Biology*, 204(14), 2469–2480.
Kemal, M. and Koçak, A.Ö. (2013) *Stamnodes depeculata* (Geometridae, Lepidoptera) in Van Province (East Turkey) with faunistical, ecological and taxonomical notes including accompanied species. *Cesa News*, 92, 15–40.
Kemp, D.J. (2000a) Contest behavior in territorial male butterflies: does size matter? *Behavioral Ecology*, 11(6), 591–596.
Kemp, D.J. (2000b) Butterfly contests: neither paradoxical nor contradictory. *Animal Behaviour*, 60, 44–46.
Kemp, D.J. (2001) Investigating the consistency of mate-locating behavior in the territorial butterfly *Hypolimnas bolina* (Lepidoptera: Nymphalidae). *Journal of Insect Behavior*, 14(1), 129–147.
Kemp, D.J. (2002a) Sexual selection constrained by life history in a butterfly. *Proceedings of the Royal Society of London B: Biological Sciences*, 269(1498), 1341–1345.
Kemp, D.J. (2002b) Butterfly contests and flight physiology: why do older males fight harder? *Behavioral Ecology*, 13(4), 456–461.
Kemp, D.J. (2002c) Visual mate-searching behaviour in the evening brown butterfly, *Melanitis leda* (L.) (Lepidoptera: Nymphalidae). *Austral Entomology*, 41(4), 300–305.

Kemp, D.J. (2003) Twilight fighting in the evening brown butterfly, *Melanitis leda* (L.) (Nymphalidae): age and residency effects. *Behavioral Ecology and Sociobiology*, 54(1), 7–13.

Kemp, D.J. (2006a) Heightened phenotypic variation and age-based fading of ultraviolet butterfly wing coloration. *Evolutionary Ecology Research*, 8(3), 515–527.

Kemp, D.J. (2006b) Contrasting lifetime patterns of territorial success in the nymphalid butterflies *Hypolimnas bolina* and *Melanitis leda*: a question of flight physiology? *Australian Journal of Zoology*, 53(6), 361–367.

Kemp, D.J. (2007a) Female butterflies prefer males bearing bright iridescent ornamentation. *Proceedings of the Royal Society of London B: Biological Sciences*, 274(1613), 1043–1047.

Kemp, D.J. (2007b) Female mating biases for bright ultraviolet iridescence in the butterfly *Eurema hecabe* (Pieridae). *Behavioral Ecology*, 19(1), 1–8.

Kemp, D.J. (2013) Contest behaviour in butterflies: fighting without weapons. In: Hardy, I.C.W. and Briffa, M. (eds), *Animal Contests*. Cambridge University Press, Cambridge, UK, pp. 134–146.

Kemp, D. (2018) Habitat selection and territoriality. In: Córdoba-Aguilar, A., González-Tokman, D. and González-Santoyo, I. (eds), *Insect Behavior: From Mechanisms to Ecological and Evolutionary Consequences*. Oxford University Press, Oxford, UK, pp. 80–97.

Kemp, D.J. and Krockenberger, A.K. (2004) Behavioural thermoregulation in butterflies: the interacting effects of body size and basking posture in *Hypolimnas bolina* (L.) (Lepidoptera: Nymphalidae). *Australian Journal of Zoology*, 52(3), 229–239.

Kemp, D.J. and Macedonia, J.M. (2006) Structural ultraviolet ornamentation in the butterfly *Hypolimnas bolina* L. (Nymphalidae): visual, morphological and ecological properties. *Australian Journal of Zoology*, 54(4), 235–244.

Kemp, D.J. and Rutowski, R.L. (2001) Spatial and temporal patterns of territorial mate locating behaviour in *Hypolimnas bolina* (L.) (Lepidoptera: Nymphalidae). *Journal of Natural History*, 35(9), 1399–1411.

Kemp, D.J. and Rutowski, R.L. (2004) A survival cost to mating in a polyandrous butterfly, *Colias eurytheme*. *Oikos*, 105(1), 65–70.

Kemp, D.J. and Rutowski, R.L. (2007) Condition dependence, quantitative genetics, and the potential signal content of iridescent ultraviolet butterfly coloration. *Evolution*, 61(1), 168–183.

Kemp, D.J. and Rutowski, R.L. (2011) The role of coloration in mate choice and sexual interactions in butterflies. *Advances in the Study of Behavior*, 43, 55–92.

Kemp, D.J. and Wiklund, C. (2001) Fighting without weaponry: a review of male-male contest competition in butterflies. *Behavioral Ecology and Sociobiology*, 49(6), 429–442.

Kemp, D.J. and Wiklund, C. (2004) Residency effects in animal contests. *Proceedings of the Royal Society of London. Series B: Biological Sciences*, 271(1549), 1707–1711.

Kemp, D.J., Rutowski, R.L. and Mendoza, M. (2005) Colour pattern evolution in butterflies: a phylogenetic analysis of structural ultraviolet and melanic markings in North American sulphurs. *Evolutionary Ecology Research*, 7(1), 133–141.

Kemp, D.J., Vukusic, P. and Rutowski, R.L. (2006) Stress-mediated covariance between nano-structural architecture and ultraviolet butterfly coloration. *Functional Ecology*, 20(2), 282–289.

Kemp, D.J., Macedonia, J.M., Ball, T.S. and Rutowski, R.L. (2008) Potential direct fitness consequences of ornament-based mate choice in a butterfly. *Behavioral Ecology and Sociobiology*, 62(6), 1017–1026.

Kemp, D.J., Herberstein, M.E. and Grether, G.F. (2011) Unraveling the true complexity of costly color signaling. *Behavioral Ecology*, 23(2), 233–236.

Kemp, D.J., Jones, D., Macedonia, J.M. and Krockenberger, A. K. (2014) Female mating preferences and male signal variation in iridescent *Hypolimnas* butterflies. *Animal Behaviour*, 87, 221–229.

Kemp, D.J., Herberstein, M.E., Fleishman, L.J., Endler, J.A., Bennett, A.T., Dyer, A.G., Hart, N.S., Marshall, J. and Whiting, M.J. (2015) An integrative framework for the appraisal of coloration in nature. *The American Naturalist*, 185(6), 705–724.

Kerkut, G.A. (ed.) (2013) *Biochemistry* (Vol. 10). Elsevier, Oxford, UK.

Kertész, K., Bálint, Z., Vértesy, Z., Márk, G.I., Lousse, V., Vigneron, J.P., Rassart, M. and Biró, L.P. (2006) Gleaming and dull surface textures from photonic-crystal-type nanostructures in the butterfly *Cyanophrys remus*. *Physical Review E*, 74(2), 021922.

Kertész, K., Molnár, G., Vértesy, Z., Koós, A.A., Horváth, Z.E., Márk, G.I., Tapasztó, L., Bálint, Z., Tamáska, I., Deparis, O. and Vigneron, J.P. (2008) Photonic band gap materials in butterfly scales: a possible source of 'blueprints'. *Materials Science and Engineering: B*, 149(3), 259–265.

Kingsolver, J.G. (1985) Butterfly thermoregulation: organismic. *Journal of Research on the Lepidoptera*, 24(1), 1–20.

Kinoshita, M. and Arikawa, K. (2000) Colour constancy in the swallowtail butterfly *Papilio xuthus*. *Journal of Experimental Biology*, 203(23), 3521–3530.

Kinoshita, S. and Yoshioka, S. (2005) Structural colors in nature: the role of regularity and irregularity in the structure. *ChemPhysChem*, 6(8), 1442–1459.

Kinoshita, M., Shimada, N. and Arikawa, K. (1999) Colour vision of the foraging swallowtail butterfly *Papilio xuthus*. *Journal of Experimental Biology*, 202(2), 95–102.

Kinoshita, S., Yoshioka, S., Fujii, Y. and Okamoto, N. (2002a) Photophysics of structural color in the *Morpho* butterflies. *Forma*, 17(2), 103–121.

Kinoshita, S., Yoshioka, S. and Kawagoe, K. (2002b) Mechanisms of structural colour in the *Morpho* butterfly: cooperation of regularity and irregularity in an iridescent scale. *Proceedings of the Royal Society of London B: Biological Sciences*, 269 (1499), 1417–1421.

Kinoshita, M., Takahashi, Y. and Arikawa, K. (2008) Simultaneous color contrast in the foraging swallowtail butterfly, *Papilio xuthus*. *Journal of Experimental Biology*, 211(21), 3504–3511.

Kinoshita, M., Takahashi, Y. and Arikawa, K. (2012) Simultaneous brightness contrast of foraging *Papilio* butterflies. *Proceedings of the Royal Society of London B: Biological Sciences*, 279(1735), 1911–1918.

Kitamura, T. and Imafuku, M. (2015) Behavioural mimicry in flight path of Batesian intraspecific polymorphic butterfly *Papilio polytes*. In *Proceedings of the Royal Society of London B: Biological Sciences*, 282(1809), 20150483

Kitching, R.L., Eastwood, R.G. and Hurley, K. (2001) Butterflies and Wallace's Line: faunistic patterns and explanatory hypotheses within the south-east Asian butterflies. In: Metcalfe, I., Smith, C.R., Morwood, M. and Davidson, I. (eds), *Faunal and Floral Migrations and Evolution in SE Asia–Australasia*. A.A. Balkema Publishers, Lisse, The Netherlands, pp. 269–286.

Kivelä, S.M. and Välimäki, P. (2008) Competition between larvae in a butterfly *Pieris napi* and maintenance of different life-history strategies. *Journal of Animal Ecology* 77, 529–539.

Klein, A.L. and De Araújo, A.M. (2010) Courtship behavior of *Heliconius erato phyllis* (Lepidoptera, Nymphalidae) towards virgin and mated females: conflict between attraction and repulsion signals? *Journal of Ethology*, 28(3), 409–420.

Klein, A.L. and de Araújo, A.M. (2013) Sexual size dimorphism in the color pattern elements of two mimetic *Heliconius* butterflies. *Neotropical Entomology*, 42(6), 600–606.

Knapton, R.W. (1985) Lek structure and territoriality in the chryxus arctic butterfly, *Oeneis chryxus* (Satyridae). *Behavioral Ecology and Sociobiology*, 17(4), 389–395.

Knight, K. (2015) Female butterflies digest sperm packages in love duel. *Journal of Experimental Biology*, 218(10), 1463–1463.

Knüttel, H. and Fiedler, K. (2001) Host-plant-derived variation in ultraviolet wing patterns influences mate selection by male butterflies. *Journal of Experimental Biology*, 204(14), 2447–2459.

Kokko, H. and Rankin, D.J. (2006) Lonely hearts or sex in the city? Density-dependent effects in mating systems. *Philosophical Transactions of the Royal Society of London B: Biological Sciences*, 361(1466), 319–334.

Kolle, M., Salgard-Cunha, P.M., Scherer, M.R., Huang, F., Vukusic, P., Mahajan, S., Baumberg, J.J. and Steiner, U. (2010) Mimicking the colourful wing scale structure of the *Papilio blumei* butterfly. *Nature Nanotechnology*, 5(7), 511.

Komae, H., Nishi, A., Hayashi, N., Wesou, C. and Kuwahara, Y. (1983) Components of the sex brand secretions of some danaid butterflies. *Agricultural and Biological Chemistry*, 47(1), 157–159.

Koshitaka, H., Kinoshita, M., Vorobyev, M. and Arikawa, K. (2008) Tetrachromacy in a butterfly that has eight varieties of spectral receptors. *Proceedings of the Royal Society of London B: Biological Sciences*, 275(1637), 947–954.

Kotiaho, J.S. (2001) Costs of sexual traits: a mismatch between theoretical considerations and empirical evidence. *Biological Reviews*, 76(3), 365–376.

Kral, K. (2016) Implications of insect responses to supernormal visual releasing stimuli in intersexual communication and flower-visiting behaviour: a review. *European Journal of Entomology*, 113, 429–437.

Krebs, R.A. (1988) The mating behaviour of *Papilio glaucus* (Papilionidae). *Journal of Research on the Lepidoptera*, 26(1–4), 27–31.

Krebs, R.A. and West, D.A. (1988) Female mate preference and the evolution of female-limited Batesian mimicry. *Evolution*, 42(5), 1101–1104.

Kreutzer, M. and Aebischer, V. (2015) The riddle of attractiveness: looking for an 'aesthetic sense' within the hedonic mind of the beholders. In: Hoquet, T. (ed.), *Current Perspectives on Sexual Selection, History, Philosophy and Theory of the Life Sciences*, 9, 263–287. doi:10.1007/978–94–017–9585–2_12

Kristensen, N.P. and Simonsen, T.J. (2012) 'Hairs' and scales. In: Kristensen, N. (ed.), *Lepidoptera, Moths and Butterflies, Vol 2: Morphology, Physiology, and Development* (Vol. 4). Walter de Gruyter, Berlin, pp. 9–22.

Kristensen, N.P., Scoble, M.J. and Karsholt, O.L.E. (2007) Lepidoptera phylogeny and systematics: the state of inventorying moth and butterfly diversity. *Zootaxa*, 1668(1), 699–747.

Kronforst, M.R., Young, L.G., Kapan, D.D., McNeely, C., O'Neill, R.J. and Gilbert, L.E. (2006) Linkage of butterfly mate preference and wing color preference cue at the genomic location of *wingless*. *Proceedings of the National Academy of Sciences USA*, 103, 6575–6580.

Kronforst, M.R., Barsh, G.S., Kopp, A., Mallet, J., Monteiro, A., Mullen, S.P., Protas, M., Rosenblum, E.B., Schneider, C.J. and Hoekstra, H.E. (2012) Unraveling the thread of nature's tapestry: the genetics of diversity and convergence in animal pigmentation. *Pigment Cell and Melanoma Research*, 25(4), 411–433.

Kruuk, H. (2003) *Niko's Nature: the Life of Niko Tinbergen and His Science of Animal Behaviour*. Oxford University Press, Oxford, UK.

Kudrna, O., Harpke, A., Lux, K., Pennerstorfer, J., Schweiger, O., Settele, J. and Wiemers, M. (2011) *Distributions Atlas of Butterflies in Europe*. Gesellschaft für Schmetterlingsschutz eV, Halle, pp. i–iv.

Kuwahara, Y. (1979) Scent scale substances of male *Pieris melete* Ménétriès (Pieridae: Lepidoptera). *Applied Entomology and Zoology*, 14(3), 350–355.

Labhart, T., Baumann, F. and Bernard, G.D. (2009) Specialized ommatidia of the polarization-sensitive dorsal rim area in the eye of monarch butterflies have non-functional reflecting tapeta. *Cell and Tissue Research*, 338(3), 391–400.

Labine, P.A. (1966) The population biology of the butterfly, *Euphydrias editha*. IV. Sperm precedence, a preliminary report. *Evolution* 20, 580–586.

Land, M.F. and Nilsson, D.E. (2012) *Animal Eyes*. Oxford University Press, Oxford, UK.

Larsdotter-Mellström, H., Eriksson, K., Janz, N., Nylin, S. and Carlsson, M.A. (2016a) Male butterflies use an anti-aphrodisiac pheromone to tailor ejaculates. *Functional Ecology*, 30(2), 255–261.

Larsdotter-Mellström, H., Eriksson, K., Liblikas, I., Wiklund, C., Borg-Karlson, A.K., Nylin, S., Janz, N. and Carlsson, M.A. (2016b) It's all in the mix: blend-specific behavioral response to a sexual pheromone in a butterfly. *Frontiers in Physiology*, 7, 68.

Larsen, T.B. (1991) The art of feigning death – thanatosis in Euploea (Danainae) and other aposematic butterflies. *The Entomologist's Record and Journal of Variation*, 103, 263–266.

Lauwers, K. and Van Dyck, H. (2006) The cost of mating with a non-virgin male in a monandrous butterfly: experimental evidence from the speckled wood, *Pararge aegeria*. *Behavioral Ecology and Sociobiology*, 60(1), 69–76.

Lawrence, C., Vukusic, P. and Sambles, R. (2002) Grazing-incidence iridescence from a butterfly wing. *Applied Optics*, 41(3), 437–441.

Lawrence, J.M. and Samways, M.J. (2001) Hilltopping in *Hyalites encedon* (Lepidoptera: Acraeinae) butterflies. *Metamorphosis*, 12(2), 41–48.

Lawrence, J.M. and Samways, M.J. (2002) Influence of hilltop vegetation type on an African butterfly assemblage and its conservation. *Biodiversity and Conservation*, 11(7), 1163–1171.

Layberry, R.A., Hall, P.W. and Lafontaine, J.D. (1998) *The Butterflies of Canada*. University of Toronto Press, Toronto, Canada.

Lea, V. (2006) Black hairstreak *Satyrium pruni* L. (Lep.: Lycaenidae) courtship behaviour. *Entomologists Record and Journal of Variation*, 118(5), 209–210.

Lederhouse, R.C. (1982) Territorial defense and lek behavior of the black swallowtail butterfly, *Papilio polyxenes*. *Behavioral Ecology and Sociobiology*, 10(2), 109–118.

Lederhouse, R.C. (1993) Territoriality along flyways as male-locating behaviour in male *Limenitis arthemis* (Nymphalidae). *Journal of the Lepidopterists' Society*, 47, 22–31.

Lederhouse, R.C. and Scriber, J.M. (1996) Intrasexual selection constrains the evolution of the dorsal color pattern of male black swallowtail butterflies, *Papilio polyxenes*. *Evolution*, 50(2), 717–722.

Lederhouse, R.C., Finke, M.D. and Scriber, J.M. (1982) The contributions of larval growth and pupal duration to protandry in the black swallowtail butterfly, *Papilio polyxenes*. *Oecologia*, 53(3), 296–300.

Lederhouse, R.C., Codella, S.G., Grossmueller, D.W. and Maccarone, A.D. (1992) Host plant-based territoriality in the white peacock butterfly, *Anartia jatrophae* (Lepidoptera: Nymphalidae). *Journal of Insect Behavior*, 5(6), 721–728.

Lehnert, M.S., Emmel, T.C. and Garraway, E. (2013) Male-male interactions in the endangered *Homerus swallowtail*, *Papilio homerus* (Lepidoptera: Papilionidae), in Jamaica. *Caribbean Journal of Science*, 47(1), 57–66.

Leimar, O., Karlsson, B. and Wiklund, C. (1994) Unpredictable food and sexual size dimorphism in insects. *Proceedings of the Royal Society of London B: Biological Sciences*, 258(1352), 121–125.

Le Roy, C., Debat, V. and Llaurens, V. (2019) Adaptive evolution of butterfly wing shape: from morphology to behaviour. *Biological Reviews*. doi:10.1111/brv.12500

Levin, M.P. (1973) Preferential mating and the maintenance of the sex-limited dimorphism in *Papilio glaucus*: evidence from laboratory matings. *Evolution*, 27(2), 257–264.

Lewington, R. (2016) *Pocket Guide to the Butterflies of Great Britain and Ireland*. Bloomsbury Publishing, London.

Ley, C. and Watt, W.B. (1989) Testing the 'mimicry' explanation for the *Colias* 'alba' polymorphism: palatability of *Colias* and other butterflies to wild bird predators. *Functional Ecology*, 3(2), 183–192.

Li, C., Wang, F., Chen, X., Zhou, C. and Yao, J. (2015) Adult Behavior of *Tirumala limniace* (Lepidoptera: Danaidae). *Journal of Insect Science*, 15(1), 76.

Li, C., Wang, H., Chen, X., Yao, J., Shi, L. and Zhou, C. (2017) Role of visual and olfactory cues in sex recognition in butterfly *Cethosia cyane cyane*. *Scientific Reports*, 7(1), 5033.

Li, Y. and Mathews, R.A. (2016) In vivo real-time monitoring of aphrodisiac pheromone release of small white cabbage butterflies (*Pieris rapae*). *Journal of Insect Physiology*, 91, 107–112.

Liao, H., Shi, L., Liu, W., Du, T., Ma, Y., Zhou, C. and Deng, J. (2017) Effects of light intensity on the flight behaviour of adult *Tirumala limniace* (Cramer) (Lepidoptera: Nymphalidae: Danainae). *Journal of Insect Behavior*, 30(2), 139–154.

Liénard, M.A., Wang, H.L., Lassance, J.M. and Löfstedt, C. (2014) Sex pheromone biosynthetic pathways are conserved between moths and the butterfly *Bicyclus anynana*. *Nature Communications*, 5, 3957.

Lindfors, V.S. (1998) Butterfly Life History and Mating Systems. Doctoral dissertation, Department of Zoology, Stockholm University, Sweden.

Lipson, R.H. and Lu, C. (2009) Photonic crystals: a unique partnership between light and matter. *European Journal of Physics*, 30(4), S33.

Liu, F., Wang, G., Jiang, L. and Dong, B. (2010) Structural colouration and optical effects in the wings of *Papilio peranthus*. *Journal of Optics*, 12(6), 065301.

Lloyd, J.E. (1979) Sexual selection in luminescent beetles. In: Blum, M. and Blum, N. (eds), *Sexual Selection and Reproductive Competition in Insects*. Academic Press, New York, pp. 293–342.

Lorenz, K. (1935) Der kumpan in der umwelt des vogels. *Journal für Ornithologie*, 83(3), 289–413.

Lorković, Z. (1993) *Leptidea reali* Reissinger, 1989 (=*lorkovicii* Real 1988), a new European species (Lepid., Pieridae). *Natura Croatica*, 2, 1–26.

Low, X.H. and Monteiro, A. (2018) Dorsal forewing white spots of male *Papilio polytes* (Lepidoptera: Papilionidae) not maintained by female mate choice. *Journal of Insect Behavior*, 31(1), 29–41.

Lucas, K.M., Windmill, J.F., Robert, D. and Yack, J.E. (2009) Auditory mechanics and sensitivity in the tropical butterfly *Morpho peleides* (Papilionoidea, Nymphalidae). *Journal of Experimental Biology*, 212(21), 3533–3541.

Lucas, K.M., Mongrain, J.K., Windmill, J F., Robert, D. and Yack, J. E. (2014) Hearing in the crepuscular owl butterfly (*Caligo eurilochus*, Nymphalidae). *Journal of Comparative Physiology A*, 200(10), 891–898.

Luke, S.M., Vukusic, P. and Hallam, B. (2009) Measuring and modelling optical scattering and the colour quality of white pierid butterfly scales. *Optics Express*, 17(17), 14729–14743.

Lundgren, L. (1977) Role of intra and interspecific male: male interactions in *Polyommatus icarus* Rott. and some other species of blues (Lycaenidae). *Journal of Research on the Lepidoptera*, 16(4), 249–264.

Lundgren, L. and Bergström, G. (1975) Wing scents and scent-released phases in the courtship behavior of *Lycaeides argyrognomon* (Lepidoptera: Lycaenidae). *Journal of Chemical Ecology*, 1(4), 399–412.

McCorkle, D.V. and Hammond, P.C. (1985) Observations on the biology of *Parnassius clodius* (Papilionidae) in the Pacific Northwest. *Journal of the Lepidopterists Society*, 39(3), 156–162.

McCulloch, K.J., Osorio, D. and Briscoe, A.D. (2016) Sexual dimorphism in the compound eye of *Heliconius erato:* a nymphalid butterfly with at least five spectral classes of photoreceptor. *Journal of Experimental Biology*, jeb-136523.

McDonald, A.K. and Nijhout, H.F. (2000) The effect of environmental conditions on mating activity of the buckeye butterfly, *Precis coenia*. *The Journal of Research on the Lepidoptera*, 35, 22–28.

Macias-Munoz, A., Smith, G. Monteiro, A. and Briscoe, A.D. (2016) Transcriptome-wide differential gene expression in *Bicyclus anynana* butterflies: female vision-related genes are more plastic. *Molecular Biology and Evolution*, 33(1), 79–92.

Macias-Munoz, A., Olguin, A.G.R. and Briscoe, A.D. (2019) Evolution of Phototransduction Genes in Lepidoptera. *BioRxiv*, 577031.

MacLean, D.B. and MacLean, B.K. (1990) *Lycaena hyllus* (Cramer) (Lycaenidae): new host and mating behavior in a drought-induced population explosion. *Journal of the Lepidopterists' Society*, 44(1), 33–34.

McMillan, W.O., Jiggins, C.D. and Mallet, J. (1997) What initiates speciation in passion-vine butterflies? *Proceedings of the National Academy of Sciences*, 94(16), 8628–8633.

McQueen, E.W. and Morehouse, N.I. (2018) Rapid divergence of wing volatile profiles between subspecies of the butterfly *Pieris rapae* (Lepidoptera: Pieridae). *Journal of Insect Science*, 18(2), 33.

Magnus, D.B.E. (1950) Beobachtungen zur Balz und Eiablage des Kaisermantels Argynnis paphia L. (Lep., Nymphalidae). *Zeitschriff für Tierpsychologie*, 7, 435–449.

Magnus, D.B.E. (1954) Experimentelle Untersuchungen am Kaisermantel zur Analyse optischer Auslösungsreize. *Deutscher Entomologentag* (Hamburg 1953), 58–75.

Magnus, D.B.E. (1958) Experimentelle Untersuchungen zur Bionomie und Ethologie des Kaisermantels *Argynnis paphia* L.(Lep. Nymph.) I. Über optische Auslöser von Anfliegereaktionen und ihre Bedeutung für das Sichfinden der Geschlechter. *Zeitschrift für Tierpsychologie*, 15(4), 397–426.

Magnus, D.B.E. (1963) Sex limited mimicry. II. Visual selection in the mate choice of butterflies. *Proceedings of the XVI International Congress of Zoology*, 4, 179–183.

Makino, K., Satoh, K., Koike, M. and Ueno, N. (1952) Sex in *Pieris rapae* L. and the pteridin content of their wings. *Nature*, 170(4335), 933.

Mallet, J. (2004) Perspectives Poulton, Wallace and Jordan: how discoveries in *Papilio* butterflies led to a new species concept 100 years ago. *Systematics and Biodiversity*, 1(4), 441–452.

Mancini, K. and Dolder, H. (2001) Ultrastructure of apyrene and eupyrene spermatozoa from the seminal vesicle of *Euptoieta hegesia* (Lepidoptera: Nymphalidae). *Tissue and Cell*, 33(3), 301–308.

Mann, F., Vanjari, S., Rosser, N., Mann, S., Dasmahapatra, K.K., Corbin, C., Linares, M., Pardo-Diaz, C., Salazar, C., Jiggins, C. and Schulz, S. (2017) The scent chemistry of *Heliconius* wing androconia. *Journal of Chemical Ecology*, 43(9), 843–857.

Marini-Filho, O.J. and Benson, W.W. (2010) Use of sound and aerial chases in sexual recognition in Neotropical *Hamadryas* butterflies (Nymphalidae). *Journal of Research on the Lepidoptera*, 42, 5–12.

Marshall, J. and Arikawa, K. (2014) Unconventional colour vision. *Current Biology*, 24(24), R1150–R1154.

Marshall, J., Roberts, N. and Cronin, T. (2014) Polarisation signals. In: Horváth, G. (ed.), *Polarized Light and Polarization Vision in Animal Sciences*. Springer, Berlin, Heidelberg, pp. 407–442. doi:10.1007/978-3-642-54718-8

Marshall, L. (1982a) Male courtship persistence in *Colias philodice* and *C. eurytheme* (Lepidoptera: Pieridae). *Journal of the Kansas Entomological Society*, 55(4), 729–736.

Marshall, L.D. (1982b) Male nutrient investment in the Lepidoptera: what nutrients should males invest? *The American Naturalist*, 120(2), 273–279.

Martínez-Lendech, N., Córdoba-Aguilar, A. and Serrano-Meneses, M.A. (2007) Body size and fat reserves as possible predictors of male territorial status and contest outcome in the butterfly *Eumaeus toxea* Godart (Lepidoptera: Lycaenidae). *Journal of Ethology*, 25(2), 195–199.

Martins, A.R.P., Faynel, C. and Robbins, R.K. (2016) Variation of male secondary sexual structures and the taxonomy of *Theritas lisus* and relatives (Lepidoptera, Lycaenidae, Eumaeini). *Proceedings of the Entomological Society of Washington*, 118(4), 555–574.

Martins, A.R.P., Duarte, M. and Robbins, R.K. (2018) Hairstreak butterflies (Lepidoptera, Lycaenidae) and evolution of their male secondary sexual organs. *Cladistics*, 0, 1–25. 10.1111/cla.12355

Matějková-Plšková, J., Shiojiri, S. and Shiojiri, M. (2009) Fine structures of wing scales in *Sasakia charonda* butterflies as photonic crystals. *Journal of Microscopy*, 236(2), 88–93.
Matějková-Plšková, J., Mika, F., Shiojiri, S. and Shiojiri, M. (2011) Fine structure of wing scales in *Chrysozephyrus* Ataxus butterflies. *Materials Transactions*, 52(3), 297–303.
Matin, T.R., Leong, M.K., Majlis, B.Y. and Gebeshuber, I.C. (2010) Correlating nanostructures with function: structural colors on the wings of a Malaysian bee. *AIP Conference Proceedings*, 1284(1), 5–14.
Matsumoto K. (1987) Mating patterns of a sphragis-bearing butterfly *Luehdorfia japonica* Leech (Lepidoptera: Papilionidae), with descriptions of mating behaviour. *Researches on Population Ecology*, 29, 97–110. doi:10.1007/BF02515428
Matsumoto, K. and Suzuki, N. (1992) Effectiveness of the mating plug in *Atrophaneura alcinous* (Lepidoptera: Papilionidae). *Behavioral Ecology and Sociobiology*, 30(3–4), 157–163. doi:10.1111/j.1600–0587.2001.tb00494.x
Matsumoto, K., Orr, A.G. and Yago, M. (2018) The occurrence and function of the sphragis in the zerynthiine genera *Zerynthia*, *Allancastria* and *Bhutanitis* (Lepidoptera: Papilionoidea: Papilionidae). *Journal of Natural History*, 52, 21–22, 1351–1376, DOI: 10.1080/00222933.2018.1464228
Matsushita, A., Awata, H., Wakakuwa, M., Takemura, S.Y. and Arikawa, K. (2012) Rhabdom evolution in butterflies: insights from the uniquely tiered and heterogeneous ommatidia of the glacial apollo butterfly, *Parnassius glacialis*. *Proceedings of the Royal Society of London B: Biological Sciences*, rspb20120475.
Matter, S.F., Reed, B., Illerbrun, K., Doyle, A., McPike, S. and Roland, J. (2012) Young love? Mating of *Parnassius smintheus* Doubleday (Papilionidae). *The Journal of the Lepidopterists' Society*, 66(2), 111–113.
Mavárez, J., Salazar, C.A., Bermingham, E., Salcedo, C., Jiggins, C.D. and Linares, M. (2006) Speciation by hybridization in *Heliconius* butterflies. *Nature*, 441(7095), 868.
Mays Jr, H.L. and Hill, G.E. (2004) Choosing mates: good genes versus genes that are a good fit. *Trends in Ecology and Evolution*, 19(10), 554–559.
Measures, D.G. (1976) *Bright Wings of Summer: Watching Butterflies*. Prentice-Hall, London.
Mega, N.O. and Araújo, A.M. (2010) Analysis of the mating behavior and some possible causes of male copulatory success in *Dryas iulia alcionea* (Lepidoptera, Nymphalidae, Heliconiinae). *Journal of Ethology*, 28, 123–132.
Meinwald, J., Meinwald, Y.C. and Mazzocchi, P.H. (1969) Sex pheromone of the queen butterfly: chemistry. *Science*, 164(3884), 1174–1175.
Mellström, H.L. and Wiklund, C. (2009) Males use sex pheromone assessment to tailor ejaculates to risk of sperm competition in a butterfly. *Behavioral Ecology*, 20(5), 1147–1151.
Mendoza-Cuenca, L., and Macías-Ordóñez, R. (2009) Female asynchrony may drive disruptive sexual selection on male mating phenotypes in a *Heliconius* butterfly. *Behavioral Ecology*, 21(1), 144–152.
Merrill, R.M., Van Schooten, B., Scott, J.A. and Jiggins, C.D. (2011) Pervasive genetic associations between traits causing reproductive isolation in *Heliconius* butterflies. *Proceedings of the Royal Society of London B: Biological Sciences*, 278(1705), 511–518.
Mérot, C., Frérot, B., Leppik, E. and Joron, M. (2015) Beyond magic traits: multimodal mating cues in *Heliconius* butterflies. *Evolution*, 69(11), 2891–2904.
Merry, J.W., Morehouse, N.I., Yturralde, K. and Rutowski, R.L. (2006) The eyes of a patrolling butterfly: visual field and eye structure in the orange sulphur, *Colias eurytheme* (Lepidoptera, Pieridae). *Journal of Insect Physiology*, 52(3), 240–248.
Meslin, C., Plakke, M.S., Deutsch, A.B., Small, B.S., Morehouse, N.I. and Clark, N.L. (2015) Digestive organ in the female reproductive tract borrows genes from multiple organ systems to adopt critical functions. *Molecular Biology and Evolution*, 32(6), 1567–1580.

Meslin, C., Cherwin, T.S., Plakke, M.S., Hill, J., Small, B.S., Goetz, B.J., Wheat, C.W., Morehouse, N.I. and Clark, N.L. (2017) Structural complexity and molecular heterogeneity of a butterfly ejaculate reflect a complex history of selection. *Proceedings of the National Academy of Sciences*, 114(27), E5406–E5413.

Meyer-Rochow, V.B. (1991) Differences in ultraviolet wing patterns in the New Zealand lycaenid butterflies *Lycaena salustius, L. rauparaha,* and *L. feredayi* as a likely isolating mechanism. *Journal of the Royal Society of New Zealand*, 21(2), 169–177.

Meyer-Rochow, V.B. and Järvilehto, M. (1997) Ultraviolet colours in *Pieris napi* from northern and southern Finland: arctic females are the brightest! *Naturwissenschaften*, 84(4), 165–168.

Michielsen, K. and Stavenga, D.G. (2008) Gyroid cuticular structures in butterfly wing scales: biological photonic crystals. *Journal of The Royal Society Interface* 5(18), 85–94.

Michielsen, K., De Raedt, H. and Stavenga, D.G. (2009) Reflectivity of the gyroid biophotonic crystals in the ventral wing scales of the green hairstreak butterfly, *Callophrys rubi*. *Journal of the Royal Society Interface*, rsif20090352.

Milam, E.L. (2010) *Looking for a Few Good Males: Female Choice in Evolutionary Biology*. The Johns Hopkins University Press, Baltimore, MD.

Miller, L.D. and Clench, H.K. (1968) Some aspects of mating behavior in butterflies. *Journal of the Lepidopterists' Society*, 22, 125–132.

Misof, B., Liu, S., Meusemann, K., Peters, R.S., Donath, A., Mayer, C., Frandsen, P.B., Ware, J., Flouri, T., Beutel, R.G. and Niehuis, O. (2014) Phylogenomics resolves the timing and pattern of insect evolution. *Science*, 346(6210), 763–767.

Mitter, K.T., Larsen, T.B., de Prins, W., De Prins, J., Collins, S., Vande Weghe, G., Safian, S., Zakharov, E.V., Hawthorne, D.J., Kawahara, A.Y. and Regier, J.C. (2011) The butterfly subfamily Pseudopontiinae is not monobasic: marked genetic diversity and morphology reveal three new species of *Pseudopontia* (Lepidoptera: Pieridae). *Systematic Entomology*, 36(1), 139–163.

Molleman, F. (2010) Puddling: from natural history to understanding how it affects fitness. *Entomologia experimentalis et applicata*, 134(2), 107–113.

Møller, A. and Jennions, M. (2001) How important are direct fitness benefits of sexual selection? *Naturwissenschaften*, 88(10), 401–415.

Monge-Nájera, J., Hernández, F., González, M.I., Soley, J., Araya, J. and Zolla, S. (1998) Spatial distribution, territoriality and sound production by tropical cryptic butterflies (Hamadryas, Lepidoptera: Nymphalidae): implications for the 'industrial melanism' debate. *Revista de Biología Tropical*, 46(2), 297–330.

Monteiro, A., Tong, X., Bear, A., Liew, S.F., Bhardwaj, S., Wasik, B.R., Dinwiddie, A., Bastianelli, C., Cheong, W.F., Wenk, M.R. and Cao, H. (2015) Differential expression of ecdysone receptor leads to variation in phenotypic plasticity across serial homologs. *PLoS Genetics*, 11(9), e1005529.

Montejo-Kovacevich, G., Goral, T. and Huertas, B. (2017) Concealed androconial scales in metalmark butterflies (Lepidoptera: Riodinidae): new insights from confocal laser scanning microscopy. *Journal of Entomological Science*, 52(4), 332–339.

Moore, S.D. (1987) Male-biased mortality in the butterfly *Euphydryas editha*: a novel cost of mate acquisition. *The American Naturalist*, 130(2), 306–309.

Morehouse, N.I. and Rutowski, R.L. (2010) In the eyes of the beholders: female choice and avian predation risk associated with an exaggerated male butterfly color. *The American Naturalist*, 176(6), 768–784.

Morehouse, N.I., Vukusic, P. and Rutowski, R. (2006) Pterin pigment granules are responsible for both broadband light scattering and wavelength selective absorption in the wing scales of pierid butterflies. *Proceedings of the Royal Society B: Biological Sciences*, 274(1608), 359–366.

Moreira, G.R. and Mielke, C.G. (2010) A new species of *Neruda* Turner, 1976 from northeast Brazil (Lepidoptera: Nymphalidae, Heiconiinae, Heliconiini). *Nachrichten des Entomologischen Vereins Apollo, NF*, 31(1/2), 85–91.

Morrow, E.H., Arnqvist, G. and Pitnick, S. (2003) Adaptation versus pleiotropy: why do males harm their mates? *Behavioral Ecology*, 14(6), 802–806.

Mueller, C.J., Wahlberg, N. and Beheregaray, L.B. (2010) 'After Africa': the evolutionary history and systematics of the genus *Charaxes* Ochsenheimer (Lepidoptera: Nymphalidae) in the Indo-Pacific region. *Biological Journal of the Linnean Society*, 100(2), 457–481.

Myers J (1972) Pheromones and courtship behavior in butterflies. *American Zoologist*, 12, 545–551.

Myers, J. and Brower, L.P. (1969) A behavioural analysis of the courtship pheromone receptors of the queen butterfly, *Danaus gilippus berenice*. *Journal of Insect Physiology*, 15(11), 2117–2130.

Nakahara, S., Zacca, T., Huertas, B., Neild, A.F., Hall, J.P., Lamas, G., Holian, L.A., Espeland, M. and Willmott, K.R. (2018) Remarkable sexual dimorphism, rarity and cryptic species: a revision of the 'aegrota species group' of the Neotropical butterfly genus *Caeruleuptychia* with the description of three new species (Lepidoptera, Nymphalidae, Satyrinae). *Insect Systematics and Evolution*, 49(2), 130–182.

Nakai, M., Honda, K., Ueda, H. and Omura, H. (2001) Is androconial secretion of *Pieris napi* (Lepidoptera, Pieridae) chemically dimorphic? *Lepidoptera Science*, 52(4), 309–312.

Nakanishi, Y., Watanabe, M. and Ito, T. (1996) Differences in lifetime reproductive output and mating frequency of two female morphs of the sulfur butterfly, *Colias erate* (Lepidoptera: Pieridae). *Journal of Research on the Lepidoptera*, 35, 1–8.

Navez, S. and Ishii, M. (2007a) Discrimination and phenology of hilltopping butterflies on two mountains on the boundary between Osaka and Nara Prefectures, central Japan. *Lepidoptera Science*, 58(1), 109–125.

Navez, S. and Ishii, M. (2007b) Comparison of behavioral patterns of hilltopping butterflies on the summit, subsummit and slope of two mountains on the boundary between Osaka and Nara Prefectures, central Japan. *Lepidoptera Science*, 58(2), 127–144.

Neild, A.F.E. and Bálint, Z. (2014) Notes on the identity of *Evenus coronata* (Hewitson, 1865) (Lepidoptera: Lycaenidae: Theclinae: Eumaeini) with the description of a remarkably overlooked sibling species. *Tropical Lepidoptera Research*, 24(2): 105–120.

Nève, G. and Singer, M.C. (2008) Protandry and postandry in two related butterflies: conflicting evidence about sex-specific trade-offs between adult size and emergence time. *Evolutionary Ecology*, 22(6), 701–709.

Ng, S.Y., Bhardwaj, S. and Monteiro, A. (2017) Males become choosier in response to manipulations of female wing ornaments in dry season *Bicyclus anynana* butterflies. *Journal of Insect Science*, 17(4), 1–6.

Nieberding, C.M. and Holveck, M.J. (2017) Laboratory social environment biases mating outcome: a first quantitative synthesis in a butterfly. *Behavioral Ecology and Sociobiology*, 71(8), 117.

Nieberding, C.M., De Vos, H., Schneider, M.V., Lassance, J.M., Estramil, N., Andersson, J., Bång, J., Hedenström, E., Löfstedt, C. and Brakefield, P.M. (2008) The male sex pheromone of the butterfly *Bicyclus anynana*: towards an evolutionary analysis. *PLoS One*, 3(7), e2751.

Nieberding, C.M., Fischer, K., Saastamoinen, M., Allen, C.E., Wallin, E.A., Hedenström, E. and Brakefield, P.M. (2012) Cracking the olfactory code of a butterfly: the scent of ageing. *Ecology Letters*, 15(5), 415–424.

Nieberding, C.M., San Martin, G., Saenko, S., Allen, C.E., Brakefield, P.M. and Visser, B. (2018) Sexual selection contributes to partial restoration of phenotypic robustness in a butterfly. *Scientific Reports* 8(1), 14315.

Nielsen, M.G. and Watt, W.B. (1998) Behavioural fitness component effects of the alba polymorphism of *Colias* (Lepidoptera, Pieridae): resource and time budget analysis. *Functional Ecology*, 12(1), 149–158.

Nielsen, M.G. and Watt, W.B. (2000) Interference competition and sexual selection promote polymorphism in *Colias* (Lepidoptera, Pieridae). *Functional Ecology*, 14(6), 718–730.

Nijhout, H.F. (1991) *The Development and Evolution of Butterfly Wing Patterns*. Smithsonian Institution Press, Washington, DC.

Nijhout, H.F. (2003) Polymorphic mimicry in *Papilio dardanus*: mosaic dominance, big effects, and origins. *Evolution and Development*, 5(6), 579–592.

Nilsson, D.E. (1989) Optics and evolution of the compound eye. In: Stavenga, D.G. and Hardie, R.C. (eds), *Facets of Vision*.Springer Science & Business Media, Berlin, Heidelberg, Germany, pp. 30–73.

Nishida, R. (2017) Chemical ecology of poisonous butterflies: model or mimic? A paradox of sexual dimorphisms in Müllerian mimicry. In: Nijhout, H.F. (ed.), *Diversity and Evolution of Butterfly Wing Patterns*.Springer, Singapore, pp. 205–220.

Nishida, R., Schulz, S., Kim, C.S., Fukami, H., Kuwahara, Y., Honda, K. and Hayashi, N. (1996) Male sex pheromone of a giant danaine butterfly, *Idea leuconoe*. *Journal of Chemical Ecology*, 22(5), 949–972.

Niu, S., Li, B., Mu, Z., Yang, M., Zhang, J., Han, Z. and Ren, L. (2015) Excellent structure-based multifunction of Morpho butterfly wings: a review. *Journal of Bionic Engineering*, 12(2), 170–189.

Niu, S., Li, B., Ye, J., Mu, Z., Zhang, J., Liu, Y. and Han, Z. (2016) Angle-dependent discoloration structures in wing scales of *Morpho menelaus* butterfly. *Science China Technological Sciences*, 59(5), 749–755.

Nukmal, N., Kanedi, M., Soekardi, H. and Lukvitasari, L. (2017) Ethogram of mating and male mate-locating strategy of *Papilio peranthus* (Lepidoptera: Papilionidae) at Gita Persada butterfly park, Lampung, Indonesia. *Journal of Entomology and Zoology Studies*, 5(1), 869–874.

Nylin, S., Wiklund, C., Wickman, P.O. and Garcia-Barros, E. (1993) Absence of trade-offs between sexual size dimorphism and early male emergence in a butterfly. *Ecology*, 74(5), 1414–1427.

Obara, Y. (1970) Studies on the mating behavior of the white cabbage butterfly, *Pieris rapae crucivora* Boisduval. *Zeitschrift für vergleichende Physiologie*, 69(1), 99–116.

Obara, Y. (1987) Avoidance of maladaptive, precocious copulation in the cabbage white butterfly, *Pieris rapae crucivora*. *Journal of Insect Physiology*, 33(6), 403–406.

Obara, Y. and Hidaka, T. (1964) Mating behavior of the cabbage white *Pieris rapae crucivora*. I. The 'flutter response' of resting male to flying males. *Zoological Magazine Tokyo*, 73, 131–135.

Obara, Y. and Hidaka, T. (1968) Recognition of the female by the male, on the basis of ultra-violet reflection, in the white cabbage butterfly, *Pieris rapae crucivora* Boisduval. *Proceedings of the Japan Academy*, 44(8), 829–832.

Obara, Y. and Majerus, M.E.N. (2000) Initial mate recognition in the British cabbage butterfly, *Pieris rapae rapae*. *Zoological Science*, 17(6), 725–730.

Obara, Y. and Majerus, M.E.N. (2009) Inaccurate mate recognition as a mating strategy of a 'pioneer male'. *Proceedings of the Japan Academy, Series B*, 85(6), 198–203.

Obara, Y., Koshitaka, H. and Arikawa, K. (2008) Better mate in the shade: enhancement of male mating behaviour in the cabbage butterfly, *Pieris rapae crucivora*, in a UV-rich environment. *Journal of Experimental Biology*, 211(23), 3698–3702.

Oberhauser, K.S. (1988) Male monarch butterfly spermatophore mass and mating strategies. *Animal Behaviour*, 36, 1384–1388.

Oberhauser, K.S. (1989) Effects of spermatophores on male and female monarch butterfly reproductive success. *Behavioral Ecology and Sociobiology*, 25(4), 237–246.

Oberhauser, K.S. (1992) Rate of ejaculate breakdown and intermating intervals in monarch butterflies. *Behavioral Ecology and Sociobiology*, 31(5), 367–373.

Oberhauser, K. and Frey, D. (1999) Coercive mating by overwintering male monarch butterflies. In: Hoth, J., Merino, L., Oberhauser, K., Pisanty, I., Price, S. and Wilkinson, T. (eds), *The 1997 North American Conference on the Monarch Butterfly*. Commission for Environmental Cooperation, Montreal, Canada, pp. 67–78.

Ockenfels, P., Boppré, M., Fischer, O. W. and Schultz, S. (1998) Origin, transfer to female and identification of a likely male anti-male pheromone in the silver-washed fritillary, *Argynnis paphia* (Lepidoptera, Nymphalidae). In: *Poster from 15th Annual Meeting of the International Society of Chemical Ecology, Ithaca, New York June 20–24, 1998*. Available at: http://www.fzi.uni-freiburg.de/pdf/p-argynnis-ISCE.pdf.

Odendaal, F.J., Iwasa, Y. and Ehrlich, P.R. (1985) Duration of female availability and its effect on butterfly mating systems. *The American Naturalist*, 125(5), 673–678.

Odendaal, F.J., Turchin, P. and Stermitz, F.R. (1988) An incidental-effect hypothesis explaining aggregation of males in a population of *Euphydryas anicia*. *The American Naturalist*, 132(5), 735–749.

Odendaal, F.J., Turchin, P. and Stermitz, F.R. (1989) Influence of host-plant density and male harassment on the distribution of female *Euphydryas anicia* (Nymphalidae). *Oecologia*, 78(2), 283–288.

Odendaal, F.J., Jones, K.N. and Stermitz, F.R. (1990) Mating behavior and male investment in *Euphydryas anicia* (Lepidoptera: Nymphalidae). *The Journal of Research on the Lepidoptera (USA)*, 28(1–2), 1–13.

Ohguchi, O. and Hidaka, T. (1988) Mate recognition in two sympatric species of butterflies, *Pieris rapae* and *P. melete*. *Journal of Ethology*, 6(1), 49–53.

Ohno, Y. and Otaki, J.M. (2015) Spontaneous long-range calcium waves in developing butterfly wings. *BMC Developmental Biology*, 15(1), 17.

Ohtani, T. (1984) Data sampling by 'single-individual trailing method' and data processing. *Kogane-mushi* (Nagasaki), 44, 27–33. (In Japanese.)

Ohtani, T. (1985) The adult behavior of the Japanese cabbage white (Lepidoptera, Pieridae) in the field: I. Behavior repertoire observed. *Lepidoptera Science*, 35(4), 161–173.

Ohtani, T. and Yamamoto, M. (1985) The adult behavior of the Japanese cabbage white (Lepidoptera, Pieridae) in the field: II. Ecological aspects of major behavior patterns. *Lepidoptera Science*, 36(2), 43–76.

Okumura, Y., Ozeki, Y., Itoh, T., Ohta, S. and Ômura, H. (2016) Volatile terpenoids from male wings lacking scent scales in *Anthocharis scolymus* (Lepidoptera: Pieridae). *Applied Entomology and Zoology*, 51(3), 385–392.

Oliver, J.C. and Monteiro, A. (2011) On the origins of sexual dimorphism in butterflies. *Proceedings of the Royal Society B*, 278(1714), 1981–1988.

Oliver, J.C., Robertson, K.A. and Monteiro, A. (2009) Accommodating natural and sexual selection in butterfly wing pattern evolution. *Proceedings of the Royal Society of London B: Biological Sciences*, 276(1666), 2369–2375.

Ômura H. and Honda, K. (2005) Chemical composition of volatile substances from adults of the swallowtail, *Papilio polytes* (Lepidoptera: Papilionidae). *Applied Entomology and Zoology*, 40, 421–427.

Ômura, H. and Honda, K. (2011) Pungent odor of the adult skipper butterfly *Erynnis montanus* (Lepidoptera: Hesperiidae). *Applied Entomology and Zoology*, 46(2), 281–286.

Ômura, H. and Yotsuzuka, S. (2015) Male-specific epicuticular compounds of the sulfur butterfly *Colias erate poliographus* (Lepidoptera: Pieridae). *Applied Entomology and Zoology*, 50(2), 191–199.

Ômura, H., Honda, K. and Hayashi, N. (2001) Identification of odoriferous compounds from adults of a swallowtail butterfly, *Papilio machaon* (Lepidoptera: Papilionidae). *Zeitschrift für Naturforschung*, 56c, 1126–1134.

Ômura, H., Yanai, N. and Honda, K. (2012) Sexual dimorphism in scent substances and cuticular lipids of adult *Papilio protenor* butterflies. *Zeitschrift für Naturforschung C*, 67(5–6), 331–341.

Ômura, H., Yakumaru, K., Honda, K. and Itoh, T. (2013) Two lactones in the androconial scent of the lycaenid butterfly *Celastrina argiolus ladonides*. *Naturwissenschaften*, 100(4), 373–377.

Ômura, H., Itoh, T., Wright, D.M., Pavulaan, H. and Schröder, S. (2015) Morphological study of alar androconia in *Celastrina* butterflies. *Entomological Science*, 18(3), 353–359.

Ômura, H., Noguchi, T. and Nehira, T. (2016) New oxygenated himachalenes in male-specific odor of the Chinese windmill butterfly, *Byasa alcinous alcinous*. *Natural Product Research*, 30(4), 406–411.

Orr, A.G. (1995) The evolution of sphragis in the Papilionidae and other butterflies. In: Scriber, J.M. Tsubaki Y. and Lederhauser R.C. (eds), *Swallowtail Butterflies: Their Ecology and Evolutionary Biology*. Scientific Publishers, Gainesville, FL, pp. 155–164

Orr, A.G. (1999a) Possible post-copulatory mate guarding in *Ornithoptera euphorion* (Gray) (Lepidoptera: Papilionidae). *Australian Entomologist*, 26(3), 71–76.

Orr, A.G. (1999b) The big greasy, *Cressida cressida* (Papilionidae). *Biology of Australian Butterflies. Monographs on Australian Lepidoptera*, 6, 115–134.

Orr, A.G. (1999c) Possible postcopulatory mate guarding in *Ornithoptera euphorion* (Gray) (Lepidoptera: Papilionidae). *The Australian Entomologist*, 26(3), 71.

Orr, A.G. (2002) The sphragis of *Heteronympha penelope* Waterhouse (Lepidoptera: Satyridae): its structure, formation and role in sperm guarding. *Journal of Natural History*, 36, 185–196.

Orr, A.G. and Kitching, R.L. (2010) *The Butterflies of Australia*. Allen & Unwin, Sydney, Australia.

Orr, A.G. and Rutowski, R.L. (1991) The function of the sphragis in *Cressida cressida* (Fab.) (Lepidoptera, Papilionidae): a visually deterrent to copulation attempts. *Journal of Natural History*, 25(3), 703–710.

Osorio, D. and Vorobyev, M. (2008) A review of the evolution of animal colour vision and visual communication signals. *Vision Research*, 48, 2042–2051.

Otis, G.W., Locke, B., McKenzie, N.G., Cheung, D., MacLeod, E., Careless, P. and Kwoon, A. (2006) Local enhancement in mud-puddling swallowtail butterflies (*Battus philenor* and *Papilio glaucus*). *Journal of Insect Behavior*, 19(6), 685–698.

Owen, D.F. (1971) *Tropical Butterflies. The Ecology and Behaviour of Butterflies in the Tropics with Special Reference to African Species*. Clarendon Press, Oxford, UK.

Owen, D.F. and Chanter, D.O. (1969) Population biology of tropical African butterflies. Sex ratio and genetic variation in *Acraea encedon*. *Journal of Zoology*, 157(3), 345–374.

Paluch, M., Casagrande, M.M. and Mielke, O.H. (2003) Tampão genital de Actinote Hübner, como caráter taxonômico (Lepidoptera, Nymphalidae, Acraeinae). *Revista Brasileira de entomologia*, 47, 573–580.

Panchen, A. (1981) Notes on the behaviour of Rajah Brooke's birdwing butterfly *Trogonoptera brookiana brookiana* (Wallace) in Sarawak. *The Entomologist's Record and Journal of Variation*, 55, 98–102.

Pantelić, D., Ćurčić, S., Savić-Šević, S., Korać, A., Kovačević, A., Ćurčić, B. and Bokić, B. (2011) High angular and spectral selectivity of purple emperor (Lepidoptera: *Apatura iris* and *A. ilia*) butterfly wings. *Optics Express*, 19(7), 5817–5826.

Papke, R.S., Kemp, D.J. and Rutowski, R.L. (2006) Multimodal signalling: structural ultraviolet reflectance predicts male mating success better than pheromones in the butterfly *Colias eurytheme* L.(Pieridae) *Animal Behaviour*, 73, 47–54.

Paré, P.W. and Tumlinson, J.H. (1999) Plant volatiles as a defense against insect herbivores. *Plant Physiology*, 121(2), 325–332.

Parker, A.R. (2000) 515 million years of structural colour. *Journal of Optics A: Pure and Applied Optics*, 2(6), R15–R28.

Parker, G.A. (1974) Assessment strategy and the evolution of fighting behaviour. *Journal of Theoretical Biology*, 47(1), 223–243.

Parker, G.A. (2014) The sexual cascade and the rise of pre-ejaculatory (Darwinian) sexual selection, sex roles, and sexual conflict. In: Rice, W.R. and Gavrilets, S. (eds), *Cold Spring Harbor Perspectives in Biology*, 6, a017509. Cold Spring Harbor Laboratory Press, pp. 1–22.

Parker, G.A. and Birkhead, T.R. (2013) Polyandry: the history of a revolution. *Philosophical Transactions of the Royal Society of London B: Biological Sciences*, 368, 20120335.

Parnell, A.J., Bradford, J.E., Curran, E.V., Washington, A.L., Adams, G., Brien, M.N., Burg, S.L., Morochz, C., Fairclough, J.P.A., Vukusic, P. and Martin, S.J. (2018) Wing scale ultrastructure underlying convergent and divergent iridescent colours in mimetic *Heliconius* butterflies. *Journal of the Royal Society Interface*, 15(141), 20170948.

Parsons, M. (1983) Notes on the courtship of *Troides oblongomaculatus papuensis* (Papilionidae) in Papua New Guinea. *Journal of the Lepidopterists' Society*, 37, 83–85.

Paul, A. (2002) Sexual selection and mate choice. *International Journal of Primatology*, 23(4), 877–904.

Pearse F.K. and Murray, N.D. (1982) Sex and variability in the common brown butterfly *Heteronympha merope merope* (Lepidoptera: Satyrinae). *Evolution*, 36, 1251–1264

Pecháček, P., Stella, D., Keil, P. and Kleisner, K. (2014) Environmental effects on the shape variation of male ultraviolet patterns in the brimstone butterfly (*Gonepteryx rhamni*, Pieridae, Lepidoptera). *Naturwissenschaften*, 101(12), 1055–1063.

Pecháček, P., Stella, D. and Kleisner, K. (2019) A morphometric analysis of environmental dependences between ultraviolet patches and wing venation patterns in *Gonepteryx* butterflies (Lepidoptera, Pieridae). *Evolutionary Ecology*, 33(1), 89–110.

Pegram, K.V., Nahm, A.C., Rutowski, R. L. and Todd, S. (2013) Warning color changes in response to food deprivation in the pipevine swallowtail butterfly, *Battus philenor*. *Journal of Insect Science*, 13(110), 1–16.

Peixoto, P.E.C. and Benson, W.W. (2009a) Seasonal effects of density on territory occupation by males of the satyrine butterfly *Paryphthimoides phronius* (Butler 1867). *Journal of Ethology*, 27(3), 489–496.

Peixoto, P.E.C. and Benson, W.W. (2009b) Daily activity patterns of two co-occurring tropical satyrine butterflies. *Journal of Insect Science*, 9(1), 1–14.

Peixoto P.E.C. and Benson W.W. (2011) Fat and body mass predict residency status in two tropical satyrine butterflies. *Ethology*, 117, 722–730.

Peixoto, P.E.C. and Benson, W.W. (2012) Influence of previous residency and body mass in the territorial contests of the butterfly *Hermeuptychia fallax* (Lepidoptera: Satyrinae). *Journal of Ethology*, 30(1), 61–68.

Peixoto, P.E.C. and Mendoza-Cuenca, L. (2013) Territorial mating systems in butterflies: what we know and what Neotropical species can show. In: Macedo, R.H. and Machado, G. (eds), *Sexual Selection: Perspectives and Models from the Neotropics*. Elsevier, London, pp. 85–114. doi:10.1016/B978-0-12-416028-6.00004-9

Peixoto, P.E.C., Medina, A.M. and Mendoza-Cuenca, L. (2014) Do territorial butterflies show a macroecological fighting pattern in response to environmental stability? *Behavioural Processes*, 109, 14–20.

Pelmyr, O. (1982) Plebeian courtship revisited: studies on the female-produced male behavior-eliciting signals in *Lycaeides idas* courtship (Lycaenidae). *Journal of Research on the Lepidoptera*, 21, 147–157.

Peña, C. and Wahlberg, N. (2008) Prehistorical climate change increased diversification of a group of butterflies. *Biology Letters*, 4(3), 274–278.

Peña, C., Wahlberg, N., Weingartner, E., Kodandaramaiah, U., Nylin, S., Freitas, A.V. and Brower, A.V. (2006) Higher level phylogeny of Satyrinae butterflies (Lepidoptera: Nymphalidae) based on DNA sequence data. *Molecular Phylogenetics and Evolution*, 40(1), 29–49.

Penney, H.D., Hassall, C., Skevington, J.H., Abbott, K.R. and Sherratt, T.N. (2012) A comparative analysis of the evolution of imperfect mimicry. *Nature*, 483(7390), 461–464.

Penz, C.M. and DeVries, P.J. (2002) Phylogenetic analysis of *Morpho* butterflies (Nymphalidae, Morphinae): implications for classification and natural history. *American Museum Novitates*, 2002(3374), 1–33.

Penz, C.M., Devries, P.J. and Wahlberg, N. (2012) Diversification of *Morpho* butterflies (Lepidoptera, Nymphalidae): a re-evaluation of morphological characters and new insight from DNA sequence data. *Systematic Entomology*, 37(4), 670–685.

Petty, R.L., Boppré, M., Schneider, D. and Meinwald, J. (1977) Identification and localization of volatile hairpencil components in male *Amauris ochlea* butterflies (Danaidae). *Experientia*, 33(10), 1324–1326.

Pfeifer, M.A., Andrick, U.R., Frey, W. and Settele, J. (2000) On the ethology and ecology of a small and isolated population of the dusky large blue butterfly *Glaucopsyche* (Maculinea) *nausithous* (Lycaenidae). *Nota Lepidopterologica*, 23(2), 147–172.

Pichaud, F., Briscoe, A. and Desplan, C. (1999) Evolution of color vision. *Current Opinion in Neurobiology*, 9(5), 622–627.

Pinzari, M. (2009) A comparative analysis of mating recognition signals in graylings: *Hipparchia statilinus* vs. *H. semele* (Lepidoptera: Nymphalidae, Satyrinae). *Journal of Insect Behaviour*, 22(3), 227–244.

Pinzari, M. and Sbordoni, V. (2013) Species and mate recognition in two sympatric grayling butterflies: *Hipparchia fagi* and *H. hermione genava* (Lepidoptera). *Ethology Ecology and Evolution*, 25(1), 28–51.

Pinzari, M., Santonico, M., Pennazza, G., Martinelli, E., Capuano, R., Paolesse, R., Di Rao, M., D'Amico, A., Cesaroni, D., Sbordoni, V. and Di Natale, C. (2018) Chemically mediated species recognition in two sympatric grayling butterflies: *Hipparchia fagi* and *Hipparchia hermione* (Lepidoptera: Nymphalidae, Satyrinae). *PloS ONE*, 13(6), .e0199997.

Pinzari, M., Pinzari, M. and Sbordoni, V. (2019) Make it simple: mating behaviour of *Euphydryas aurinia* provincialis (Lepidoptera: Nymphalidae). *The European Zoological Journal*, 86(1), 220–232.

Pirih, P., Wilts, B.D. and Stavenga, D.G. (2011) Spatial reflection patterns of iridescent wings of male pierid butterflies: curved scales reflect at a wider angle than flat scales. *Journal of Comparative Physiology A*, 197(10), 987–997.

Pivnick, K.A. and McNeil, J.N. (1985) Mate location and mating behavior of *Thymelicus lineola* (Lepidoptera: Hesperiidae). *Annals of the Entomological Society of America*, 78(5), 651–656.

Pivnick, K.A., Lavoie-Dornik, J.A.N.Y. and McNeil, J.N. (1992) The role of the androconia in the mating behaviour of the European skipper, *Thymelicus lineola*, and evidence for a male sex pheromone. *Physiological Entomology*, 17(3), 260–268.

Plakke, M.S., Deutsch, A.B., Meslin, C., Clark, N.L. and Morehouse, N.I. (2015) Dynamic digestive physiology of a female reproductive organ in a polyandrous butterfly. *Journal of Experimental Biology*, 218(10), 1548–1555.

Platt, A.P. and Allen, J.F. (2001) Sperm precedence and competition in doubly-mated *Limenitis arthemis-astyanax* butterflies (Rhopalocera: Nymphalidae). *Annals of the Entomological Society of America*, 94(5), 654–663.

Platt, A.P., Rawson, G.W. and Balogh, G. (1978) Inter-specific hybridization involving *Limenitis archippus* and its congeneric species. *Journal of the Lepidopterists' Society*, 32(4), 289–303.

Plattner, L. (2004) Optical properties of the scales of *Morpho rhetenor* butterflies: theoretical and experimental investigation of the back-scattering of light in the visible spectrum. *Journal of the Royal Society Interface*, 1, 49–59.

Pliske, T.E. (1975a) Courtship behavior and use of chemical communication by males of certain species of ithomiine butterflies (Nymphalidae: Lepidoptera). *Annals of the Entomological Society of America*, 68(6), 935–942.

Pliske, T.E. (1975b) Courtship behavior of the monarch butterfly, *Danaus plexippus*. *Annals of the Entomological Society of America*, 68(1), 143–151.

Pliske, T.E. (1975c) Attraction of Lepidoptera to plants containing pyrrolizidine alkaloids. *Environmental Entomology*, 4(3), 455–473.

Pliske, T.E. and Eisner, T. (1969) Sex pheromone of the queen butterfly: biology. *Science*, 164(3884), 1170–1172.

Pliske, T.E. and Salpeter, M.M. (1971) The structure and development of the hairpencil glands in males of the queen butterfly, *Danaus gilippus berenice*. *Journal of Morphology*, 134(2), 215–241.

Pliske, T.E., Edgar, J.A. and Culvenor, C.C. (1976) The chemical basis of attraction of ithomiine butterflies to plants containing pyrrolizidine alkaloids. *Journal of Chemical Ecology*, 2(3), 255–262.

Pohl, N.B., Van Wyk, J. and Campbell, D.R. (2011) Butterflies show flower colour preferences but not constancy in foraging at four plant species. *Ecological Entomology*, 36(3), 290–300.

Pola, M. and García-París, M. (2005) Marine puddling in *Papilio polytes* (Lepidoptera: Papilionidae). *Florida Entomologist*, 88(2), 211–213.

Poladian, L., Wickham, S., Lee, K. and Large, M.C. (2009) Iridescence from photonic crystals and its suppression in butterfly scales. *Journal of the Royal Society Interface*, 6(suppl_2), S233–S242.

Potyrailo, R.A., Bonam, R.K., Hartley, J.G., Starkey, T.A., Vukusic, P., Vasudev, M., Bunning, T., Naik, R.R., Tang, Z., Palacios, M.A. and Larsen, M. (2015) Towards outperforming conventional sensor arrays with fabricated individual photonic vapour sensors inspired by *Morpho* butterflies. *Nature Communications*, 6, 7959.

Poulton, E.B. (1890) *The Colours of Animals: Their Meaning and Use, Especially Considered in the Case of Insects*. International Scientific Series, Vol. LXVII. D. Appleton, New York, pp. 286–286.

Pouya, C. (2012) Investigating naturally occurring 3-dimensional photonic crystals. PhD thesis, University of Exeter, Exeter, UK.

Powell, J.A. (1968) A study of area occupation and mating behavior in *Incisalia iroides* (Lepidoptera: Lycaenidae). *Journal of the New York Entomological Society*, 1, 47–57.

Preston-Mafham, R. and Preston-Mafham, K. (1988) *Butterflies of the World*. Blandford Press, London.

Price, T. (1998) Sexual selection and natural selection in bird speciation. *Philosophical Transactions of the Royal Society of London B: Biological Sciences*, 353(1366), 251–260.

Prieto, C. and Dahners, H.W. (2006) Eumaeini butterflies (Lepidoptera: Lycaenidae) from San Antonio hill: richness dynamic and hilltopping behaviour. *Revista Colombiana de Entomología*, 32(2), 179–190.

Prieto, C. and Dahners, H.W. (2009) Resource utilization and environmental and spatio-temporal overlap of a hilltopping Lycaenid butterfly community in the Colombian Andes. *Journal of Insect Science*, 9, 1–12.

Prudic, K.L., Jeon, C., Cao, H. and Monteiro, A. (2011) Developmental plasticity in sexual roles of butterfly species drives mutual sexual ornamentation. *Science*, 331(6013), 73–75.

Prudic, K.L., Stoehr, A.M., Wasik, B.R. and Monteiro, A. (2015) Eyespots deflect predator attack increasing fitness and promoting the evolution of phenotypic plasticity. *Proceedings of the Royal Society of London B: Biological Sciences*, 282, 20141531.

Prum, R.O. and Torres, R.H. (2013) Fourier blues: structural coloration of biological tissues. In *Excursions in Harmonic Analysis, Volume 2: The February Fourier Talks at the Norbert Wiener Center*. Birkhäuser, Boston, MA, pp. 401–421.

Prum, RO., Torres, R.H., Williamson, S. and Dyck, J. (1998) Coherent light scattering by blue feather barbs. *Nature*, 396(6706), 28–29.

Prum R.O., Quinn, T. and Torres, R.H. (2006) Anatomically diverse butterfly scales all produce structural colours by coherent scattering. *Journal of Experimental Biology*, 209, 748–765.

Pyrcz, T.W., Garlacz, R., Kertész, K., Biró, L.P. and Bálint, Z. (2018) An imperfect imago? Post-mating loss of iridescent scales in *Cheimas* butterflies may change female from attractive to cryptic (Lepidoptera: Nymphalidae: Satyrinae). *Journal of Natural History*, 52(19–20), 1333–1350.

Qiu, X. and Arikawa, K. (2003) Polymorphism of red receptors: sensitivity spectra of proximal photoreceptors in the small white butterfly *Pieris rapae crucivora*. *The Journal of Experimental Biology*, 206, 2787–2793.

Qiu, X., Vanhoutte, K.A., Stavenga, D.G. and Arikawa, K. (2002) Ommatidial heterogeneity in the compound eye of the male small white butterfly, *Pieris rapae crucivora*. *Cell and Tissue Research*, 307(3), 371–379.

Quigley, C. and Fusani, L. (2018) Courtship and mating. In: Skinner M.K. (ed.), *Encyclopedia of Reproduction*, Vol. 6, Elsevier, Oxford, pp. 67–71.

Racevska, E. and Roberts, S. H. (2018) Sexual selection. In: Vonk, J. and Shackelford, T.K. (eds), *Encyclopedia of Animal Cognition and Behavior*. Springer International Publishing AG, Basel, Switzerland, pp. 1–11.

Rajyaguru, P.K., Pegram, K.V., Kingston, A.C. and Rutowski, R.L. (2013) Male wing color properties predict the size of nuptial gifts given during mating in the pipevine swallowtail butterfly (*Battus philenor*). *Naturwissenschaften*, 100(6), 507–513.

Ramos, R.R. and Francini, R.B. (2014) Intergeneric mating attempt: males of *Tegosa claudina* Eschscholtz and of *Chlosyne lacinia saundersi* Doubleday. *News of the Lepidopterists Society*, 56(4), 170–174.

Rauser, C.L. and Rutowski, R.L. (2003) Male-specific structures on the wings of the Gulf fritillary butterfly, *Agraulis vanillae* (Nymphalidae). *Journal of the Lepidopterists' Society*, 57(4), 279–283.

Ravenscroft, N.O. (1994) Environmental influences on mate location in male chequered skipper butterflies, *Carterocephalus palaemon* (Lepidoptera: Hesperiidae). *Animal Behaviour*, 47(5), 1179–1187.

Ray, P. and Dutta, W. (2016) A report on intergeneric mating among butterfly species in nature. *Proceedings of the Zoological Society*, 69(2), 259–261.

Réal, P. (1988) Lépidoptères nouveaux principalement jurassiens. *Mém. Comité de Liaison Rech. Ecofaunist. Jura*, 4, 1–28. (Liaison Committee for ecofaunistic research in the Jura).

Rhainds, M. (2010) Female mating failures in insects. *Entomologia Experimentalis et Applicata*, 136(3), 211–226.

Ribarič, D. and Gogala, M. (1996) Acoustic behaviour of some butterfly species of the genus *Erebia* (Lepidoptera: Satyridae). *Acta Entomologica Slovenica*, 4(1), 5–12.

Ridley, M. (1989a) The incidence of sperm displacement in insects: four conjectures, one corroboration. *Biological Journal of the Linnean Society*, 38, 349–367.

Ridley, M. (1989b) The timing and frequency of mating in insects. *Animal Behaviour*, 37, 535–545.

Rister, J. and Desplan, C. (2011) The retinal mosaics of opsin expression in invertebrates and vertebrates. *Developmental Neurobiology*, 71(12), 1212–1226.

Robbins, R.K. (1982) How many butterfly species. *News of the Lepidopterists' Society*, 24(3), 40–41.

Robbins, R.K. (1990) Systematic implications of butterfly leg structures that clean the antennae. *Psyche*, 96(3–4), 209–222.

Robbins, R.K. and Opler, P.A. (1997) Butterfly diversity and a preliminary comparison with bird and mammal diversity. In: Reaka-Kudla, M.L., Wilson, D.E. and Wilson, E.O. (eds), *Biodiversity II, Understanding and Protecting our Biological Resources*. Joseph Henry Press, Washington DC, pp. 68–82.

Robbins, R.K., Martins, A.R.P., Busby, R.C. and Duarte, M. (2012) Loss of male secondary sexual structures in allopatry in the Neotropical butterfly genus *Arcas* (Lycaenidae: Theclinae: Eumaeini). *Insect Systematics and Evolution*, 43, 35–65.

Robbins, R.K., Heredia, M.D. and Busby, R.C. (2015) Male secondary sexual structures and the systematics of the *Thereus oppia* species group (Lepidoptera, Lycaenidae, Eumaeini). *ZooKeys*, 520, 109–130.

Robertson, K.A. and Monteiro, A. (2005) Female *Bicyclus anynana* butterflies choose males on the basis of their dorsal UV-reflective eyespot pupils. *Proceedings of the Royal Society of London B: Biological Sciences*, 272, 1541–1546.

Rodríguez, R.L. (2015) Mating is a give-and-take of influence and communication between the sexes. In: Peretti, A.V. (ed.), *Cryptic Female Choice in Arthropods*. Springer, Cham, Switzerland, pp. 479–496.

Rogers, S.H. and Wells, H. (1984) The structure and function of the bursa copulatrix of the monarch butterfly (*Danaus plexippus*). *Journal of Morphology*, 180(3), 213–221.

Rosenberg, R.H. (1989) Behavior of the territorial species *Limenitis weidemeyerii* (Nymphalidae) within temporary feeding areas. *Journal of the Lepidopterists' Society*, 43(2), 102–107.

Rosenberg, R.H. and Enquist, M. (1991) Contest behaviour in Weidemeyer's admiral butterfly *Limenitis weidemeyerii* (Nymphalidae): the effect of size and residency. *Animal Behaviour*, 42(5), 805–811.

Ross, G.N. (2002) Life history of the Seminole crescent, *Anthanassa texana seminole* (Lepidoptera: Nymphalidae). *Holarctic Lepidoptera*, 9(1–2).

Rossato, D.O., Boligon, D., Fornel, R., Kronforst, M.R., Gonçalves, G.L. and Moreira, G.R. (2018) Subtle variation in size and shape of the whole forewing and the red band among co-mimics revealed by geometric morphometric analysis in *Heliconius* butterflies. *Ecology and Evolution*, 8(6), 3280–3295.

Rothschild, M. (1978) Hell's angels. *Antenna*, 2(2), 38–39.

Rothschild, M. and Marsh, N. (1978) Some peculiar aspects of danaid/plant relationships. *Entomologia experimentalis et applicata*, 24(3), 637–650.

Ruse, M. (2015) Sexual selection: why does it play such a large role in the descent of man? In: Hoquet, T. (ed.), *Current Perspectives on Sexual Selection*. Springer, Dordrecht, Germany, pp. 3–17.

Rutowski, R.L. (1977a) Chemical communication in the courtship of the small sulphur butterfly *Eurema lisa* (Lepidoptera: Pieridae). *Journal of Comparative Physiology*, 115, 75–85.

Rutowski, R.L. (1977b) The use of visual cues in sexual and species discrimination by males of the small sulphur butterfly *Eurema lisa* (Lepidoptera, Pieridae). *Journal of Comparative Physiology*, 115(1), 61–74.

Rutowski, R.L. (1978) The form and function of ascending flights in *Colias* butterflies. *Behavioral Ecology and Sociobiology*, 3(2), 163–172.

Rutowski, R.L. (1979) Courtship behavior of the checkered white, *Pieris protodice* (Pieridae). *Journal of the Lepidopterists' Society*, 33, 42–49.

Rutowski, R.L. (1980a) Courtship solicitation by females of the checkered white butterfly, *Pieris protodice*. *Behavioral Ecology and Sociobiology*, 7(2), 113–117.

Rutowski, R.L. (1980b) Male scent-producing structures in *Colias* butterflies. *Journal of Chemical Ecology*, 6(1), 13–26.

Rutowski, R.L. (1981a) Courtship behaviour of the dainty sulfur butterfly, *Nathalis iole* with a description of a new, faculative male display (Pieridae). *Journal of Research on the Lepidoptera*, 20(3), 161–169.

Rutowski, R.L. (1981b) Sexual discrimination using visual cues in the checkered white butterfly (*Pieris protodice*). *Zeitschrift für Tierpsychologie*, 55(4), 325–334.

Rutowski, R.L. (1982a) Mate choice and lepidopteran mating behavior. *The Florida Entomologist*, 65(1), 72–82.

Rutowski, R.L. (1982b) Epigamic selection by males as evidenced by courtship partner preferences in the checkered white butterfly (*Pieris protodice*). *Animal Behaviour*, 30(1), 108–112.
Rutowski, R.L. (1983a) Courtship leading to copulation in the cloudless sulphur, *Phoebis sennae* (Pieridae). *Journal of Research on the Lepidoptera*, 22(4), 249–253.
Rutowski, R.L. (1983b) The wing-waving display of *Eurema daira* males (Lepidoptera: Pieridae): its structure and role in successful courtship. *Animal Behaviour*, 31(4), 985–989.
Rutowski, R.L. (1984a) Courtship leading to copulation in the cloudless sulphur, *Phoebis sennae* (Pieridae). *Journal of Research on the Lepidoptera*, 22(4), 249–253.
Rutowski, R.L. (1984b) Courtship behavior of the Gulf fritillary, *Agraulis vanillae* (Nymphalidae). *Journal of the Lepidopterists' Society*, 38(1), 23–31.
Rutowski, R.L. (1984c) Sexual selection and the evolution of butterfly mating behaviour. *Journal of Research on the Lepidoptera*, 23(2), 125–142.
Rutowski, R.L. (1984d) Production and use of secretions passed by males at copulation in *Pieris protodice* (Lepidoptera, Pieridae). *Psyche: A Journal of Entomology*, 91(1–2), 141–152.
Rutowski, R.L. (1985) Evidence for mate choice in a sulphur butterfly (*Colias eurytheme*) *Zeitschrift für Tierpsychologie*, 70(2), 103–114.
Rutowski, R.L. (1989) Mating strategies in butterflies, *Scientific American*, 279(1), 64–69.
Rutowski, R.L. (1991) The evolution of male mate-locating behaviour in butterflies. *American Naturalist*, 138, 1121–1139.
Rutowski, R.L. (1992) Male mate-locating behavior in the common eggfly, *Hypolimnas bolina* (Nymphalidae). *Journal of the Lepidopterists' Society*, 46(1), 24–38.
Rutowski, R.L. (1994) Questions about butterfly behaviour: the case of the Empress Leilia. *American Butterflies*, May, 20–23.
Rutowski, R.L. (1997a) When butterflies meet. *American Butterflies*, Fall, 18–23.
Rutowski, R.L. (1997b) Sexual dimorphism, mating systems and ecology in butterflies. In: Choe, J.C. and Crespi, B.J. (eds), *The Evolution of Mating Systems in Insects and Arachnids*. Cambridge University Press, Cambridge, UK, pp. 257–268.
Rutowski, R.L. (1998) Mating strategies in butterflies. *Scientific American*, 279(1), 64–69.
Rutowski, R.L. (2000a) Postural changes accompany perch location changes in male butterflies (*Asterocampa leilia*) engaged in visual mate searching. *Ethology*, 5, 453–466.
Rutowski, R.L. (2000b) Variation of eye size in butterflies: inter-and intraspecific patterns. *Journal of Zoology*, 252(02), 187–195.
Rutowski, R.L. (2003) Visual ecology of adult butterflies. In: Boggs CL, Watt, W.B. and Ehrlich, P.R. (eds), *Butterflies. Ecology and Evolution Taking Flight*. University of Chicago Press, Chicago, IL, pp. 9–25.
Rutowski, R.L. and Gilchrist, G.W. (1986) Copulation in *Colias eurytheme* (Lepidoptera: Pieridae): patterns and frequency. *Journal of Zoology*, 209(1), 115–124.
Rutowski, R.L. and Gilchrist, G.W. (1987) Courtship, copulation and oviposition in the Chalcedon checkerspot, *Euphydryas chalcedona* (Lepidoptera: Nymphalidae). *Journal of Natural History*, 21(5), 1109–1117.
Rutowski, R.L. and Kemp, D.J. (2017) Female iridescent colour ornamentation in a butterfly that displays mutual ornamentation: is it a sexual signal? *Animal Behaviour*, 126, 301–307.
Rutowski, R.L. and Rajyaguru, P.K. (2013) Male-specific iridescent coloration in the pipevine swallowtail (*Battus philenor*) is used in mate choice by females but not sexual discrimination by males. *Journal of Insect Behavior*, 26(2), 200–211.
Rutowski, R.L. and Schaefer, J. (1984) Courtship behavior of the gulf fritillary, *Agraulis vanillae* (Nymphalidae). *Journal of the Lepidopterists' Society*, 38, 23–31.
Rutowski, R L. and Warrant, E.J. (2002) Visual field structure in the Empress Leilia, *Asterocampa leilia* (Lepidoptera, Nymphalidae): dimensions and regional variation in acuity. *Journal of Comparative Physiology A*, 188(1), 1–12.

Rutowski, R.L., Long, C.E., Marshall, L.D. and Vetter, R.S. (1981) Courtship solicitation by *Colias* females (Lepidoptera: Pieridae). *American Midland Naturalist*, 105, 334–340.

Rutowski, R.L., Newton, M. and Schaefer, J. (1983) Interspecific variation in the size of the nutrient investment made by male butterflies during copulation. *Evolution*, 37(4), 708–713.

Rutowski, R.L., Gilchrist, G.W. and Terkanian, B. (1987) Female butterflies mated with recently mated males show reduced reproductive output. *Behavioral Ecology and Sociobiology*, 20(5), 319–322.

Rutowski, R.L., Alcock, J. and Carey, M. (1989) Hilltopping in the pipevine swallowtail butterfly (*Battus philenor*). *Ethology*, 82(3), 244–254.

Rutowski, R.L., Demlong, M.J. and Leffingwell, T. (1994) Behavioural thermoregulation at mate encounter sites by male butterflies (*Asterocampa leilia,* Nymphalidae). *Animal Behaviour*, 48(4), 833–841.

Rutowski, R.L., Demlong, M.J. and Terkanian, B. (1996) Seasonal variation in mate-locating activity in the desert hackberry butterfly (*Asterocampa leilia*; Lepidoptera: Nymphalidae). *Journal of Insect Behavior*, 9(6), 921–931.

Rutowski, R. L., Terkanian, B., Eitan, O. and Knebel, A. (1997) Male mate-locating behavior and yearly population cycles in the snout butterfly, *Libytheana bachmanii* (Libytheidae). *Journal of the Lepidopterists' Society*, 51(3), 197–207.

Rutowski, R.L., McCoy, L. and Demlong, M.J. (2001) Visual mate detection in a territorial male butterfly (*Asterocampa leilia*): effects of distance and perch location. *Behaviour*, 138(1), 31–43.

Rutowski, R.L., Macedonia, J.M., Morehouse, N. and Taylor-Taft, L. (2005) Pterin pigments amplify iridescent ultraviolet signal in males of the orange sulphur butterfly, *Colias eurytheme*. *Proceedings of the Royal Society of London B: Biological Sciences*, 272(1578), 2329–2335.

Rutowski, R.L., Macedonia, J.M., Kemp, D.J. and Taylor-Taft, L. (2007a) Diversity in structural ultraviolet coloration among female sulphur butterflies (Coliadinae, Pieridae). *Arthropod Structure and Development*, 36(3), 280–290.

Rutowski, R.L., Macedonia, J.M., Merry, J.W., Morehouse, N.I., Yturralde, K., Taylor-Taft, L., Gaalema, D., Kemp, D.J. and Papke, R.S. (2007b) Iridescent ultraviolet signal in the orange sulphur butterfly (*Colias eurytheme*): spatial, temporal and spectral properties. *Biological Journal of the Linnean Society*, 90(2), 349–364.

Rutowski, R.L., Gislén, L. and Warrant, E.J. (2009) Visual acuity and sensitivity increase allometrically with body size in butterflies. *Arthropod Structure and Development*, 38(2), 91–100.

Rutowski, R.L., Nahm, A.C. and Macedonia, J.M. (2010) Iridescent hindwing patches in the pipevine swallowtail: differences in dorsal and ventral surfaces relate to signal function and context. *Functional Ecology*, 24(4), 767–775.

Saba, M., Wilts, B.D., Hielscher, J. and Schröder-Turk, G.E. (2014) Absence of circular polarisation in reflections of butterfly wing scales with chiral gyroid structure. *Materials Today: Proceedings*, 1, 193–208.

Saccheri, I.J., Brakefield, P.M. and Nichols, R.A. (1996) Severe inbreeding depression and rapid fitness rebound in the butterfly *Bicyclus anynana* (Satyridae). *Evolution*, 50(5), 2000–2013.

Saccheri, I., Kuussaari, M., Kankare, M., Vikman, P., Fortelius, W. and Hanski, I. (1998) Inbreeding and extinction in a butterfly metapopulation. *Nature*, 392(6675), 491–494.

Sáfián, S. (2013) Observation of hill-topping behaviour by the giant African swallowtail – *Papilio antimachus* Drury, 1782 and other recent records from Liberia (West Africa) (Lepidoptera: Papilionidae). *SHILAP Revista de Lepidopterología*, 41, 163.

Sáfián, S. (2014) Hill-topping in some African *Celaenorrhinus* Hübner, 1819. *Metamorphosis*, 25, 1–2.

Sáfián, S. (2015) Behaviour and development of *Pseudopontia gola* Sáfián & Mitter, 2011 (Lepidoptera: Pieridae). *SHILAP Revista de Lepidopterología*, 43(169), 85–89.

Sahoo, R.K., Warren, A.D., Collins, S.C. and Kodandaramaiah, U. (2017) Hostplant change and paleoclimatic events explain diversification shifts in skipper butterflies (Family: Hesperiidae). *BMC Evolutionary Biology*, 17(1), 174.

Sánchez, V. and Cordero, C. (2014) Sexual coevolution of spermatophore envelopes and female genital traits in butterflies: evidence of male coercion? *PeerJ*, 2, e247.

Sánchez, V., Hernández-Baños, B.E. and Cordero, C. (2011) The evolution of a female genital trait widely distributed in the Lepidoptera: comparative evidence for an effect of sexual coevolution. *PLoS ONE*, 6, e22642.

San Martin, G., Bacquet, P. and Nieberding, C.M. (2011) Mate choice and sexual selection in a model butterfly species, *Bicyclus anynana*: state of the art. *Proceedings of the Netherlands Entomological Society Meeting*, 22, 9–22.

Sappington, T.W. and Taylor, O.R. (1990a) Disruptive sexual selection in *Colias eurytheme* butterflies. *Proceedings of the National Academy of Sciences*, 87(16), 6132–6135.

Sappington, T.W. and Taylor, O.R. (1990b) Developmental and environmental sources of pheromone variation in *Colias eurytheme* butterflies. *Journal of Chemical Ecology*, 16(9), 2771–2786.

Sasaki, N. and Watanabe, M. (2015) Alternative ejaculate allocation tactics in relation to male mating history of the swallowtail butterfly, *Papilio xuthus* L. (Lepidoptera: Papilionidae). *Journal of Research on the Lepidoptera*, 48, 1–8.

Sasaki, N., Konagaya, T., Watanabe, M. and Rutowski, R.L. (2015a) Indicators of recent mating success in the pipevine swallowtail butterfly (*Battus philenor*) and their relationship to male phenotype. *Journal of Insect Physiology*, 83, 30–36.

Sasaki, N., Todoriki, S. and Watanabe, M. (2015b) Effect of sperm ejection by females on male fertilization success in the swallowtail butterfly, *Papilio xuthus* L. (Lepidoptera: Papilionidae). *Journal of Research on the Lepidoptera*, 48, 39–44.

Sasaki, N., Konagaya, T., Watanabe, M. and Rutowski, R.L. (2017) Estimating the mating success of male butterflies in the field. In: Sekimura, T. and Nijhout, H.F. (eds), *Diversity and Evolution of Butterfly Wing Patterns*. Springer, Singapore, pp. 255–268.

Sauman, I., Briscoe, A.D., Zhu, H., Shi, D., Froy, O., Stalleicken, J., Yuan, Q., Casselman, A. and Reppert, S.M. (2005) Connecting the navigational clock to sun compass input in monarch butterfly brain. *Neuron*, 46(3), 457–467.

Schneider, D., Boppré, M., Schneider, H., Thompson, W.R., Boriack, C.J., Petty, R.L. and Meinwald, J. (1975) A pheromone precursor and its uptake in male *Danaus* butterflies. *Journal of Comparative Physiology*, 97(3), 245–256.

Schröder-Turk, G.E., Wickham, S., Averdunk, H., Brink, F., Gerald, J.F., Poladian, L., Large, M.C.J. and Hyde, S.T. (2011) The chiral structure of porous chitin within the wing-scales of *Callophrys rubi*. *Journal of Structural Biology*, 174(2), 290–295.

Schulz, S. (1998) Insect-plant interactions – metabolism of plant compounds to pheromones and allomones by Lepidoptera and leaf beetles. *European Journal of Organic Chemistry*, 1998, 13–20.

Schulz, S. and Nishida, R. (1996) The pheromone system of the male danaine butterfly, *Idea leuconoe*. *Bioorganic and Medicinal Chemistry*, 4(3), 341–349.

Schulz, S., Francke, W., Edgar, J. and Schneider, D. (1988) Volatile compounds from androconial organs of danaine and ithomiine butterflies. *Zeitschrift für Naturforschung C*, 43(1–2), 99–104.

Schulz, S., Boppré, M. and Vane-Wright, R. (1993) Specific mixtures of secretions from male scent organs of African milkweed butterflies (Danainae). *Philosophical Transactions of the Royal Society of London B: Biological Sciences*, 342 (1300), 161–181.

Schulz, S., Yildizhan, S. Stritzke, K. Estrada C. and L.E. Gilbert. (2007) Macrolides from the scent glands of the tropical butterflies *Heliconius cydno* and *Heliconius pachinus*.*Organic and Biomolecular Chemistry*, 5, 3434–3441.

Schulz, S., Estrada, C., Yildizhan, S., Boppré, M. and Gilbert, L.E. (2008) An antiaphrodisiac in *Heliconius melpomene* butterflies. *Journal of Chemical Ecology*, 34(1), 82–93.
Schurian, K.G. and Fiedler, K. (1996) . *Nachrichten des Entomologischen Vereins Apollo, NF*, 16(4), 329–343.
Schwab, I.R. (2012) *Evolution's Witness: How Eyes Evolved*. Oxford University Press, Oxford, UK.
Scott, J.A. (1970) Hilltopping as a mating mechanism to aid the survival of low density species. *Journal of Research on the Lepidoptera*, 7, 191–204.
Scott, J. (1973a) Mating of butterflies. *Journal of Research on the Lepidoptera*, 11, 99–127.
Scott, J.A. (1973b) Adult behavior and population biology of two skippers (Hesperiidae) mating in contrasting topographic sites. *Journal of Research on the Lepidoptera*, 12, 181–196.
Scott, J.A. (1973c) Convergence of population biology and adult behaviour in two sympatric butterflies, *Neominois ridingsii* (Papilionoidea: Nymphalidae) and *Amblyscirtes simius* (Hesperioidea: Hesperiidae). *The Journal of Animal Ecology*, 42(3), 663–672.
Scott, J.A. (1974a) Mate-locating behavior of butterflies. *American Midland Naturalist*, 91, 103–117.
Scott, J.A. (1974b) Population biology and adult behavior of *Lycaena arota* (Lycaenidae). *Journal of the Lepidopterists' Society*, 28(1), 64–72.
Scott, J.A. (1974c) Adult behavior and population biology of *Poladryas minuta* (Nymphalidae) and the relationship of the Texas and Colorado populations. *Pan Pacific Entomologist*, 50(1), 9–22
Scott, J.A. (1974d) The interaction of behavior, population biology, and environment in *Hypaurotis crysalus* (Lepidoptera). *American Midland Naturalist*, 383–394.
Scott, J. (1975a) *Pyrgus xanthus* (Hesperiidae): systematics, foodplants and behavior. *Journal of the Lepidopterists' Society*, 29, 213–220.
Scott, J.A. (1975b) Variability of courtship of the buckeye butterfly, *Precis coenia* (Nymphalidae). *Journal of Research on the Lepidoptera*, 35, 22–28.
Scott, J.A. (1975c) Movements of *Euchloe ausonides*. *Journal of the Lepidopterists' Society*, 29, 24–31.
Scott, J.A. (1983) Mate-locating behavior of western North American butterflies. II. New observations and morphological adaptations. *Journal of Research on the Lepidoptera*, 21(3), 177–187.
Scott, J. (1986a) *The Butterflies of North America*. Stanford University Press, Stanford, CA.
Scott, J.A. (1986b) The courtship of *Phyciodes*, and the relationship between *Phyciodes tharos tharos* and *Phyciodes tharos morpheus* (= *pascoensis*) in Colorado. *Papilio. New Series*, 5, 1–8.
Scott, J.A. (1992) *The Butterflies of North America: a Natural History and Field Guide*. Stanford University Press, Stanford, CA.
Scott, J.A. (2006) Territoriality in butterflies. *Papilio New Series*, 14, 67–74.
Scott, J.A. (2010) New terminology for describing mate-locating behavior of butterflies (and moths), with examples in Colorado. *News of the Lepidopterists' Society*, 52, 58–62.
Scott, J.A. (2014) Lepidoptera of North America 13. Flower visitation by Colorado butterflies (40,615 records) with a review of the literature on pollination of Colorado plants and butterfly attraction (Lepidoptera: Hesperioidea and Papilionoidea). *Lepidoptera of North America*. Contributions of the C.P. Gillette Museum of Arthropod Diversity Colorado State University.
Scott, J.A., Guppy, C.S., Pelham, J.P. and Calhoun, J.V. (2012) *Callophrys* green hairstreaks regain traditional names. *News of the Lepidopterists' Society*, 54(2), 40.
Scriber, J.M. (2002) A female *Papilio canadensis* (Lepidoptera: Papilionidae) puddles with males. *The American Midland Naturalist*, 147(1), 175–178.
Sellier, R. (1972) Etude ultrastructurale en microscopie electronique a balayage des poils androconiaux alaires chez les lepidopteres diurnes. *Comptes Rendus de lilcadkmie des Sciences*, Strie D, 275, 2239–2242.

Shapiro, A.M. (1965) Ecological and behavioral notes on *Hesperia metea* and *Atrytonopsis hianna* (Hesperiidae). *Journal of the Lepidopterists' Society*, 19(4), 215–221.
Shapiro, A.M. (1970) The role of sexual behavior in density-related dispersal of pierid butterflies. *The American Naturalist*, 104(938), 367–372.
Shapiro, A.M. (1973) An attempted interfamilial mating (Lycaenidae, Nymphalidae). *Journal of the Lepidopterists' Society*, 27(2), 159.
Shapiro, A.M. (1975) Ecological and behavioral aspects of coexistence in six crucifer-feeding pierid butterflies in the central Sierra Nevada. *American Midland Naturalist*, 424–433.
Shapiro, I.D. (1977) Interaction of population biology and mating behavior of the fiery skipper, *Hylephila phylaeus* (Hesperiidae). *American Midland Naturalist*, 85–94.
Shapiro, A.M. (1988) Homosexual pseudocopulation in *Eucheira socialis* (Pieridae). *The Journal of Research on the Lepidoptera (USA)*, 27(3–4), 262.
Shapiro, A.M. and Porter, A.H. (1989) The lock-and-key hypothesis: evolutionary and biosystematic interpretation of insect genitalia. *Annual Review of Entomology*, 34(1), 231–245.
Shields, O. (1967) Hilltopping. *Journal of Research on the Lepidoptera*, 6(2), 69–178.
Shields, O. (1989) World numbers of butterflies. *Journal of the Lepidopterists' Society*, 43(3), 178–183.
Shields and Emmel, J.F. (1973) A review of carrying pair and mating times in butterflies. *Journal of Research on the Lepidoptera*, 12, 25–64.
Shreeve, T.G. (1984) Habitat selection, mate location, and microclimatic constraints on the activity of the speckled wood butterfly *Pararge aegeria*. *Oikos*, 42, 371–377.
Shreeve, T.G., Dennis, R.L. and Wakeham-Dawson, A. (2000) Phylogenetic, habitat, and behavioural aspects of possum behaviour in European Lepidoptera. *Journal of Research on the Lepidoptera*, 39, 80–85.
Shuker, D.M. (2010) Sexual selection: endless forms or tangled bank? *Animal Behaviour*, 79(3), e11–e17.
Shull, E.M. (1978) Mating behavior of butterflies (Papilionoidea) and skippers (Hesperioidea) in Indiana. *Proceedings of the Indiana Academy of Science*, 88, 200–208.
Shyy, W., Kang, C.K., Chirarattananon, P., Ravi, S. and Liu, H. (2016) Aerodynamics, sensing and control of insect-scale flapping-wing flight. *Proceedings of the Royal Society A: Mathematical, Physical and Engineering Sciences*, 472(2186), 20150712.
Siddique, R.H., Gomard, G. and Hölscher, H. (2015) The role of random nanostructures for the omnidirectional anti-reflection properties of the glasswing butterfly. *Nature Communications*, 6, 6909.
Siddique, R.H., Vignolini, S., Bartels, C., Wacker, I. and Hölscher, H. (2016) Colour formation on the wings of the butterfly *Hypolimnas salmacis* by scale stacking. *Scientific Reports*, 6, 36204.
Silberglied, R.E. (1977) Communication in the Lepidoptera. In: Sebeok, T.A. (ed.), *How Animals Communicate*. Indiana University Press, Bloomington, IN, pp. 362–402.
Silberglied, R.E. (1979) Communication in the ultraviolet. *Annual Review of Ecology and Systematics*, 10(1), 373–398.
Silberglied R.E. (1984) Visual communication and sexual selection among butterflies. In: Vane-Wright, R.I. and Ackery, P.R. (eds), *The Biology of Butterflies* (Symposia of the Royal Entomological Society of London, 11). Academic Press, London, pp. 207–223.
Silberglied, R.E. and Taylor, O.T. (1973) Ultraviolet differences between the sulphur butterflies, *Colias eurytheme* and *C. philodice*, and a possible isolating mechanism. *Nature*, 241, 406–408.
Silberglied, R.E. and Taylor, O.R. (1978) Ultraviolet reflection and its behavioral role in the courtship of the sulfur butterflies *Colias eurytheme* and *C. philodice* (Lepidoptera, Pieridae). *Behavioral Ecology and Sociobiology*, 3(3), 203–243.
Silberglied, R.E., Aiello, A. and Lamas, G. (1979) Neotropical butterflies of the genus *Anartia*: systematics, life histories and general biology (Lepidoptera: Nymphalidae). *Psyche: A Journal of Entomology*, 86(2–3), 219–260.

Silberglied, R.E., Shepherd, J.G. and Dickinson, J.L. (1984) Eunuchs: the role of apyrene sperm in Lepidoptera? *The American Naturalist*, 123(2), 255–265.

Siler, C.D. and Welton, L.J. (2010) Geographic variation in Philippine mimicry system: hypothesized widespread coral snake (*Hemibungarus calligaster*) mimicry by Lepidopteran larvae (*Bracca* sp.) on Luzon Island, Philippines. *Herpetological Review*, 41(4), 427.

Simmons, L.W. (2014) Sexual selection and genital evolution. *Austral Entomology*, 53(1), 1–17.

Singer, M.C. (1982) Sexual selection for small size in male butterflies. *The American Naturalist*, 119(3), 440–443.

Sison-Mangus, M.P. and Briscoe, A.D. (2009) Molecular and physiological innovations of the lycaenid butterfly eyes. In: Goldsmith M.R. and Marec, F. (eds), *Molecular Biology and Genetics of the Lepidoptera*. CRC Press, University of California, Irvine, CA.

Sison-Mangus, M.P., Bernard, G.D., Lampel, J. and Briscoe, A.D. (2006) Beauty in the eye of the beholder: the two blue opsins of lycaenid butterflies and the opsin gene-driven evolution of sexually dimorphic eyes. *Journal of Experimental Biology*, 209(16), 3079–3090.

Smith, D.A.S. (1976) Phenotypic diversity, mimicry and natural selection in the African butterfly *Hypolimnas misippus* L. (Lepidoptera: Nymphalidae). *Biological Journal of the Linnean Society*, 8(3), 183–204.

Smith, D.A. (2014) *African Queens and Their Kin: a Darwinian Odyssey*. Brambleby Books, Taunton, UK.

Smith, J.M. and Parker, G.A. (1976) The logic of asymmetric contests. *Animal Behaviour*, 24(1), 159–175.

Snell-Rood, E.C. and Papaj, D.R. (2009) Patterns of phenotypic plasticity in common and rare environments: a study of host use and color learning in the cabbage white butterfly *Pieris rapae*. *The American Naturalist*, 173(5), 615–631.

Sohn, J.C., Labandeira, C.C., Davis, D.R. and Mitter, C. (2012) An annotated catalog of fossil and subfossil Lepidoptera (Insecta: Holometabola) of the world. *Zootaxa*, 3286, 1–132.

Sohn, J.C., Labandeira, C.C. and Davis, D.R. (2015) The fossil record and taphonomy of butterflies and moths (Insecta, Lepidoptera): implications for evolutionary diversity and divergence-time estimates. *BMC Evolutionary Biology*, 15(1), 1.

Solensky, M.J. (2004) The effect of behavior and ecology on male mating success in overwintering monarch butterflies (*Danaus plexippus*). *Journal of Insect Behavior* 17(6), 723–743.

Solensky, M.J. and Oberhauser, K.S. (2004) Behavioral and genetic components of male mating success in monarchs. In: Oberhauser, K. and Solensky, M. (eds), *The Monarch Butterfly: Biology and Conservation*. Cornell University Press, Ithaca, NY, pp. 61–68.

Solensky, M.J. and Oberhauser, K.S. (2009) Male monarch butterflies, *Danaus plexippus*, adjust ejaculates in response to intensity of sperm competition. *Animal Behaviour*, 77, 465–472.

Sourakov, A. (2008) Pupal mating in zebra longwing (*Heliconius charithonia*): photographic evidence. *News of the Lepidopterists' Society*, 50, 26–32.

Sourakov, A. and Emmel, T.C. (1997) Mating habits in the genus *Acraea*, with a possible explanation for monosexual populations. *Tropical Lepidoptera*, 8(3), 33–35.

South, A. and Lewis, S.M. (2011) The influence of male ejaculate quantity on female fitness: a meta-analysis. *Biological Reviews*, 86, 299–309.

Southcott, L. and Kronforst, M.R. (2018) Female mate choice is a reproductive isolating barrier in *Heliconius* butterflies. *Ethology*, 124(12), 862–869.

Sperling, F.A.H. (1990) Natural hybrids of *Papilio* (Insecta: Lepidoptera): poor taxonomy or interesting evolutionary problem? Canadian *Journal of Zoology* (Ottawa), 6, 1790–1799.

Spitzer, L., Beneš, J., Vrba, P. and Zlatnık, M. (2010) Three observation of interspecific mating attempts by males of the meadow brown (*Maniola jurtina* (Linnaeus, 1758)) in the wild

(Lepidoptera, Nymphalidae: Satyrinae, Heliconiinae). *Nachrichten des Entomologischen Vereins Apollo*, 31(3), 166–168.
Srygley, R.B. (1994) Shivering and its cost during reproductive behaviour in Neotropical owl butterflies, *Caligo* and *Opsiphanes* (Nymphalidae: Brassolinae). *Animal Behaviour*, 47(1), 23–32.
Srygley, R.B. and Penz, C.M. (1999) Lekking in Neotropical owl butterflies, *Caligo illioneus* and *C. oileus* (Lepidoptera: Brassolinae). *Journal of Insect Behavior*, 12(1), 81–103.
Stallings, D.B., Stallings, V.N., Turner, J.R. and Turner, B.R. (1985) Courtship and oviposition patterns of two *Agathymus* (Megathymidae). *Journal of the Lepidopterists' Society (USA)*, 39(3), 171–176.
Stavenga, D.G. (1979) Pseudopupils of compound eyes. In: Autrum, H. (ed.), *Comparative Physiology and Evolution of Vision in Invertebrates*. Springer, Berlin, Heidelberg, pp. 357–439.
Stavenga, D.G. (2014) Thin film and multilayer optics cause structural colors of many insects and birds. *Materials Today: Proceedings*, 1, 109–121.
Stavenga, D.G. and Arikawa, K. (2006) Evolution of color and vision of butterflies. *Arthropod Structure and Development*, 35(4), 307–318.
Stavenga, D.G. and Hardie, R.C. (eds) (1989) *Facets of Vision*. Springer Science & Business Media, Berlin, Heidelberg, Germany.
Stavenga, D.G. and Leertouwer, H.L. (2007) Colourful butterfly wings: scale stacks, iridescence and sexual dichromatism of Pieridae. *Entomologische Berichten-Nederlandsche Entomologische Vereenigung*, 67(5), 158.
Stavenga, D.G., Foletti, S., Palasantzas, G. and Arikawa, K. (2006a) Light on the moth-eye corneal nipple array of butterflies. *Proceedings of the Royal Society of London B: Biological Sciences*, 273(1587), 661–667.
Stavenga, D.G., Giraldo, M. A. and Hoenders, B.J. (2006b) Reflectance and transmittance of light scattering scales stacked on the wings of pierid butterflies. *Optics Express*, 14(11), 4880–4890.
Stavenga, D.G., Leertouwer, H.L., Pirih, P. and Wehling, M.F. (2009) Imaging scatterometry of butterfly wing scales. *Optics Express*, 17(1), 193–202.
Stavenga, D.G., Giraldo, M.A. and Leertouwer, H.L. (2010) Butterfly wing colors: glass scales of *Graphium sarpedon* cause polarized iridescence and enhance blue/green pigment coloration of the wing membrane. *Journal of Experimental Biology*, 213(10), 1731–1739. doi:10.1242/jeb.041434
Stavenga, D.G., Matsushita, A., Arikawa, K., Leertouwer, H.L. and Wilts, B.D. (2012) Glass scales on the wing of the swordtail butterfly *Graphium sarpedon* act as thin film polarizing reflectors. *Journal of Experimental Biology*, 215, 657–662.
Stavenga, D.G., Leertouwer, H.L. and Wilts, B.D. (2014a) The colouration toolkit of the pipevine swallowtail butterfly, *Battus philenor*: thin films, papiliochromes, and melanin. *Journal of Comparative Physiology A*, 200(6), 547–561.
Stavenga, D.G., Leertouwer, H.L. and Wilts, B.D. (2014b) Coloration principles of nymphaline butterflies – thin films, melanin, ommochromes and wing scale stacking. *Journal of Experimental Biology*, 217(12), 2171–2180.
Stavenga, D.G., Matsushita, A. and Arikawa, K. (2015) Combined pigmentary and structural effects tune wing scale coloration to color vision in the swallowtail butterfly *Papilio xuthus*. *Zoological Letters*, 1(1), 1.
Steiger, S. and Stökl, J. (2014) The role of sexual selection in the evolution of chemical signals in insects. *Insects*, 5(2), 423–438.
Stella, D., Faltýnek Fric, Z., Rindoš, M., Kleisner, K. and Pecháček, P. (2018a) Distribution of ultraviolet ornaments in *Colias* butterflies (Lepidoptera: Pieridae). *Environmental Entomology*, 47(5), 1344–1354.

Stella, D., Pecháček, P., Meyer-Rochow, V. B. and Kleisner, K. (2018b) UV reflectance is associated with environmental conditions in Palaearctic *Pieris napi* (Lepidoptera: Pieridae). *Insect science*, 25(3), 508–518.

Stevens, M. (2005) The role of eyespots as anti-predator mechanisms, principally demonstrated in the Lepidoptera. *Biological Reviews*, 80(4), 573–588.

Stevens, M. (2007) Predator perception and the interrelation between different forms of protective coloration. *Proceedings of the Royal Society of London B: Biological Sciences*, 274(1617), 1457–1464.

Stevens, M. (2015) Anti-predator coloration and behaviour: a longstanding topic with many outstanding questions. *Current Zoology*, 61, 703–707.

Stillwell, R.C., Blanckenhorn, W.U., Teder, T., Davidowitz, G. and Fox, C.W. (2010) Sex differences in phenotypic plasticity affect variation in sexual size dimorphism in insects: from physiology to evolution. *Annual Review of Entomology*, 55, 227–245.

Stjernholm, F. and Karlsson, B. (2000) Nuptial gifts and the use of body resources for reproduction in the green-veined white butterfly *Pieris napi*. *Proceedings of the Royal Society of London B: Biological Sciences*, 267(1445), 807–811.

Stoddard, M.C. (2012) Mimicry and masquerade from the avian visual perspective. *Current Zoology*, 58(4), 630–648.

Stoehr, A. M., Hayes, K. and Wojan, E.M. (2016) Assessing the role of wing spots in intraspecific communication in the cabbage white butterfly using a simple device to increase butterfly responses. *Journal of Insect Behavior*, 29(3), 243–255.

Stride, G.O. (1956) On the courtship behaviour of *Hypolimnas misippus* L. (Lepidoptera, Nymphalidae), with notes on the mimetic association with *Danaus chrysippus* L. (Lepidoptera, Danaidae). *The British Journal of Animal Behaviour*, 4(2), 52–68.

Stride, G.O. (1957) Investigations into the courtship behaviour of the male of *Hypolimnas misippus* L. (Lepidoptera, Nymphalidae), with special reference to the role of visual stimuli. *The British Journal of Animal Behaviour*, 5(4), 153–167.

Stride, G.O. (1958) Further studies on the courtship behaviour of African mimetic butterflies. *Animal Behaviour*, 6(3–4), 224–230.

Stüning, D. (2006) The genus *Bracca* Hübner in the Oriental and Australian tropics: distribution patterns and the phenomenon of strikingly different island races (Geometridae, Ennominae). *Spixiana*, 29/3, 207–208.

Stutt, A.D. and Willmer, P. (1998) Territorial defence in speckled wood butterflies: do the hottest males always win? *Animal Behaviour*, 55, 1341–1347.

Sugawara, T. (1979) Stretch reception in the bursa copulatrix of the butterfly, *Pieris rapae crucivora*, and its role in behaviour. *Journal of Comparative Physiology*, 130, 191–199.

Sun, G., Fang, Y., Cong, Q. and Ren, L.Q. (2009) Anisotropism of the non-smooth surface of butterfly wing. *Journal of Bionic Engineering*, 6(1), 71–76.

Sun, J., Bhushan, B. and Tong, J. (2013) Structural coloration in nature. *Research Advances*, 3(35), 14862–14889.

Suzuki, N. and Matsumoto, K. (1990) Pair-clinging behavior by males of *Atrophaneura alcinous* (Lepidoptera: Papilionidae). *Journal of Ethology*, 8(2), 45–51.

Suzuki, N. and Matsumoto, K. (1992) Lifetime mating success of males in a natural population of the papilionid butterfly, *Atrophaneura alcinous* (Lepidoptera: Papilionidae). *Researches on Population Ecology*, 34(2), 397–407.

Suzuki, Y. (1976) So-called territorial behaviour of the small copper, *Lycaena phlaeas daimio* Seitz (Lepidoptera, Lycaenidae) 昆蟲, 44(2), 193–204. (*Kontyû*: Japanese Journal of Entomology, Tokyo)

Suzuki, Y. (1979) Mating frequency in females of the small cabbage white, *Pieris rapae crucivora* Boisduval (Lepidoptera: Pieridae) 昆蟲, 47(3), 335–339. (*Kontyû*: Japanese Journal of Entomology, Tokyo)

Suzuki, Y. (1980) The forming process of male-biased sex ratio at an emergence site in *Pieris rapae crucivora* Boisduval (Lepidoptera, Pieridae). *Kontyû*, 48(2), 299–306.

Suzuki, Y., Nakanishi, A., Shima, H., Yata, O. and Saigusa, T. (1977) Mating behaviour of four Japanese species of the genus *Pieris* (Lepidoptera, Pieridae) 昆蟲, 45(2), 300–313. (*Kontyû*: Japanese Journal of Entomology, Tokyo).

Svärd, L. (1985) Paternal investment in a monandrous butterfly, *Pararge aegeria*. *Oikos*, 45, 66–70.

Svärd, L. and Wiklund, C. (1986) Different ejaculate delivery strategies in first versus subsequent matings in the swallowtail butterfly *Papilio machaon* L. *Behavioral Ecology and Sociobiology*, 18(5), 325–330.

Svärd, L. and Wiklund, C. (1988a) Prolonged mating in the monarch butterfly *Danaus plexippus* and nightfall as a cue for sperm transfer. *Oikos*, 52, 351–354.

Svärd, L. and Wiklund, C. (1988b) Fecundity, egg weight and longevity in relation to multiple matings in females of the monarch butterfly. *Behavioral Ecology and Sociobiology*, 23(1), 39–43.

Svärd, L. and Wiklund, C. (1989) Mass and production rate of ejaculates in relation to monandry/polyandry in butterflies. *Behavioral Ecology and Sociobiology*, 24(6), 395–402.

Svärd, L. and Wiklund, C. (1991) The effect of ejaculate mass on female reproductive output in the European swallowtail butterfly, *Papilio machaon* (L.) (Lepidoptera: Papilionidae). *Journal of Insect Behavior*, 4(1), 33–41.

Swallow, J.G., and Wilkinson, G.S. (2002) The long and short of sperm polymorphisms in insects. *Biological Reviews*, 77(2), 153–182.

Sweeney, A., Jiggins, C. and Johnsen, S. (2003) Insect communication: polarized light as a butterfly mating signal. *Nature*, 423(6935), 31.

Tabashnik, B.E. (1980) Population structure of pierid butterflies. *Oecologia*, 47(2), 175–183.

Takemura, S.Y. and Arikawa, K. (2006) Ommatidial type-specific interphotoreceptor connections in the lamina of the swallowtail butterfly, *Papilio xuthus*. *The Journal of Comparative Neurology*, 494, 663–672.

Takemura, S.Y., Stavenga, D.G. and Arikawa, K. (2007) Absence of eye shine and tapetum in the heterogeneous eye of *Anthocharis* butterflies (Pieridae). *The Journal of Experimental Biology*, 210, 3075–3081.

Takeuchi, T. (2006) The effect of morphology and physiology on butterfly territoriality. *Behaviour*, 143(3), 393–403.

Takeuchi, T. (2010) Mate-locating behavior of the butterfly *Lethe diana* (Lepidoptera: Satyridae): do males diurnally or seasonally change their mating strategy? *Zoological Science*, 27(10), 821–825.

Takeuchi, T. (2012) Cost of reproduction in males of a satyrine butterfly *Lethe diana*. *Physiological Entomology*, 37(2), 171–176.

Takeuchi T. (2016) Early comers occupy popular territories in a lycaenid butterfly, *Chrysozephyrus smaragdinus* (Lepidoptera: Lycaenidae). *Entomological Science*, 19, 42–48.

Takeuchi, T. (2017) Agonistic display or courtship behavior? A review of contests over mating opportunity in butterflies. *Journal of Ethology*, 35(1), 3–12.

Takeuchi T. and Honda, K. (2009) Early comers become owners: effect of residency experience on territorial contest dynamics in a lycaenid butterfly. *Ethology*, 115, 767–773.

Takeuchi, T. and Imafuku, M. (2005a) Territorial behavior of a green hairstreak *Chrysozephyrus smaragdinus* (Lepidoptera: Lycaenidae): site tenacity and wars of attrition. *Zoological Science*, 22(9), 989–994.

Takeuchi, T. and Imafuku, M. (2005b) Territorial behavior of *Favonius taxila* (Lycaenidae): territory size and persistency. *Journal of Research on the Lepidoptera*, 38, 59–66.

Takeuchi, Y., Arikawa, K. and Kinoshita, M. (2006) Color discrimination at the spatial resolution limit in a swallowtail butterfly, *Papilio xuthus*. *The Journal of Experimental Biology*, 209, 2873–2879.

Takeuchi, T., Yabuta, S. and Tsubaki, Y. (2016) The erroneous courtship hypothesis: do insects really engage in aerial wars of attrition? *Biological Journal of the Linnean Society*, 118(4), 970–981.

Talla, V., Suh, A., Kalsoom, F., Dincă, V., Vila, R., Friberg, M., Wiklund, C. and Backström, N. (2017) Rapid increase in genome size as a consequence of transposable element hyperactivity in wood-white (*Leptidea*) butterflies. *Genome Biology and Evolution*, 9(10), 2491–2505.

Tam, H.L., Cheah, K.W., Goh, D.T. and Goh, J.K. (2013) Iridescence and nano-structure differences in *Papilio* butterflies. *Optical Materials Express*, 3(8), 1087–1092.

Tang-Martinez, Z. and Ryder, T. B. (2005) The problem with paradigms: Bateman's worldview as a case study. *Integrative and Comparative Biology*, 45(5), 821–830.

Taylor Jr, O.R. (1972) Random vs. non-random mating in the sulfur butterflies, *Colias eurytheme* and *Colias philodice* (Lepidoptera: Pieridae). *Evolution*, 26(3), 344–356.

Taylor Jr, O.R. (1973a) A non-genetic 'polymorphism' in *Anartia fatima* (Lepidoptera: Nymphalidae). *Evolution*, 27(1), 161–164.

Taylor Jr, O.R. (1973b) Reproductive isolation in *Colias eurytheme* and *C. philodice* (Lepidoptera: Pieridae): use of olfaction in mate selection. *Annals of the Entomological Society of America*, 66(3), 621–626.

Tennent, W.J. (1987) A note on the apparent lowering of moral standards in the Lepidoptera. *Entomologist's Record and Journal of Variation*, 99, 81–83.

Thornhill, R. and Alcock, J. (1983) *The Evolution of Insect Mating Systems*. Harvard University Press, Cambridge, MA.

Thurman, T.J. and Seymoure, B.M. (2016). A bird's eye view of two mimetic tropical butterflies: coloration matches predator's sensitivity. *Journal of Zoology*, 298(3), 159–168.

Thurman, T.J., Brodie, E., Evans, E. and McMillan, W.O. (2017) Facultative pupal mating in *Heliconius erato*: implications for mate choice, female preference, and speciation. *Ecology and Evolution*, 8(3), 1882–1889.

Thurman, T.J., Brodie, E., Evans, E. and McMillan, W.O. (2018). Facultative pupal mating in *Heliconius erato*: implications for mate choice, female preference, and speciation. *Ecology and Evolution*, 8(3), 1882–1889.

Tigreros, N., Mowery, M.A. and Lewis, S.M. (2014) Male mate choice favors more colorful females in the gift-giving cabbage butterfly. *Behavioral Ecology and Sociobiology*, 68(9), 1539–1547.

Tinbergen, N. (1941) Ethologische Beobachtungen am Samtfalter, *Satyrus semele* L. *Journal of Fuer Ornithologie*, 89(3), 133–144.

Tinbergen, N. (1948) Social releasers and the experimental method required for their study. *The Wilson Bulletin*, 6–51.

Tinbergen, N. (1972) *The Animal in Its World (Explorations of an Ethologist, 1932–1972): Field Studies* (Vol. 84). Harvard University Press, Harvard, CT.

Tinbergen, N., Meeuse, B.J.D. and Varossieau, L.K. (1942) Die Balz des Samtfalters, *Eumenis (= Satyrus) semele* (L.). *Ethology*, 5(1), 182–226.

Tinbergen N., Meeuse B.J.D., Boerema L.K. and Varossieau W. (1972) The courtship of the grayling *Eumenis* (= *Satyrus*) *semele*. In: Tinbergen N. (ed.), *The Animal in its World*. Allen & Unwin, London, pp. 197–249.

Tinghitella, R.M., Weigel, E.G., Head, M. and Boughman, J.W. (2013) Flexible mate choice when mates are rare and time is short. *Ecology and Evolution*, 3(9), 2820–2831.

Tokita, C.K., Oliver, J.C. and Monteiro, A. (2013) A survey of eyespot sexual dimorphism across Nymphalid butterflies. *International Journal of Evolutionary Biology*, 2013, 926702. doi:10.1155/2013/926702

Torres-Vila, L.M. and Jennions, M.D. (2005) Male mating history and female fecundity in the Lepidoptera: do male virgins make better partners? *Behavioral Ecology and Sociobiology*, 57(4), 318–326.

Torres-Vila, L.M., Rodríguez-Molina, M.C. and Jennions, M.D. (2004) Polyandry and fecundity in the Lepidoptera: can methodological and conceptual approaches bias outcomes? *Behavioral Ecology and Sociobiology*, 55(4), 315–324.

Tregenza, T. and Wedell, N. (2000) Genetic compatibility, mate choice and patterns of parentage: invited review. *Molecular Ecology*, 9(8), 1013–1027.

Tregenza, T., Wedell, N. and Chapman, T. (2006) Introduction. Sexual conflict: a new paradigm? *Philosophical Transactions of the Royal Society of London B: Biological Sciences*, 361, 229–234.

Treusch, H.W. (1967) Bisher unbekanntes gezieltes Duftanbieten paarungsbereiter Argynnis paphia-Weibchen. *Naturwissenschaften*, 54(22), 592–592.

Trigo, J.R. and Motta, P.C. (1990) Evolutionary implications of pyrrolizidine alkaloid assimilation by danaine and ithomiine larvae (Lepidoptera: Nymphalidae). *Experientia*, 46(3), 332–334.

Trigo, J.R., Brown, K.S., Witte, L., Hartmann, T., Ernst, L. and Barata, L.E.S. (1996) Pyrrolizidine alkaloids: different acquisition and use patterns in Apocynaceae and Solanaceae feeding ithomiine butterflies (Lepidoptera: Nymphalidae). *Biological Journal of the Linnean Society*, 58(1), 99–123.

Trujano-Ortega, M., García-Vázquez, U.O., Callaghan, C. J., Ávalos-Hernández, O., Luis-Martínez, M.A. and Llorente-Bousquets, J.E. (2018). Two new genera of metalmark butterflies of North and Central America (Lepidoptera, Riodinidae). *ZooKeys*, (729), 61–85.

Trzeciak, T.M., Wilts, B.D., Stavenga, D.G. and Vukusic, P. (2012) Variable multilayer reflection together with long-pass filtering pigment determines the wing coloration of papilionid butterflies of the nireus group. *Optics Express*, 20, 8877–8890.

Tschudi-Rein, K. and Benz, G. (1990) Mechanisms of sperm transfer in female *Pieris brassicae* (Lepidoptera: Pieridae). *Annals of the Entomological Society of America*, 83(6), 1158–1164.

Tsubaki, Y. and Matsumoto, K. (1998) Fluctuating asymmetry and male mating success in a sphragis-bearing butterfly *Luehdorfia japonica* (Lepidoptera: Papilionidae). *Journal of Insect Behavior*, 11(4), 571–582.

Tuomaala, M., Kaitala, A. and Rutowski, R.L. (2012) Females show greater changes in wing colour with latitude than males in the green-veined white butterfly, *Pieris napi* (Lepidoptera: Pieridae). *Biological Journal of the Linnean Society*, 107(4), 899–909.

Turlure, C. and Van Dyck, H. (2009) On the consequences of aggressive male mate-locating behaviour and micro-climate for female host plant use in the butterfly *Lycaena hippothoe*. *Behavioral Ecology and Sociobiology*, 6, 1581–1591.

Turlure, C., Schtickzelle, N., Van Dyck, H., Seymoure, B. and Rutowski, R. (2016) Flight morphology, compound eye structure and dispersal in the bog and the cranberry fritillary butterflies: an inter- and intraspecific comparison. *PLoS ONE*, 11(6), e0158073. doi:10.1371/journal.pone.0158073

Turner, J.R. (1978) Why male butterflies are non-mimetic: natural selection, sexual selection, group selection, modification and sieving. *Biological Journal of the Linnean Society*, 10(4), 385–432.

Uchiyama, H., Awata, H., Kinoshita, M. and Arikawa, K. (2013) Rough eyes of the northeast-Asian wood white, *Leptidea amurensis*. *Journal of Experimental Biology*, 216(18), 3414–3421.

Umebachi, Y. (1977) *Distribution of Papiliochrome in Papilionid Butterflies. The Science Reports of the Kanazawa University*, 22, 187–195

Vahed, K. (1998) The function of nuptial feeding in insects: a review of empirical studies. *Biological Reviews*, 73(1), 43–78.

Välimäki, P., Kaitala, A. and Kokko, H. (2006) Temporal patterns in reproduction may explain variation in mating frequencies in the green-veined white butterfly *Pieris napi*. *Behavioral Ecology and Sociobiology*, 61(1), 99–107.

Välimäki, P., Kivelä, S.M. and Mäenpää, M.I. (2011) Mating with a kin decreases female remating interval: a possible example of inbreeding avoidance. *Behavioral Ecology and Sociobiology*, 65(11), 2037–2047.

Van Dyck, H. (2003) Mate-location, a matter of design? Adaptive morphological variation in the speckled wood butterfly. In: Boggs L.C., Watt, W.B. and Ehrlich, P.R. (eds), *Butterflies: Ecology and Evolution Taking Flight*. University of Chicago Press, Chicago, IL, pp. 353–366.

Van Dyck, H. and Matthysen, E. (1998) Thermoregulatory differences between phenotypes in the speckled wood butterfly: hot perchers and cold patrollers? *Oecologia*, 114(3), 326–334.

Van Dyck, H., Matthysen, E. and Dhondt, A.A. (1997a) The effect of wing colour on male behavioural strategies in the speckled wood butterfly. *Animal Behaviour*, 53, 39–51.

Van Dyck, H., Matthysen, E. and Dhondt, A.A. (1997b) Mate-locating strategies are related to relative body length and wing colour in the speckled wood butterfly *Pararge aegeria*. *Ecological Entomology*, 22, 116–120.

van Eldijk, T.J., Wappler, T., Strother, P.K., van der Weijst, C.M., Rajaei, H., Visscher, H. and van de Schootbrugge, B. (2018) A Triassic-Jurassic window into the evolution of Lepidoptera. *Science Advances*, 4(1), e1701568.

Van Hooijdonk, E., Barthou, C., Vigneron, J.P. and Berthier, S. (2012) Angular dependence of structural fluorescent emission from the scales of the male butterfly *Troïdes magellanus* (Papilionidae). *JOSA B*, 29(5), 1104–1111.

van Nieukerken, E.J., Kaila, L., Kitching, I.J., Kristensen, N.P., Lees, D.C., Minet, J., Mitter, C., Mutanen, M., Regier, J.C., Simonsen, T.J. and Wahlberg, N. (2011) Order Lepidoptera Linnaeus, 1758. *Zootaxa*, 3148, 212–221.

Vane-Wright, D. (2004) Entomology: butterflies at that awkward age. *Nature*, 428(6982), 477–480.

Vane-Wright, R.I. (1972) Pre-courtship activity and a new scent organ in butterflies. *Nature*, 239(5371), 338.

Vane-Wright, R.I. and Boppré, M. (1990) The unknown male of *Tiradelphe schneideri* (Lepidoptera, Danainae): missing piece in a butterfly puzzle. *Lepidoptera Science*, 41(3), 193–199.

Vane-Wright, R.I. and Boppré, M. (1993) Visual and chemical signalling in butterflies: functional and phylogenetic perspectives. *Philosophical Transactions of the Royal Society of London B: Biological Sciences*, 340(1292), 197–205.

Vane-Wright, R.I. and Boppré, M. (2004) Adult morphology and the higher classification of *Bia* Hübner (Lepidoptera: Nymphalidae). *Bonner zoologische Beiträge* (Bonn zoological Bulletin), 53, 235–254.

Vane-Wright, R.I., Boppré, M. and Ackery, P.R. (2002) Miriamica, a new genus of milkweed butterflies with unique androconial organs (Lepidoptera: Nymphalidae). *Zoologischer Anzeiger-A Journal of Comparative Zoology*, 241(3), 255–267.

Vanjari, S., Mann, F., Merrill, R.M., Schulz, S. and Jiggins, C.D. (2015a) Male sex pheromone components in the butterfly *Heliconius melpomene*. *bioRxiv*, 033506. doi:https://doi.org/10.1101/033506

Vanjari, S., Mann, F., Merrill, R.M., Schulz, S. and Jiggins, C.D. (2015b) Male sex pheromone components in the butterfly *Heliconius melpomene*. *Bior*χ*ivbeta*, 033506.

Velde, L.V. (2011) Alternative mate location strategies in the speckled wood butterfly (*Pararge aegeria*): an evolutionary ecological approach. PhD thesis, UCL-Université Catholique de Louvain, Belgium.

Velde, L.V., Damiens, D. and Van Dyck, H. (2011) Spermatophore and sperm allocation in males of the monandrous butterfly *Pararge aegeria*: the female's perspective. *Ethology*, 117(7), 645–654.

Velde, L.V., Silvestre, P., Damiens, D. and Van Dyck, H. (2012) Male reproductive investment relative to age and flight behaviour in the monandrous butterfly *Pararge aegeria*. *Behavioral Ecology and Sociobiology*, 66(3), 347–359.

Vértesy, Z., Bálint, Z., Kertész, K., Vigneron, J.P., Lousse, V. and Biró, L.P. (2006) Wing scale microstructures and nanostructures in butterflies – natural photonic crystals. *Journal of Microscopy*, 224(1), 108–110.

Vetter, R.S. and Rutowski, R.L. (1978) External sex brand morphology of three sulphur butterflies (Lepidoptera: Pieridae). *Psyche: A Journal of Entomology*, 85(4), 383–393.

Vieira, M.C. and Peixoto, P.E. (2013) Winners and losers: a meta-analysis of functional determinants of fighting ability in arthropod contests. *Functional Ecology*, 27(2), 305–313.

Vigneron, J.P. and Simonis, P. (2010) Structural colours. *Advances in Insect Physiology*, 38, 181–218.

Vigneron, J.P., Kertész, K., Vértesy, Z., Rassart, M., Lousse, V., Bálint, Z. and Biró, L.P. (2008) Correlated diffraction and fluorescence in the backscattering iridescence of the male butterfly *Troides magellanus* (Papilionidae). *Physical Review E*, 78(2), 021903.

Visser, B., Dublon, I. A., Heuskin, S., Laval, F., Bacquet, P.M., Lognay, G. and Nieberding, C.M. (2018) Common practice solvent extraction does not reflect actual emission of a sex pheromone during butterfly courtship. *bioRxiv*, 270462. http://dx.doi.org/10.1101/270462

Vlasanek, P. and Konvicka, M. (2009) Sphragis in *Parnassius mnemosyne* (Lepidoptera: Papilionidae): male-derived insemination plugs loose efficiency with progress of female flight. *Biologia*, 64(6), 1206–1211.

Vlasanek, P., Hauck, D. and Konvicka, M. (2009) Adult sex ratio in the *Parnassius mnemosyne* butterfly: effects of survival, migration, and weather. *Israel Journal of Ecology and Evolution*, 55(3), 233–252.

Vogel R. 1912. Uber die Chordotonalorgane in der Wurzel der Schmetterlingsflugel. *Zeitschrift für wissenschaftliche Zoologie*, 100, 210–244.

von Frisch, K. (1914) Der Farbensinn und Formensinn der Biene. *Zoologische Jahrbücher. Abteilung für allgemeine Zoologie und Physiologie der Tiere*, 35, 1–188.

Vukusic, P. (2011) Structural colour: elusive iridescence strategies brought to light. *Current Biology*, 21(5), R187–R189.

Vukusic, P. and Sambles, J.R. (2003) Photonic structures in biology. *Nature*, 424(6950), 852.

Vukusic, P., Sambles, J.R., Lawrence, C.R. and Wootton, R.J. (1999) Quantified interference and diffraction in single Morpho butterfly scales. *Proceedings of the Royal Society of London B: Biological Sciences*, 266(1427), 1403–1411.

Vukusic, P., Sambles, J.R. and Lawrence, C.R. (2000) Structural colour: colour mixing in wing scales of a butterfly. *Nature*, 404(6777), 457–457.

Vukusic, P., Sambles, R., Lawrence, C. and Wakely, G. (2001a) Sculpted-multilayer optical effects in two species of *Papilio* butterfly. *Applied Optics*, 40(7), 1116–1125.

Vukusic, P., Sambles, J.R., Lawrence, C.R. and Wootton, R.J. (2001b) Structural colour: now you see it – now you don't. *Nature*, 410(6824), 36.

Vukusic, P., Sambles, J.R., Lawrence, C.R. and Wootton, R.J. (2002) Limited-view iridescence in the butterfly *Ancyluris meliboeus*. *Proceedings of the Royal Society of London B: Biological Sciences*, 269(1486), 7–14.

Vukusic, P., Sambles, J.R. and Lawrence, C.R. (2004) Structurally assisted blackness in butterfly scales. *Proceedings of the Royal Society of London B: Biological Sciences*, 271(Suppl 4), S237–S239.

Vukusic, P., Kelly, R. and Hooper, I. (2009) A biological sub-micron thickness optical broadband reflector characterized using both light and microwaves. *Journal of the Royal Society Interface*, 6(Suppl 2), S193–S201.

Wago, H. (1977) Mating behavior of *Zizeeria maha argia* in connection with the visual stimuli from *Zizina otis*. *Applied Entomology and Zoology*, 12(2), 203–205.

Wago, H. (1978) Studies on the mating behavior of the pale grass blue, *Zizeeria maha argia* (Lepidoptera: Lycaenidae) III. Olfactory cues in sexual discrimination by males. *Applied Entomology and Zoology*, 13, 283–289.

Wago, H., Unno, K. and Suzuki, Y. (1976) Studies on the mating behavior of the pale grass blue, *Zizeeria maha argia* (Lepidoptera: Lycaenidae): I. Recognition of conspecific individuals by flying males. *Applied Entomology and Zoology*, 11(4), 302–311.

Wahlberg, N. (2006) That awkward age for butterflies: insights from the age of the butterfly subfamily Nymphalinae (Lepidoptera: Nymphalidae). *Systematic Biology*, 55(5), 703–714.

Wahlberg, N., Braby, M.F., Brower, A.V., de Jong, R., Lee, M.M., Nylin, S., Pierce, N.E., Sperling, F.A., Vila, R., Warren, A.D. and Zakharov, E. (2005) Synergistic effects of combining morphological and molecular data in resolving the phylogeny of butterflies and skippers. *Proceedings of the Royal Society B: Biological Sciences*, 272(1572), 1577–1586.

Wahlberg, N., Leneveu, J., Kodandaramaiah, U., Peña, C., Nylin, S., Freitas, A. V. and Brower, A. V. (2009) Nymphalid butterflies diversify following near demise at the Cretaceous/Tertiary boundary. *Proceedings of the Royal Society of London B: Biological Sciences*, rspb20091303.

Wakakuwa, M., Stavenga, D. G., Kurasawa, M. and Arikawa, K. (2004) A unique visual pigment expressed in green, red and deep-red receptors in the eye of the small white butterfly, *Pieris rapae crucivora*. *Journal of Experimental Biology*, 207(16), 2803–2810.

Wakeham-Dawson, A., Kudrna, O. and Dennis, R.L. (2007) Description of androconia in the Palaearctic Asian *Pseudochazara baldiva* (Moore, 1865) butterfly species-group (Nymphalidae: Satyrinae) with designation of two lectotypes and reference to type and other material in the Natural History Museum, London. *Nota Lepidopterologica*, 30(2), 211.

Walker, W.F. (1980) Sperm utilization in nonsocial insects. *American Naturalist*, 115, 780–799.

Wallace, A.R. (1865) I. On the phenomena of variation and geographical distribution as illustrated by the Papilionidæ of the Malayan Region. *Transactions of the Linnean Society of London*, 25(1), 1–71.

Walters, J.R., Stafford, C., Hardcastle, T.J. and Jiggins, C.D. (2012) Evaluating female remating rates in light of spermatophore degradation in *Heliconius* butterflies: pupal-mating monandry versus adult-mating polyandry. *Ecological Entomology*, 37(4), 257–268.

Wang, H.L., Brattström, O., Brakefield, P.M., Francke, W. and Löfstedt, C. (2014) Identification and biosynthesis of novel male specific esters in the wings of the tropical butterfly, *Bicyclus martius sanaos*. *Journal of Chemical Ecology*, 40(6), 549–559.

Warren, A.D., Ogawa, J.R. and Brower, A.V.Z. (2009) Revised classification of the Hesperiidae (Lepidoptera: Hesperioidea) based on combined molecular and morphological data. *Systematic Entomology*, 34, 467–523.

Wasik, B.R., Liew, S. F., Lilien, D.A., Dinwiddie, A.J., Noh, H., Cao, H. and Monteiro, A. (2014) Artificial selection for structural color on butterfly wings and comparison with natural evolution. *Proceedings of the National Academy of Sciences*, 111(33), 12109–12114.

Watanabe, M. (1988) Multiple matings increase the fecundity of the yellow swallowtail butterfly, *Papilio xuthus* L., in summer generations. *Journal of Insect Behavior*, 1(1), 17–29.

Watanabe, M. (2016a) Historical perspective. *Sperm Competition in Butterflies*, Ecological Research Monographs, Springer, Tokyo, pp. 7–32.

Watanabe, M. (2016b) Avoidance of sperm competition in males. In: Watanabe, M. (ed.) *Sperm Competition in Butterflies*. Ecological Research Monographs. Springer, Tokyo, pp. 121–133.

Watanabe, M. and Ando, S. (1993) Influence of mating frequency on lifetime fecundity in wild females of the small white *Pieris rapae* (Lepidoptera, Pieridae). *Japanese Journal of Entomology*, 61, 691–696.

Watanabe, M. and Hirota, M. (1999) Effects of sucrose intake on spermatophore mass produced by male swallowtail butterfly *Papilio xuthus* L. *Zoological Science*, 16(1), 55–61.

Watanabe, M. and Irie, S. (2011) Female abdomen position signals receptivity during courtship in the Japanese sulfur butterfly, *Colias erate* (Pieridae). *The Journal of the Lepidopterists' Society*, 65(4), 259–263.

Watanabe, M. and Kobayashi, T. (2006) Total sperm ejaculation in monandrous (*Papilio machaon*) and polyandrous (*P. xuthus*) swallowtail butterflies (Lepidoptera: Papilionidae) restricted to larval stage-derived nutrients. *Journal of Research on the Lepidoptera*, 39, 1–7.

Watanabe, M. and Nakanishi, Y. (1996) Population structure and dispersals of the sulfur butterfly *Colias erate* (Lepidoptera: Pieridae) in an isolated plain located in a cool temperate zone of Japan. *Japanese Journal of Applied Entomology and Zoology*, 64(1), 17–29.

Watanabe, M. and Nishimura, M. (2001) Reproductive output and egg maturation in relation to mate-avoidance in monandrous females of the small copper, *Lycaena phlaeas* (Lycaenidae). *Journal of the Lepidopterists' Society*, 54(3), 83–87.

Watanabe, M. and Sato, K. (1993) A spermatophore structured in the bursa copulatrix of the small white *Pieris rapae* (Lepidoptera, Pieridae) during copulation, and its sugar content. *Journal of Research on the Lepidoptera*, 32, 26–36.

Watanabe, M. and Sasaki, N. (2010) Pattern of sperm storage and migration in the reproductive tract of the swallowtail butterfly *Papilio xuthus*: cryptic female choice after second mating. *Physiological Entomology*, 35, 328–333.

Watanabe, M., Nakanishi, Y. and Bon'no, M. (1997) Prolonged copulation and spermatophore size ejaculated in the sulfur butterfly, *Colias erate* (Lepidoptera: Pieridae) under selective harassments of mated pairs by conspecific lone males. *Journal of Ethology*, 15(1), 45–54.

Watanabe, M., Bon'no, M. and Hachisuka, A. (2000) Eupyrene sperm migrates to spermatheca after apyrene sperm in the swallowtail butterfly, *Papilio xuthus* L. (Lepidoptera: Papilionidae). *Journal of Ethology*, 18(2), 91–99.

Watt, W.B. (1973) Adaptive significance of pigment polymorphisms in Colias butterflies. III. Progress in the study of the 'alba' variant. *Evolution*, 27(4), 537–548.

Watt, W.B., Carter, P.A. and Donohue, K. (1986) Females' choice of 'good genotypes' as mates is promoted by an insect mating system. *Science*, 233, 1187–1190.

Webb, H.K., Crawford, R.J. and Ivanova, E.P. (2014) Wettability of natural superhydrophobic surfaces. *Advances in Colloid and Interface Science*, 210, 58–64.

Wedell, N. (1996) Mate quality affects reproductive effort in a paternally investing species. *The American Naturalist*, 148, 1075–1088.

Wedell, N. (2001) Female remating in butterflies: Interaction between female genotype and non-fertile sperm. *Journal of Evolutionary Biology*, 14, 746–754.

Wedell, N. (2005) Sperm competition in butterflies and moths. In: Fellowes, M., Holloway, G. and Rolff, J. (eds), *Insect Evolutionary Ecology: Proceedings of the Royal Entomological Society's 22nd Symposium*. CAB International, Wallingford, UK, pp. 49–82.

Wedell, N. (2010) Variation in male courtship costs in butterflies. *Behavioral Ecology and Sociobiology*, 64(9), 1385–1391.

Wedell, N. and Cook, P.A. (1998) Determinants of paternity in a butterfly. *Proceedings of the Royal Society of London B: Biological Sciences*, 265(1396), 625–630.

Wedell, N. and Cook, P.A. (1999a) Butterflies tailor their ejaculate in response to sperm competition risk and intensity. *Proceedings of the Royal Society of London B: Biological Sciences*, 266, 1033–1039.

Wedell, N. and Cook, P.A. (1999b) Strategic sperm allocation in the small white butterfly *Pieris rapae* (Lepidoptera: Pieridae). *Functional Ecology*, 13(1), 85–93.

Wedell, N. and Karlsson, B. (2003) Paternal investment directly affects female reproductive effort in an insect. *Proceedings of the Royal Society of London B: Biological Sciences*, 270(1528), 2065–2071.

Wedell, N., Wiklund, C. and Cook, P.A. (2002) Monandry and polyandry as alternative lifestyles in a butterfly. *Behavioral Ecology*, 13(4), 450–455.

Wedell, N., Kvarnemo, C. and Tregenza, T. (2006) Sexual conflict and life histories. *Animal Behaviour*, 71(5), 999–1011.

Wedell, N., Wiklund, C. and Bergström, J. (2009) Coevolution of non-fertile sperm and female receptivity in a butterfly. *Biology Letters*, 5, 678–681.

Wells, C.N., Spencer, L. and Simons, D. (2011) Reproductive behavior of *Speyeria diana* (Nymphalidae) in Arkansas. *The Journal of the Lepidopterists' Society*, 65(1), 51–54.

Westerman, E.L., Hodgins-Davis, A., Dinwiddie, A. and Monteiro, A. (2012) Biased learning affects mate choice in a butterfly. *Proceedings of the National Academy of Sciences*, 109(27), 10948–10953.

Westerman, E.L., Drucker, C. B. and Monteiro, A. (2014) Male and female mating behavior is dependent on social context in the butterfly *Bicyclus anynana*. *Journal of Insect Behavior*, 27(4), 478–495.

Westerman, E.L., Letchinger, R., Tenger-Trolander, A., Massardo, D., Palmer, D. and Kronforst, M.R. (2018) Does male preference play a role in maintaining female limited polymorphism in a Batesian mimetic butterfly?. *Behavioural Processes*, 150, 47–58.

White, T.E., Macedonia, J., Birch, D., Dawes, J. and Kemp, D.J. (2012) The nanoanatomical basis of sexual dimorphism in iridescent butterfly colouration. *Australian Journal of Zoology*, 60(2), 101–107.

White, T.E., Zeil, J. and Kemp, D.J. (2014) Signal design and courtship presentation coincide for highly biased delivery of an iridescent butterfly mating signal. *Evolution*, 69(1), 14–25.

Whitehorn, P.R., Tinsley, M.C. and Goulson, D. (2009) Kin recognition and inbreeding reluctance in bumblebees. *Apidologie*, 40(6), 627–633.

Whiteley, D. (1992) An atlas of Sheffield area butterflies. *Sorby Record*, 29, 19–56.

Wickman, P.O. (1985a) The influence of temperature on the territorial and mate locating behaviour of the small heath butterfly, *Coenonympha pamphilus* (L.) (Lepidoptera: Satyridae). *Behavioral Ecology and Sociobiology*, 16(3), 233–238.

Wickman, P.O. (1985b) Territorial defence and mating success in males of the small heath butterfly, *Coenonympha pamphilus* L. (Lepidoptera: Satyridae). *Animal Behaviour*, 33, 1162–1168.

Wickman, P.O. (1985c) Male determined mating duration in butterflies? *Journal of the Lepidopterists' Society*, 39, 341–342.

Wickman, P.O. (1986) Courtship solicitation by females of the small heath butterfly, *Coenonympha pamphilus* (L.) (Lepidoptera, Satyridae) and their behavior in relation to male territories before and after copulation. *Animal Behaviour* 34, 153–157.

Wickman, P.O. (1988) Dynamics of mate-searching behaviour in a hilltopping butterfly, *Lasiommata megera* (L.): the effects of weather and male density. *Zoological Journal of the Linnean Society*, 93(4), 357–377.

Wickman, P.O. (1992) Mating systems of *Coenonympha* butterflies in relation to longevity. *Animal Behaviour*, 44(1), 141–148.

Wickman, P.O. and Jansson, P. (1997) an estimate of female mate searching costs in the lekking butterfly *Coenonympha pamphilus*. *Behavioral Ecology and Sociobiology*, 40(5), 321–328.

Wickman, P.O. and Rutowski, R.L. (1999) The evolution of mating dispersion in insects. *Oikos*, 84, 463–472.

Wickman, P.O. and Wiklund, C. (1983) Territorial defence and its seasonal decline in the speckled wood butterfly (*Pararge aegeria*). *Animal Behaviour*, 31, 1206–1216.

Wickham, S., Large, M.C., Poladian, L. and Jermiin, L.S. (2006) Exaggeration and suppression of iridescence: the evolution of two-dimensional butterfly structural colours. *Journal of the Royal Society Interface*, 3(6), 99–109.

Wiernasz, D.C. (1989) Female choice and sexual selection of male wing melanin pattern in *Pieris occidentalis* (Lepidoptera). *Evolution*, 43(8), 1672–1682.

Wiernasz, D.C. (1995) Male choice on the basis of female melanin pattern in *Pieris* butterflies. *Animal Behaviour*, 49(1), 45–51.

Wijnen, B., Leertouwer, H.L. and Stavenga, D.G. (2007) Colors and pterin pigmentation of pierid butterfly wings. *Journal of Insect Physiology*, 53(12), 1206–1217.

Wiklund, C. (1977) Courtship behaviour in relation to female monogamy in *Leptidea sinapis* (Lepidoptera). *Oikos*, 29(2), 275–283.

Wiklund, C. (1982) Behavioural shift from courtship solicitation to mate avoidance in female ringlet butterflies (*Aphantopus hyperanthus*) after copulation. *Animal Behaviour*, 30(3), 790–793.

Wiklund, C. (2003) Sexual selection and the evolution of butterfly mating systems. In: Boggs L.C., Watt, W.B. and Ehrlich, P.R. (eds), *Butterflies: Ecology and Evolution Taking Flight*. University of Chicago Press, Chicago, IL, pp. 67–90.

Wiklund, C. and Fagerström, T. (1977) Why do males emerge before females? *Oecologia*, 31(2), 153–158.

Wiklund, C. and Forsberg, J. (1985) Courtship and male discrimination between virgin and mated females in the orange tip butterfly *Anthocharis cardamines*. *Animal Behaviour*, 34(2), 328–332.

Wiklund, C. and Forsberg, J. (1991) Sexual size dimorphism in relation to female polygamy and protandry in butterflies: a comparative study of Swedish Pieridae and Satyridae. *Oikos*, 60(3), 373–381.

Wiklund, C. and Kaitala, A. (1995) Sexual selection for large male size in a polyandrous butterfly: the effect of body size on male versus female reproductive success in *Pieris napi*. *Behavioral Ecology*, 6(1), 6–13.

Wiklund, C. and Karlsson, B. (1988) Sexual size dimorphism in relation to fecundity in some Swedish satyrid butterflies. *The American Naturalist*, 131(1), 132–138.

Wiklund, C. and Solbreck, C. (1982) Adaptive versus incidental explanations for the occurrence of protandry in butterfly, *Leptidea sinapis* L. *Evolution*, 36(1), 56–62.

Wiklund, C., Kaitala, A., Lindfors, V. and Abenius, J. (1993) Polyandry and its effect on female reproduction in the green-veined white butterfly (*Pieris napi* L.). *Behavioral Ecology and Sociobiology*, 33(1), 25–33.

Wiklund, C., Karlsson, B. and Leimar, O. (2001) Sexual conflict and cooperation in butterfly reproduction: a comparative study of polyandry and female fitness. *Proceedings of the Royal Society of London B: Biological Sciences*, 268(1477), 1661–1667.

Wiklund, C., Gotthard, K. and Nylin, S. (2003) Mating system and the evolution of sex-specific mortality rates in two nymphalid butterflies. *Proceedings of the Royal Society of London B: Biological Sciences*, 270(1526), 1823–1828.

Wilts, B.D., Leertouwer, H.L. and Stavenga, D.G. (2008) Imaging scatterometry and microspectrophotometry of lycaenid butterfly wing scales with perforated multilayers. *Journal of The Royal Society Interface*, 6(Suppl. 2), S185–S192.

Wilts, B.D., Michielsen, K., De Raedt, H. and Stavenga, D.G. (2011) Iridescence and spectral filtering of the gyroid-type photonic crystals in *Parides sesostris* wing scales. *Interface Focus*, 2(5), 681–687.

Wilts, B.D., Trzeciak, T.M., Vukusic, P. and Stavenga, D.G. (2012) Papiliochrome II pigment reduces the angle dependency of structural wing colouration in nireus group papilionids. *Journal of Experimental Biology*, 215(5), 796–805.

Wilts, B.D., Pirih, P., Arikawa, K. and Stavenga, D.G. (2013) Shiny wing scales cause spec(tac) ular camouflage of the angled sunbeam butterfly, *Curetis acuta*. *Biological Journal of the Linnean Society*, 109(2), 279–289.

Wilts, B.D., IJbema, N. and Stavenga, D.G. (2014) Pigmentary and photonic coloration mechanisms reveal taxonomic relationships of the cattlehearts (Lepidoptera: Papilionidae: Parides). *BMC Evolutionary Biology*, 14(1), 160.

Wilts, B.D., Matsushita, A., Arikawa, K. and Stavenga, D.G. (2015) Spectrally tuned structural and pigmentary coloration of birdwing butterfly wing scales. *Journal of the Royal Society Interface*, 12(111), 20150717.

Wilts, B.D., Zubiri, B.A., Klatt, M.A., Butz, B., Fischer, M.G., Kelly, S.T., Spiecker, E., Steiner, U. and Schröder-Turk, G.E. (2017a) Butterfly gyroid nanostructures as a time-frozen glimpse of intracellular membrane development. *Science Advances*, 3(4), e1603119.

Wilts, B.D., Wijnen, B., Leertouwer, H.L., Steiner, U. and Stavenga, D.G. (2017b) Extreme refractive index wing scale beads containing dense pterin pigments cause the bright colors of pierid butterflies. *Advanced Optical Materials*, 5(3), 1600879.

Winter, D. (1985) Misdirected monarch mating behavior (Danaidae: *Danaus plexippus*) or noblesse oblige? *Journal of the Lepidopterists' Society*, 39(4) 334.

Woo, S., Lee, C.E. and Park, H.C. (1990) A microscopic study on the bursa copulatrix of Korean Pieridae (Lepidoptera). *Esakia*, Special Issue No. 1, 167–172.

Woronik, A. (2017) A functional genomic investigation of an alternative life history strategy: The Alba polymorphism in *Colias croceus*. Doctoral dissertation, Department of Zoology, Stockholm University, Sweden.

Woronik, A., Stefanescu, C., Käkelä, R., Wheat, C.W. and Lehmann, P. (2018) Physiological differences between female limited, alternative life history strategies: the *Alba* phenotype in the butterfly *Colias croceus*. *Journal of Insect Physiology*, 107, 257–264.

Wuest, J. (1998) The pheromone dispersing apparatus in some Hesperiinae (Lepidoptera, Hesperiidae, Hesperiinae). *Revue Suisse de Zoologie*, 105(4), 813–822.

Wyatt, T.D. (2009) Fifty years of pheromones. *Nature*, 457(7227), 262.

Yack, J.E. (2004) The structure and function of auditory chordotonal organs in insects. *Microscopy Research and Technique*, 63(6), 315–337.

Yack, J.E., Otero, L.D., Dawson, J.W., Surlykke, A. and Fullard, J.H. (2000) Sound production and hearing in the blue cracker butterfly *Hamadryas feronia* (Lepidoptera, Nymphalidae) from Venezuela. *Journal of Experimental Biology*, 203(24), 3689–3702.

Yagi, N. (1954) Note of electron microscopic research on pterin pigments in the scales of Pierid butteflies. 日本動物学彙報, 27(3), 113–114. (Japanese Zoological Dictionary the Zoological Society of Japan)

Yamamoto, M. (1981) Adult duration of the Japanese cabbage white, *Pieris rapae crucivora* (Lepidoptera: Pieridae). *Applied Entomology and Zoology*, 16(3), 310–312.

Yata, O., Abe, M. and Nakai, M. (1986) Relationship between androconial scale and the male wing odour in *Pieris melete* Menetries (Lepidoptera, Pieridae). *Sieboldia Acta Biologica*, 5(1), 131–136.

Yildizhan, S., van Loon, J., Sramkova, A., Ayasse, M., Arsene, C., ten Broeke, C. and Schulz, S. (2009). Aphrodisiac pheromones from the wings of the small cabbage white and large cabbage white butterflies, *Pieris rapae* and *Pieris brassicae*. *ChemBioChem*, 10(10), 1666–1677.

Yoshida, A., Noda, A., Yamana, A. and Numata, H. (2000) Arrangement of scent scales in the male wings of the small white cabbage butterfly (Lepidoptera: Pieridae). *Entomological Science*, 3(2), 345–349.

Yoshioka, S. and Kinoshita, S. (2004) Wavelength-selective and anisotropic light-diffusing scale on the wing of the Morpho butterfly. *Proceedings of the Royal Society of London B: Biological Sciences*, 271(1539), 581–587.

Yoshioka, S., Fujita, H., Kinoshita, S. and Matsuhana, B. (2014a) Alignment of crystal orientations of the multi-domain photonic crystals in *Parides sesostris* wing scales. *Journal of the Royal Society Interface*, 11, 20131029.

Yoshioka, S., Matsuhana, B. and Fujita, H. (2014b) Polarization-dependent tessellated pattern of the wing scale of the *Parides sesostris* butterfly. *Materials Today: Proceedings*, 1, 186–192.

Yuan, F., Bernard, G.D., Le, J. and Briscoe, A.D. (2010) Contrasting modes of evolution of the visual pigments in *Heliconius* butterflies. *Molecular Biology and Evolution*, 27(10), 2392–2405.

Yuvaraj, J.K. (2017) Evolution of olfaction in Lepidoptera and Trichoptera: gene families and antennal morphology. Doctoral thesis, Lund University, Lund, Sweden.

Yuvaraj, J.K., Andersson, M.N., Anderbrant, O. and Löfstedt, C. (2018) Diversity of olfactory structures: a comparative study of antennal sensilla in Trichoptera and Lepidoptera. *Micron*, 111, 9–18.

Zaccardi, G., Kelber, A., Sison-Mangus, M.P. and Briscoe, A.D. (2006) Color discrimination in the red range with only one long-wavelength sensitive opsin. *Journal of Experimental Biology*, 209(10), 1944–1955.

Zagrobelny, M., Bak, S. and Møller, B.L. (2008) Cyanogenesis in plants and arthropods. *Phytochemistry*, 69(7), 1457–1468.

Zahavi, A. (1975) Mate selection – a selection for a handicap. *Journal of Theoretical Biology*, 53(1), 205–214.

Zakharov, E. (2005) Synergistic effects of combining morphological and molecular data in resolving the phylogeny of butterflies and skippers. *Proceedings of the Royal Society of London B: Biological Sciences*, 272(1572), 1577–1586.

Zhan, S., Merlin, C., Boore, J. L. and Reppert, S.M. (2011) The monarch butterfly genome yields insights into long-distance migration. *Cell*, 147(5), 1171–1185.

Zhao, Q.B., Guo, X.M., Fan, T.X., Ding, J., Zhang, D. and Guo, Q.X. (2011) Art of blackness in butterfly wings as natural solar collector. *Soft Matter*, 7, 11433.

Ziemba, K.S. and Rutowski, R.L. (2000) Sexual dimorphism in eye morphology in a butterfly (*Asterocampa leilia*; Lepidoptera, Nymphalidae). *Psyche*, 103(1–2), 25–36.

Zonneveld, C. (1996) Being big or emerging early? Polyandry and the trade-off between size and emergence in male butterflies. *The American Naturalist*, 147(6), 946–965.

Index

Note: bold page numbers indicate figures; numbers in brackets preceded by *n* refer to chapter endnotes.

This book is published by **CABI**, an international not-for-profit organisation that improves people's lives worldwide by providing information and applying scientific expertise to solve problems in agriculture and the environment.

CABI is also a global publisher producing key scientific publications, including world renowned databases, as well as compendia, books, ebooks and full text electronic resources. We publish content in a wide range of subject areas including: agriculture and crop science / animal and veterinary sciences / ecology and conservation / environmental science / horticulture and plant sciences / human health, food science and nutrition / international development / leisure and tourism.

The profits from CABI's publishing activities enable us to work with farming communities around the world, supporting them as they battle with poor soil, invasive species and pests and diseases, to improve their livelihoods and help provide food for an ever growing population.

CABI is an international intergovernmental organisation, and we gratefully acknowledge the core financial support from our member countries (and lead agencies) including:

UKaid from the British people

Ministry of Agriculture People's Republic of China

Australian Government Australian Centre for International Agricultural Research

Agriculture and Agri-Food Canada

Ministry of Foreign Affairs of the Netherlands

Schweizerische Eidgenossenschaft Confédération suisse Confederazione Svizzera Confederaziun svizra Swiss Agency for Development and Cooperation SDC

Discover more

To read more about CABI's work, please visit: **www.cabi.org**

Browse our books at: **www.cabi.org/bookshop**,
or explore our online products at: **www.cabi.org/publishing-products**

Interested in writing for CABI? Find our author guidelines here:
www.cabi.org/publishing-products/information-for-authors/